9/24
5—

ENDEAVOURING
BANKS

ENDEAVOURING BANKS

Exploring collections from
The *Endeavour* Voyage
1768–1771

Neil Chambers

With contributions by

Anna Agnarsdóttir, Sir David Attenborough,
Jeremy Coote, Philip J. Hatfield and John Gascoigne

PAUL HOLBERTON PUBLISHING

UNIVERSITY OF WASHINGTON PRESS

NEWSOUTH PUBLISHING

Copyright © 2016
Texts copyright © the authors

All rights reserved. No part of this publication may be transmitted in any
form or by any means, electronic or mechanical, including photocopy,
recording or any storage or retrieval system, without the prior permission
in writing from the copyright holder and publisher.

Published in the United States of America by
University of Washington Press, Seattle
www.washington.edu/uwpress

ISBN 978-0-295-99811-4

Library of Congress Control Number: 2016932086

First edition published in the UK and Europe by Paul Holberton Publishing,
in the US and Canada by University of Washington Press
and in Australia by NewSouth Publishing.

Designed and typeset in Sirenne Text MVB by Laura Parker,
Paul Holberton publishing, London

Printed and bound in China by 1010 Printing International Ltd

FRONT COVER: *The Fly Catching Macaroni*, 1772 (cat. 135), detail
BACK COVER: *Artocarpus altilis* (Parkinson) Fosberg (cat. 40), detail
FRONTISPIECE: Benjamin West RA, *Joseph Banks*, 1771–2 (cat. 143), detail

CONTENTS

Foreword 6
SIR DAVID ATTENBOROUGH

Acknowledgements 9

Introduction 10

Background to the Endeavour *Voyage* 21
JOHN GASCOIGNE

THE VOYAGE: AIMS AND ORGANIZATION 27

THE ATLANTIC STAGE 51

Dressing Up, Taking Over and Passing On: 80
Joseph Banks and Artificial Curiosities from the Endeavour *Voyage*
JEREMY COOTE

THE SOCIETY ISLANDS 86

The Material History of the Endeavour: *Joseph Banks at the British Library* 154
PHILIP J. HATFIELD

NEW ZEALAND 160

Exploring Collections from the Endeavour *Voyage* 200
NEIL CHAMBERS

THE EAST COAST OF AUSTRALIA 219

HOMEWARD BOUND 264

After the Endeavour: *What Next for Joseph Banks?* 269
ANNA AGNARSDÓTTIR

AFTERMATH 275

Notes 294
Bibliography and Sources 298
Photographic Credits 304

FOREWORD

SIR DAVID ATTENBOROUGH

WE TEND TO TAKE IT FOR GRANTED THESE DAYS that an expedition setting out for a little-known part of the world will bring back detailed observations and representative collections of all the new things it encounters – the animals, the plants, the rocks, the landscapes as well as the artefacts produced by the human inhabitants. But that has not always been so.

The first expedition to do such a thing on any scale was that which in 1768 set out for the Pacific in HM Bark *Endeavour*. It was commanded by Lieutenant James Cook RN. His orders were to make astronomical observations of the transit of the planet Venus across the face of the Sun. These, when compared with similar observations made simultaneously around the known world, would enable astronomers to calculate the distance of the Earth from the Sun. That, in turn, would make it possible to establish the distance of the Earth from the Moon and the planets, and ultimately, the stars.

After that had been done, he was instructed to sail southwards in order to establish the existence, predicted by geographers, of a vast unknown continent lying across the other end of the earth. If he discovered it, he had to survey it and bring back samples of its rocks and any plants and animals that it might contain. If he failed to find any such place, then he was to return home, on the way mapping such new islands as he might encounter and claiming those hitherto unknown to European explorers in the name of His Britannic Majesty.

To achieve these aims, Cook took with him Charles Green, assistant to astronomer royal Nevil Maskelyne. But he also had a companion – an extremely wealthy, twenty-five-year-old landowner from Lincolnshire named Joseph Banks, who, with the help of the Royal Society, had persuaded the lords of the Admiralty to let him join the expedition.

Banks was a passionate naturalist. As a youth, he had become obsessed by plants and had compiled a flora of his own home county. At university in Oxford, he had engaged a private tutor from Cambridge to teach him something about the science of botany. He had then joined a fishery protection vessel as a naturalist on a voyage to Labrador and Newfoundland. The voyage to the Pacific, however, was a challenge on an incomparably greater scale. For that he would take on the responsibility of gathering specimens and compiling scientific accounts of every natural thing the *Endeavour* encountered on the other side of the globe. And he would do all of this at his own expense.

To assist him, he recruited the Swedish naturalist Daniel Solander and, as a clerk and assistant artist, another Swede, Herman Diedrich Spöring. He also engaged two artist-draughtsmen to make images of the animals and plants. To look after his personal welfare, he had four servants. And finally, completing his entourage, he brought two of his favourite greyhounds. He was indeed an exceedingly rich young man.

It took the expedition eight months to sail across the Atlantic, round the southern tip of South America and then voyage across the vast emptiness of the Pacific to reach Tahiti. Only two European ships had preceded them there, one under an Englishman, Captain Samuel Wallis, who had landed there two years earlier, and another commanded by a Frenchman, Louis Antoine de Bougainville, who had arrived nine months after Wallis. While the *Endeavour's* astronomers set up their observatory, Banks and his helpers set about collecting whatever they could of every aspect of the island – its birds, its plants and the implements, weapons and costumes produced by its inhabitants.

Eight weeks later, when the astronomers had finished their work, the expedition left, heading south as instructed to search for the notional Great Southern Continent. They went sufficiently far south for Cook to assure himself that no habitable land lay beyond. Having now been away for a year and with his two major tasks completed, he started on the return journey.

But it was to be a circuitous one. In compliance with his instructions, he continued to sail westwards. He headed for the islands now called New Zealand, and sailed around them, making periodic contacts with the Māori inhabitants. Sydney Parkinson drew their portraits. Banks collected their artefacts. And Cook made a running survey of the coasts of both islands. He then continued westwards until he sighted the east coast of Australia.

No European ship had ever charted this part of the largely unknown continent before. Nor had it yet acquired its name of Australia. Joseph Banks, when he went ashore, was so thrilled by the new and wonderful plants he found that Cook named that part of the coast Botany Bay. The expedition then sailed on northwards, keeping relatively close to the coast so that it could be charted, and battled its way for a thousand miles along the labyrinthine length of the Great Barrier Reef, before finally being able to turn westwards through the Torres Straits and into the archipelagos of what is now Indonesia. And so back home.

The *Endeavour* finally docked in Dover in July 1771. On board, she had an astonishing quantity of specimens and meticulous records, both written and painted, of where and how they had been collected. There were over a thousand zoological specimens, mostly skins or skeletons. Among them were examples gathered in Australia of strange pouched animals that had never been seen before in Europe. There were 30,000 pressed and dried botanical specimens, including 1,400 species that were new to science. And a great range of islanders' artefacts, 'natural curiosities' as they were termed, comprising implements, weapons and costumes.

All these things, however, had been gathered at a cost. Even before *Endeavour* entered the Pacific, two of Bank's servants had died of exposure when a small shore party led by Banks had become stranded in Tierra del Fuego, the desolate southern tip of South America. In Tahiti, soon after landing, one of his draughtsmen, John Buchan, had died during an epileptic fit. And when at last *Endeavour* left the Pacific, and lay anchored in Java for several weeks undergoing much needed repairs, thirty of the ship's company died of dysentery and malaria, including Herman Spöring and his fellow artist, the young botanical illustrator Sydney Parkinson.

But the treasures that the *Endeavour* brought back in her hold were an astonishment to all who saw them, as wonderful and exotic to European eyes in the eighteenth century as the small rocks brought back by the twentieth century's explorers from the surface of the moon – and assuredly far more varied and beautiful.

Banks had been responsible for collecting them and now he determined their disposal. Some of the ethnographic objects he sent to scholarly institutions, such as the University of Oxford and the British Museum, and others he gave away to colleagues with more of an interest in them than he. The bulk of the plant specimens, Banks's own particular obsession, he kept and eventually bequeathed to the British Museum, from whence they later passed to the Natural History Museum in South Kensington. So, ultimately the collections were widely dispersed.

As a consequence, it has not been possible since that time for anyone to see the full range and richness of the *Endeavour's* collections in one place. Nearly two hundred and fifty years later, however, that was put right. In 2014, a great range of them were re-assembled from various collections for a celebratory exhibition in Lincoln, Banks's home city. And here is its richly illustrated catalogue. In it you will find reproductions of some of Parkinson's exquisite botanical drawings, together with coloured engravings of them printed from the plates that Banks commissioned, but which were never printed in his lifetime. Here too is a selection of the ethnographic objects, which show the great range and depth of Bank's interests. And how beautiful most of them are, with a perfection of form and detail all the more remarkable to European eyes in that they were all fashioned without the use of metal tools.

No expedition before had brought back collections of such a size and importance to such a range of the natural sciences. And Joseph Banks, who was still only twenty-eight years old, had been directly responsible for their assembly. Benjamin West, a fashionable artist of the day, painted Banks's portrait, standing surrounded by a small sample of the artefacts he had gathered – a Tahitian stone-bladed adze, a Māori paddle and spear and a Tahitian headdress. An open volume of plant drawings lies at his feet. On his shoulders he wears a *kaitaka*, a Māori cloak woven from the fibres of phormium, the New Zealand flax. As he steps forward towards the spectator, he proudly points a finger to its fringe of highly prized near-sacred dog-hair tassels. Such garments decorated in this way were reserved by the Māori for the most senior members of their nobility and young Joseph Banks wears it with visible pride. As well he might.

ACKNOWLEDGEMENTS

THE SUBJECT OF THIS VOLUME is the historic voyage of HM Bark *Endeavour* from August 1768 to July 1771, and especially the contribution of Joseph Banks and the team that he took on the mission to gather natural history and ethnographic specimens as well as to illustrate the people and places encountered. The volume is intended to help open up for further exploration the world of collecting that was so memorably enlarged on this seminal mission by examining surviving material from a range of disciplines as gathered at each of the locations that were visited. Much of this material is now separated, being held in various institutions devoted to each of these disciplines, and this and the unique value of anything connected with *Endeavour* means that most of what remains is rarely if ever seen together in public today. An exhibition to try to overcome this problem, and to bring more of these hidden treasures to light in a venue outside London, where the majority now reside, was mounted in early 2014 at the Collection, Lincoln's main museum. Lincolnshire was where Banks's main country estates were located, from which he drew the income that he deployed in the cause of science and discovery, and the Collection holds the massive post-voyage portrait of Banks by Benjamin West showing the young naturalist as a pioneering collector in various fields (frontispiece and number 143). West's portrait provided an impressive centrepiece for the exhibition. The Heritage Lottery Fund provided a generous grant to make this event possible. The exhibition was curated by the author, and it set a record for attendance at the museum, thus achieving one of its main objectives. This would not have been possible without the essential support of the institutions that loaned objects, their staff and others. It is a great pleasure to acknowledge these here.

Lending institutions, including those staff with whom the author had direct contact, were: Boston Borough Council; the British Library, Arnold Hunt, Andrea Clarke, Barbara O'Connor and Tom Harper; the British Museum, Jill Hasell and Rachael Murphy; the Captain Cook Memorial Museum, Whitby; Derbyshire Record Office, Becky Sheldon, Karen Millhouse and Mark Smith; the Goodwood Collection, Goodwood House; the Linnean Society of London; the National Archives, Kew; the National Maritime Museum, Greenwich, Christine Riding and Nigel Rigby; the National Museum of Scotland, Alison Morrison-Low; the Natural History Museum, London, Roberto Portela Miguez, Andrea Hart, Armando Mendez, Paul Cooper and Goulven Keineg; the Pitt Rivers Museum, University of Oxford, Jeremy Coote; the Royal Collection, by permission of Her Majesty Queen Elizabeth II; the Royal College of Surgeons, Sam Alberti and Martyn Cooke; the Royal Horticultural Society, Charlotte Brooks; the Royal Society, London; the School of Oriental and African Studies, London, Joanne Anthony, Sujan Nandanwar and Winifred Assan; the Science Museum, London; the Collection, Lincoln, Jules Rich, William Mason, Andrea Martin and Dawn Heywood.

Known *Endeavour*-voyage material from one other institution that did not lend objects to the original exhibition has been used in this volume in order to broaden its scope, with thanks: the Museum of Archaeology and Anthropology, University of Cambridge, Rachel Hand.

Other institutions and individuals were consulted or otherwise assisted in this work: Mark Eldridge, Australian Museum, Sydney; Patricia Wallace, University of Canterbury, New Zealand; Etnografiska Museet, Stockholm, Lars-Erik Barkman, Magnus Johansson and Rose-Marie Westling; the Museum of New Zealand (Te Papa Tongarewa), Wellington, Puawai Cairns, Dougal Austin, Rhonda Paku and Claire Regnault; the National Library of Australia, Pip Manifold. Particular thanks are due to Joanna Kapusta, David Hibberd and Pamela Carr for their stalwart patience and valuable expertise not only in the mounting of the exhibition, but also in the preparation of the present volume arising from it. This volume develops some of the themes of the exhibition, broadens its availability and marks too the imminent 250th anniversary of the launch of HM Bark *Endeavour*, when the impulses that led to the mission as well as its enormous impact and enduring legacy will undoubtedly be more fully re-examined.

My gratitude to Paul Holberton and Laura Parker for their friendly professionalism. Warm thanks are due to the contributors of essays for their comments on the text and tremendous scholarship. In order of appearance: Professor John Gascoigne; Jeremy Coote; Dr Phil Hatfield and Professor Anna Agnarsdóttir. Warm thanks are also due to Sir David Attenborough, who not only opened the exhibition during a memorable evening on 14 February 2014, but also kindly volunteered the present foreword.

Introduction

JOSEPH BANKS was born into a wealthy but untitled family descended from Yorkshire stock, whose main estates were by the late eighteenth century concentrated in Lincolnshire, with the family seat being situated at Revesby Abbey, near Horncastle. As established landowners the Banks family held a respectable position in county society and even had minor status within London's political and learned circles – Joseph Banks's grandfather was a fellow of the Royal Society and more than one forebear sat in parliament. Banks was born at 30 Argyll Street, London, on 13 February 1743, and was educated at Harrow School and then Eton College, where he found the traditional focus on classical studies uninspiring and in his leisure time was drawn instead to investigation of the natural world. He was an undergraduate at Christ Church, Oxford, but went down before taking a degree like many gentlemen of his class. His father died young in 1761, and so Banks was for a while under the guardianship of his uncle, Robert Banks-Hodgkinson, a considerate man with family property in Derbyshire and through marriage also in Wales, the former of which Banks eventually inherited.

After leaving Oxford, Banks settled into life in London, becoming a familiar figure among the various societies and clubs that formed its social and intellectual scene. In particular, he studied the natural collections and library at the recently founded British Museum, and there in 1764 made the acquaintance of the great friend of his early years, Daniel Solander. From Sweden and a favourite pupil of the great naturalist and professor at Uppsala University Carl Linnaeus, Solander was employed cataloguing the natural collections at the museum. For this he used his master's system, which forms the basis of modern systematic classification in natural history. In ensuing years Banks and Solander would both promote use of the Linnaean system in the organization of London's existing natural collections as well as those that were increasingly being gathered from across the globe. Banks collected plants abroad for the first time in 1766, sailing with his old Eton contemporary Constantine John Phipps to the coasts of Newfoundland and Labrador on a fisheries patrol. Before departing Banks was elected to the Society of Antiquaries and while away to the Royal Society, two of London's key intellectual hubs, in both of which he would later play a leading role. This early overseas collecting trip helped strengthen Banks's standing among naturalists in London and, no less importantly, it provided valuable preparation for greater voyages to come.

Historic opportunities to travel and collect abroad opened up during the years immediately following Banks's 1766 trip. At this time Europe's academies were alive with talk about the forthcoming transit of Venus, observation of which would enable astronomers to determine the distance between the Earth and Sun, and hence obtain an agreed unit of measure for use in their study of the solar system. The British astronomer Edmond Halley had earlier in the century suggested overseas missions to observe periodic transits of Venus in order to find this distance, and it was a matter not only of scientific importance but also of national prestige that Britain should now play its part alongside many other nations in the international effort to observe the next one. The transit of 1761 (such transits come in pairs eight years apart) provided a sort of rehearsal for observation of the ensuing transit, but the latter event needed to be successfully observed since there would not be another transit for more than a hundred years. Consequently, the Royal Society applied to the Crown for support in launching a special mission to the Pacific, to which George III gave £4,000, with a Royal Navy vessel also being provided for the voyage. The vessel would be called HM Bark *Endeavour*, and she would sail under the experienced naval commander and marine surveyor Lieutenant James Cook.

A subsidiary aim of the mission was to probe for an undiscovered southern continent, *Terra Australis Incognita*, that many thought lay in the southern oceans, a matter of strategic as well as geographical importance given the resources and trade that such a landmass might yield. Banks saw the opportunity that the mission would provide to increase knowledge of plant and animal life in the areas to be visited. Working through his contacts at the Royal Society and the Admiralty, he was able to secure a place at his own expense as a supernumerary on the voyage. He took, too, a party of collectors and illustrators, including Daniel Solander, thereby expanding the scientific range of the expedition to include natural history and laying down, albeit unknowingly, a framework for future British voyages of discovery that led, ultimately, to the voyage of Darwin in the *Beagle*. British and French missions that were launched into the Pacific at this time (Byron, Wallis, Carteret, Bougainville) heralded a vast expansion of geographical, natural and human discovery, and they had an immeasurable impact on the peoples of the Pacific as also on European art and understanding. The subject of this volume is one of these historic voyages, that of HM Bark *Endeavour* from August 1768 to July 1771, and especially Joseph Banks's contribution to its achievements as seen in expedition specimens, artwork and records surviving today in Great Britain and elsewhere.

I Portrait, Sir Joseph Banks, *aet* 71

By Thomas Phillips RA. Commissioned by the Corporation of Boston, Lincolnshire, 1813 and completed by February 1814. Exhibited at the Royal Academy, 1814 (52).

Boston Guildhall, Boston Borough Council. BOSGM 1964.6. Unframed 140 × 109 cm. Oil on canvas.

In a letter to his friend, the botanist and antiquary Dawson Turner, dated 6 September 1813, Thomas Phillips explained: 'I have painted portraits of 2 important personages since you left town – Lord Byron and another of Sir Jos. Banks, with whom I had the gratification of living for a fortnight, and you will be glad to hear that he is entirely recovered as to bodily health, but unfortunately is not yet able to walk'. Phillips was a leading portrait painter by the time the Corporation of Boston commissioned him to depict Banks in 1813 at a cost of 100 guineas. He would during his career produce more than 700 portraits and he had, as indicated in his comment to Turner, painted Banks before (see below). Phillips is especially noted for his depictions of natural philosophers (as scientists would then have been known), explorers, writers and other thinkers of his day. He painted no less than five presidents of the Royal Society and, taking an interest in the sciences, was himself elected a society fellow in 1819. His paintings of Romantic intellectuals are particularly admired, and during the summer period that he painted Banks for the Corporation he also produced two fine if different portraits of Lord Byron: one a half-length view of the poet, confidently posed in an open white shirt and wrapped in a dark cloak; the other a three-quarter length painting in which Byron is dramatically attired in Albanian dress, giving him the appearance of an exotic traveller. Byron had recently returned from a Mediterranean grand tour, and in these portraits Phillips captured his sitter at the height of his energy and allure.

By contrast Banks, the ageing veteran of James Cook's first great circumnavigation, was incapacitated by severe gout and unable to leave London in 1813. The present oil portrait of him as a county gentleman and grandee was therefore probably painted at his home at 32 Soho Square, which would explain why Phillips stayed with his subject for two weeks and was able to report on the state of his health to Turner, a mutual friend. The painting shows Banks calmly glancing to one side, his powerful gaze averted from the viewer. The debility of the sitter is concealed. Instead, as in his other portraits, what Phillips sought to convey were the interests and concerns of his subject. Since this one was in recognition of Banks's many services to the borough and neighbourhood of Boston, where he acted as recorder from 1809 to his death in 1820, it was his county connections that Phillips chose to highlight.

Banks's main country seat was situated in Lincolnshire at Revesby Abbey, and his extensive estates provided him with much of the wealth that he devoted to science and collecting in the cause of learning. An improving landowner, he managed his properties and tenants with care, and was also, as this portrait demonstrates, an influential figure in county affairs. In 1797 he had briefly been a Lieutenant-Colonel in the Northern Battalion of the Lincolnshire Supplemental Militia, after helping to resolve local difficulties raising the militia during the wars with France. It may, therefore, be the uniform of this unit that Banks is shown wearing, and, if so, it was a post that he soon relinquished due to his increasingly poor physical condition. On his breast is the star and, just visible beneath his coat, the red sash of the Order of the Bath, a prestigious award made to Banks in 1795 for his services to learning. Phillips also depicts Banks holding the plan of a fen drainage scheme with a parliamentary act by his elbow, illustrating the fact that Banks devoted much of his time in the county to navigation and drainage projects intended to enhance communication and agricultural production. In these ways, then, Phillips stressed Banks's Lincolnshire activities, but an earlier commission resulted in a quite different treatment.

This was the iconic portrait of Banks as president of the Royal Society, started in 1808 and finished in 1809 for the Spanish mathematician and astronomer José de Mendoza y Rios. In it Phillips shows the naturalist dominating the canvas in his full presidential regalia, with ceremonial mace and an inkstand before him. Banks firmly returns the viewer's gaze, rather than looking to one side as in the corporation painting. This, the first version, now resides in the Dixson Galleries, New South Wales, and was thrice copied by Phillips in and shortly after 1810 (copies at the Linnean Society of London, the National Portrait Gallery and the Caroline Simpson Collection, Historic Houses Trust of New South Wales). These copy versions differ in certain details of dress and setting, sometimes adhering more closely to the first version and sometimes not, but the forward-facing pose of Banks remains basically the same. Five years later Phillips reproduced

the 1809 portrait, this time for the Royal Society, with Banks again seen in his presidential dress, and that version is still held by the society. A year after Banks's death in 1820, a final copy was completed for the Royal Horticultural Society, with Banks shown soberly dressed in grey in his Soho Square study (see number 142). Thus the 1809 portrait proved a good model for institutional and other commissions, featuring as it does the direct Banks stare so familiar during this period at committee and other meetings both in London and Lincolnshire. Phillips also appears to have produced a copy of the less imposing Boston Corporation portrait, presumably in the years immediately following completion of the original shown here. In 2011 this copy version was purchased by the National Portrait Gallery, Canberra. A later copy was made, in oils of the head and shoulders only, by Charles Ernest White, probably in the 1920s. This is currently held by the Sir Joseph Banks Society in Horncastle, Lincolnshire, on loan from the Environment Agency.

2 John Gerard, *The Herball or Generall Historie of Plantes* (London: John Norton, 1597)

The British Library, London. 449.k.4. Rebound, brown leather spine and brown cloth boards. Title, place and date of publication, gilt lettering on spine. Author given. 33.5 × 24.2 cm (closed).

In the summer of 1757, while on vacation from Eton College, Banks reputedly found a first-edition copy of Gerard's *Herball* in his mother's dressing room and this helped stimulate his interest in plants and the natural world. Until then he had struggled to engage with the classical curriculum at Eton, his housemaster Edward Young writing to Banks's father earlier that same year to report on his deficiencies as a student and seek a parental reprimand. The young Joseph Banks was by all accounts a pleasant boy, who, in his master's view, enjoyed play and the outdoors rather too much and the study of Latin and Greek rather too little. Following the awakening of his interest in the natural world, however, Banks's time and energy were increasingly devoted to collecting plants and insects. He roamed in search of specimens, searching along the Thames in term time and among the fens of his Lincolnshire homeland during vacations. While at Eton he is said to have paid sixpence to the local women who collected plants for apothecaries and druggists in return for each useful piece of information about the names and uses of plants that they gave him – an early example of the combination of wealth and initiative for which he would later become famous.

This copy of Gerard's *Herball*, first published by the queen's printer John Norton in 1597, is from Banks's own library. The English botanist and herbalist John Gerard's illustrated *Herball* was the best known botany book in English in the seventeenth century, although much of its content was probably derived from translations of the 1554 herbal *Cruydeboeck* by the Flemish botanist and physician Rembert Dodoens. Of the more than 1,800 woodcuts that illustrate Gerard's work only sixteen are original, the remainder being printed from blocks used to illustrate Dodoen's and other earlier Continental herbals. Gerard's herbal also contained many of his own remarks and, notwithstanding various errors and repetitions of folklore, it still remains one of the best-known English works of its type. A new and enlarged edition by the apothecary Thomas Johnson was published in 1633. Banks's library copy is here open at page 787, at the Marsh Mallow (now *Althaea officinalis* L.), a plant that Banks would have found both along the Thames valley and in damp places in Lincolnshire. This plant is well known to have medicinal and culinary uses and would almost certainly have been one of the herbs collected by the women paid by Banks to gather plants while he was at Eton (see number 16).

Note that the 1597 edition was dedicated to William Cecil, 1st Baron Burghley, whose gardens in London and Hertfordshire were supervised by Gerard, and that Banks's family was connected to the Burghley line through the marriage of his mother's aunt to Brownlow Cecil, 8th earl of Exeter. Moreover, Baron Burghley once owned the Revesby estate later purchased by Joseph Banks's great-grandfather in 1714, and Joseph Banks's parents, William and Sarah, were married in the chapel at Burghley House, Stamford, in September 1741.

HISTORIE OF PLANTS.

and if a man be first annointed with the leaues stamped with a little oile, he shall not be stung at all as *Dioscorides* saith.

The decoction of Mallowes with their rootes drunken, are good against all venome and poison, B if it be incontinently taken after the poison, so that it be vomited vp againe.

The leaues of Mallowes boiled till they be soft and applied, do mollifie tumours and harde swel- C lings of the mother, if they do withall sit ouer the fume thereof, and bathe themselues therewith.

The decoction vsed in glisters is good against the roughnes and fretting of the guts, bladder, D and fundament.

The rootes of the Veruaine Mallowe do heale the bloudie flixe and inward burstings, if they be E drunke with wine and water, as *Dioscorides* and *Paulus AEgineta* testifie.

Of Marshe Mallowe. Chap. 338.

✱ *The kindes.*

There be diuers sorts of Marsh Mallowes, differing very notably as shall be declared.

1 *Althæa Ibiscus.*
Marsh Mallowe.

2 *Althæa palustris.*
Water Mallowe.

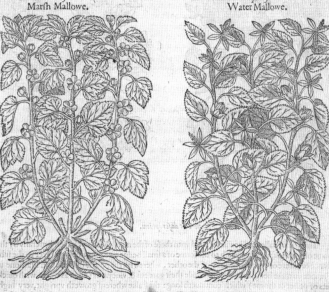

✱ *The description.*

1 MArsh Mallowe is also a certaine kinde of wilde Mallowe: it hath broade leaues, small toward the point, soft, white, and freezed or cottoned, and slightly nicked about the edges: the stalkes be rounde and straight, three or fower foote high, of a whitish graie colour: whereon do grow flowers like vnto those of the wilde Mallowes, yet not red as they are, but commonly white, or of a very light purple colour out of a white; the knoppe or round button wherein

3 & 4 Portrait medallions of Joseph Banks and Daniel Solander

By Wedgwood & Bentley. Modelled by John Flaxman, 1775.

BANKS
The British Museum, London. 1887,0307,I.61. Height 10.06, width 7.38 and depth 1.28 cm. Jasper ware (stoneware) oval cameo; white relief bust in profile, left side; moulded and applied to deep blue ground; 'BANKS' impressed beneath profile; gilt metal frame; maker's mark on reverse.

SOLANDER
The British Museum, London. 1909,1201.141. Height 8.4 cm. Jasper ware oval cameo; white relief bust in profile, right side; moulded and applied to deep blue ground; 'SOLANDER' impressed beneath profile; gilt metal frame; maker's mark on reverse.

Josiah Wedgwood manufactured ornamental and useful wares with great commercial success, and experimented endlessly to improve and widen the range of his products. Based in Staffordshire, he was the master potter of his age, and his plates, cups, saucers, pots, vases and medallions were much in demand both in England and abroad. Many of these were made using techniques that Wedgwood himself developed or refined – creamware, black basalt and jasper – and were decorated with glazes and patterns that were highly fashionable and much admired for their elegance and finish. Wedgwood's designs were mainly neo-classical in style, in keeping with the prevailing taste of the age. He promoted his goods inventively using showrooms, illustrated catalogues and the presentation of samples to potential customers such as Queen Charlotte, to whom he sent a cream-coloured earthenware tea set, thereafter successfully marketing this product as Queen's Ware. He also sought by various means to increase the speed and efficiency of production in his factories, and to improve the means of transporting goods to market. More than simply a leading potter and an inventor, Wedgwood was very much a pioneering modern industrialist.

Portraiture of all kinds grew in popularity in England during the second half of the eighteenth century. With the introduction of Wedgwood's jasper medallions it was possible to make cameo portraits of eminent persons quickly and in sufficient numbers to meet demand. The portraits were produced in bas-relief and then fixed to a coloured medallion before being fired. Such medallions were less expensive than most other forms of portrait, thereby enabling a greater number of people to afford them. Some were produced to flatter patrons or as private commissions, while many featured popular or edifying figures for wider public consumption. The first portrait medallions by Wedgwood were produced in basalt ware in 1771. However, with the introduction of jasper in the mid-1770s sharper detailing could be worked into the white bas-relief portraits. By this time Wedgwood was in partnership with Thomas Bentley and the famous Etruria factory where all their ornamental wares were manufactured was fully operational. The English worthies featured in portrait medallions produced by Wedgwood and Bentley date back as far as Chaucer and hundreds of these portraits were made. In this pair, Banks and Solander, both personally known to Wedgwood, are shown in head-and-shoulders profile on classic blue-and-white jasper medallions. Wedgwood employed many skilful modellers, foremost among them being the sculptor John Flaxman jnr, who started work for Wedgwood in 1775. Both the present portraits are by his hand, dating to that year.

Daniel Solander was a Swedish botanist, a favourite pupil of the great Carl Linnaeus at Uppsala University and the close friend and travelling companion of Joseph Banks on the *Endeavour* voyage. At university Solander distinguished himself under the tutelage of Linnaeus, who treated him almost as a son. Solander went on collecting expeditions in 1753 and 1755, visiting districts around the northern shores of the Gulf of Bothnia, including that of his birthplace at Piteå. Before completing his degree he travelled to England for what was intended to be a temporary visit, arriving in June 1760. In England he was welcomed by British naturalists, particularly the zoologist John Ellis, and he actively promoted his master's sexual classification system among them. He also sent back seeds and plants to the Hortus Upsalienis. He did not, however, return home to Sweden.

England, and particularly London, offered excellent opportunities for the study of natural history and a social life that the gregarious Solander evidently enjoyed. He declined a professorship at St Petersburg secured for him by Linnaeus, and then the offer of his master's own chair at Uppsala, which was in the gift of its incumbent. This and the marriage in 1764 to another man of Linnaeus's eldest daughter, whom Solander had once hoped to wed, brought contacts with Linnaeus to a virtual close. In 1763 Solander was employed to catalogue the natural history collections displayed in the public rooms of the British Museum, and in 1765 he started cataloguing the plants contained in the Sloane herbarium. He entered his descriptions and notes on slips of paper 6 inches by 4 inches in size, and thereby formed his manuscript slip catalogue, the slips being arranged according to the Linnaean system. These were then filed in small boxes that became known as 'Solander cases'. The cases gave their name to Solander boxes, containers that are

3

4

today used in archives and libraries for the storage and preservation of old or finely bound books, manuscripts, maps, prints and other documents. Solander boxes are in the form of a book constructed of wood or thick card and have a hinged lid connected to a base. Solander later used his slips to describe collections made on Banks's voyages, most notably those gathered on *Endeavour*, apparently with a view to producing a new edition of Linneaus's *Systema Naturae*, but this work was never finished.

It was probably at the British Museum in 1764 that Solander first met Joseph Banks, and they formed a strong friendship. Both men were devoted to botany, each was at home in London's learned circles and together they would promote the use of Linnaeus's classificatory system across a period when travel and collecting abroad became global in range and scope. Solander was in the vanguard of the Swedish disciples of Linnaeus who employed this system on foreign expeditions, becoming the first Swede to circumnavigate the world after Banks invited him to join the *Endeavour* mission in 1768. A fine naturalist, he contributed greatly to the systematic description and classification of the plant and animal specimens that were gathered during the voyage. He also accompanied Banks on a trip to Iceland in 1772 following abortive plans to join Cook on his second Pacific mission. Once settled back in London, Solander divided his time between work on Banks's collections and those at the British Museum, where in 1773 he was promoted to the post of under-librarian or, in modern parlance, 'keeper' of the department of Natural and Artificial Curiosities. He was a regular visitor to Banks's London residences at New Burlington Street and then Soho Square, where he acted as the first of Banks's librarian curators, the next being the Swede Jonas Dryander and, after him, the Scottish botanist and voyager Robert Brown. Solander was a key figure in the preparation of Banks's planned *Florilegium*, a fourteen-volume work comprising the new plant species collected during the *Endeavour* voyage, to which both Banks and Solander would have put their names as authors had it been published. Indeed, Solander's premature death in 1782 at Soho Square from a cerebral haemorrhage was doubtless a contributory factor in the failure to publish this work. It caused, too, the usually stoic Banks to write a moving tribute to his deceased friend (see Chambers, ed., *Select Letters*, letter 23).

Much of Solander's work remained unpublished during his lifetime, which served to obscure the quality and extent of his contribution. However, the *Florilegium* finally appeared in full in the 1980s, and Solander's British Museum catalogue is now recognized as a formidable work of scholarship in its own right. He was, moreover, responsible for much of the scientific content of the first edition of the *Hortus Kewensis*, a catalogue of the plants cultivated at Kew Gardens, although not credited as such in that work. Recent publication of Solander's correspondence and a biography of him by Dr Edward Duyker have helped to increase understanding of the importance of this previously underrated naturalist. Solander was an intelligent, genial and pioneering naturalist, who carried the Linnaean classficatory system into new fields of natural history collecting at a time when major discoveries were being made of lands previously unknown to European science. He is commemorated in the names of many plants, including *Banksia solandri* R. Br.

5 Portrait, Sydney Parkinson

The Natural History Museum, London. 4264. 28 × 18 cm. Oil on board.

This is an oil painting, possibly a self-portrait, of Sydney Parkinson. Parkinson was born in Edinburgh to Quaker parents, but his father, a brewer, died when he was still young, leaving the family in poverty. After finishing school Parkinson was apprenticed to a woollen draper, perhaps following in the footsteps of his brother Stanfield, who went on to become an upholsterer. Parkinson's talents as a natural history artist were already apparent, and, though the link is not confirmed, it may be that he studied under William de la Cour, an artist of French descent who in 1760 opened in Edinburgh the first publicly maintained art school in Great Britain. De la Cour was a talented painter who worked in a variety of fields, including portraits and landscapes, and he even produced a well-executed illustration of some locusts that apparently fell in England in 1748, which was engraved for sale. Parkinson later excelled in depicting insects and other small invertebrates, but it is his scenic work on the *Endeavour* that most clearly shows a sense of artistic design perhaps attributable to early training under the Frenchman in Edinburgh. At exhibitions there or in London, possibly at the Free Society of Artists where he exhibited flower paintings in 1765–6, Parkinson could also have studied contemporary landscapes and other works that influenced his subsequent approach in the Pacific. His primary artistic focus remained, however, natural history and within it the illustration of plants in particular.

Following a family move to London, Parkinson was engaged by a fellow Scottish Quaker, James Lee, to draw plants at the Lee and Kennedy Vineyard Nursery and to instruct his daughter. In 1767 Lee introduced Parkinson to Banks, for whom the young illustrator painted watercolours of zoological specimens from Banks's expedition to Labrador and Newfoundland of the previous year, as well as copying paintings of Indian mammals and birds belonging to Gideon Loten, a former Dutch governor of Ceylon (Sri Lanka) then living in London. Though Banks reserved work on a limited number of his Newfoundland plants for the eminent and by this time ageing German painter Georg Dionysius Ehret, contact with a figure of Ehret's stature would clearly have been beneficial to Parkinson's development. The Parkinson illustrations from Banks's Newfoundland expedition are held at the British Museum along with animal drawings by a number of other artists that Banks employed over the years, amounting to more than 500 items, while those by Ehret are at the Natural History Museum, London. The zoological specimens from the expedition were mostly dispersed or lost, but many plant specimens do still survive at the Natural History Museum and probably also in collections on the Continent. For an authoritative survey of Banks's Newfoundland expedition, see A.M. Lysaght, *Joseph Banks in Newfoundland and Labrador, 1766: His Diary, Manuscripts and Collections* (London, 1971).

Capable and diligent, Parkinson was recruited by Banks to join the party that he took on the *Endeavour* voyage at a total cost, the zoologist John Ellis said, of £10,000. Parkinson was to produce the natural history illustrations for Banks. Landscapes and figures were to be the work of Alexander Buchan, but his death at Tahiti from a fit meant that these tasks also fell to Parkinson, who was by then fully occupied with the burgeoning natural collections. Fortunately, Banks's secretary, a Swede called Herman Diedrich Spöring, was able to assist in the additional duties being shouldered by Parkinson, who all but gave up producing finished plant and animal drawings during the Pacific stage of the voyage, and instead rapidly laid down sketches annotated with details as to their colouring for later completion. It should be added, too, that between them Buchan, Spöring and Parkinson were responsible for all the coastal profiles drawn on the voyage bar one. This was a considerable workload, but the heaviest burden in terms of visual art fell on Parkinson. Thus Banks wrote, undoubtedly of Parkinson, not long after departing the aptly named Botany Bay: 'This evening we finishd Drawing the plants got in the last harbour, which had been kept fresh till this time by means of tin chests and wet cloths. In 14 days just, one draughtsman has made 94 sketch drawings, so quick a hand has he acquird by use' (*Banks's Journal*, vol. 2, p. 62). Like Buchan, however, neither Parkinson nor Spöring survived the mission, both dying in January 1771 within two days of each other from fevers contracted amid the filth of Batavia (Jakarta) as *Endeavour* made its way home.

Parkinson behaved exceptionally well throughout a long and demanding voyage, and during that time

far exceeded Banks's expectations in terms of the quantity and range of work that he produced. He drew more than 1,300 illustrations, largely of plants, but also of marine creatures, some birds and insects, as well as important drawings of figures and landscapes. These provide a visual record of the discoveries made and the places visited on *Endeavour*, many illustrated by him for the first time in the history of Western science. Parkinson also collected vocabularies, just as Banks did, and he wrote a lively and observant journal account of the voyage. Although not today counted one of Britain's foremost botanical illustrators in this period (among whom the Bauer brothers are thought to be preeminent – Ferdinand Bauer followed in Parkinson's footsteps as natural history artist on the *Investigator* mission to Australia, 1801–5), Parkinson possessed considerable flair and grace in his botanical work, and his overall skill matured rapidly during an arduous mission. Had he survived to complete his expedition illustrations, and to continue to develop as he was doing aboard *Endeavour*, it seems likely that he would have achieved comparable standing to the leading natural history artists of his day and certainly a prodigious career output. Of the Parkinson illustrations that survive in various repositories, the main concentrations are to be found in London, at the Natural History Museum and the British Library, the former holding much of the natural history and the latter the ethnographic works that he drew on *Endeavour*. The British Museum also holds Parkinson illustrations formerly belonging to Banks, chiefly of different animals completed before this voyage.

Yet Parkinson's *Endeavour* artwork remained unpublished, since the 14-volume *Florilegium* that Banks planned on returning to England was never finished. Using the illustrations and specimens that were available, however, Banks advanced the work a considerable way in concert with Solander and a succession of paid artists and engravers (see numbers 122–6). No less regrettably, there was an ugly disagreement with Stanfield Parkinson over the publication of his late brother's voyage account. As Parkinson's employer Banks took control of the artist's work and so access to it could only be gained through him. While he lay dying Sydney granted James Lee perusal of the account, but confusion over whether this included full ownership of it only served to fan increasingly heated feelings. Acrimonious exchanges took place between Stanfield and Banks, with Stanfield insisting that he and his sister were being unfairly deprived of their legal right to their brother's papers under a will made before he departed England. Stanfield also contended that various drawings and artefacts were owed to the Parkinson family. Banks was not the sort of person to take lightly being crossed or accused of dishonest dealings, and Stanfield was an unstable character who ultimately went insane and died in St Luke's Hospital for Lunatics, London. The preconditions for a clash were all there, and it is perhaps no surprise that one duly took place. One point is clear, however. Banks was at this time preoccupied with planning for another Pacific mission under Cook. He therefore surrendered his own journal to be incorporated with Cook's in the official *Endeavour* voyage account edited by Dr John Hawkesworth, and only reluctantly agreed to supervise engravings of illustrations made by his own artists for that work. It is unlikely that he harboured any serious concerns in respect of the production of Hawkesworth's account and still less so of publishing a narrative himself.

The respected Quaker physician and botanist John Fothergill, a friend to the Parkinson family, attempted to mediate in the dispute. He helped secure an agreement whereby Banks gained uncontested ownership of Sydney's papers and any specimens wanting in Banks's own collections for £500. Banks had already determined to pay the outstanding salary owed to Sydney as well as a small consideration for any clothing and minor items that were not brought back from the voyage, a sum amounting to not much more than £151. This enhanced offer was intended finally to settle matters. Banks also loaned Sydney's journal notes to Stanfield to read (a complete journal thought to have existed was not located), but under the condition that he make no improper use of them. Feeling, nevertheless, that he had been duped regarding some residual objects from his brother's collections that were allegedly retained by Banks, Stanfield commenced arrangements for publication of his brother's account. Hawkesworth managed to obtain a temporary injunction against Stanfield to prevent the forthcoming official account being pre-empted, but following its appearance in June 1773, and lifting of the injunction, Stanfield published a single-volume account, including in it a preface denouncing Banks's conduct. A hack writer, Dr William Kenrick, ghosted the preface for a fee and assisted in production of the work, entitled *A Journal of a Voyage to the South Seas in his Majesty's Ship, The Endeavour* (London 1773). Appalled at this, Fothergill purchased some 400 remaining copies after Stanfield's death, and obtained the rights to the book intending to reissue it with a measured counter to Stanfield's controversial preface. He died before he could complete the task, but his friend John Coakley Lettsom fulfilled this plan in 1784 in what was effectively a second edition.

Eventually, in the 1980s, Alecto Historical Editions lavishly printed 100 sets of the 738 extant engraved copper plates of Parkinson's illustrations, as prepared by Banks for his *Florilegium*. The plates are held at the Natural History Museum, London, along with Parkinson's botanical sketches and completed drawings. The latter were formerly bound in eighteen volumes, which also included the work of the artists employed by Banks in London to complete Parkinson's draft sketches for engraving, along with a set of original uncoloured proof prints from the plates. In addition, there were three further volumes containing 299 zoological drawings and paintings by Parkinson, Buchan and Spöring, of which 268 are by Parkinson. These volumes were all disbound as part of work on the collection in the 1980s. Their contents and Parkinson's edited account remain extremely valuable records of his contribution to this seminal voyage – as also of the complexities involved in publishing its results.

Background to the Endeavour *Voyage*

JOHN GASCOIGNE
University of New South Wales

A NUMBER OF DIFFERENT IMPULSES COMBINED to propel the *Endeavour* to the distant Pacific on its epochal voyage of 1768 to 1771. The most overt reason for the voyage was scientific – participation in the worldwide observation of the transit of Venus of 1769. It was an astronomical event which fascinated all of Europe, and leadership in planning for observing it came, to Britain's embarrassment, from the French. This made it all the more important for Britain not to be seen to be falling behind – thus facilitating the royal largesse that made the *Endeavour* voyage possible. Why did the transit of Venus of 1769 loom so large in the European consciousness? In part because, like all the planet's transits, it came as a pair eight years apart with an earlier one in 1761 (the last such transit had been seen in 1639). In that sense the European scientific community had had a chance for a dress rehearsal, with 1769 being the occasion to learn from the lessons of 1761.

One lesson had been the desirability of a vantage point in the South Seas so that the overall accuracy of the observations made around the globe would be enhanced by a Southern Hemisphere perspective. Such a desideratum was an instance of the global approach the transit prompted, with observations being made around the world the better to determine more accurately the distance from the Earth to the Sun. This brought scientific advantage but it was also hoped that it might yield practical navigational fruit as sailors commonly determined their position relative to the Sun and other heavenly bodies. Then, as now, scientific institutions such as Britain's Royal Society, the chief promoter of the expedition, were anxious to persuade government that utilitarian as well as scientific advantages could be gained from the expenditure of state money on research.

Such considerations prompted the Council of the Royal Society on 15 February 1768 to send a memorial to King George III arguing forcibly 'That the passage of the Planet Venus over the Disc of the Sun, which will happen on the 3rd of June in the year 1769, is a Phaenomenon that must, if the same be accurately observed in proper places, contribute greatly to the improvement of Astronomy on which Navigation so much depends'.[1] It was an issue that this august body had been pondering for some time, giving particular attention to the possibilities of a point of observation in the southern seas. Thus, in November 1767 its 'Committee for the Transit' had argued for the desirability of sending two observers to that part of the world, suggesting as locations the but dimly known islands of the Marquesas (sighted by the Spaniard Mendaña in 1595) or the Rotterdam or Amsterdam Islands (the Tongan islands of Nomuka and Tongatabu, sighted by the Dutch Tasman in 1643).[2] In the event, the location chosen for the observations was Tahiti, which had been added to the European map of the world by Samuel Wallis in 1767. His return to England in May 1768 came just in time to reorientate the *Endeavour* to the island that was to shape European perceptions of an Edenic South Seas (see number 6).

As the Royal Society's references to these long distant voyages of Mendaña and Tasman suggest, the geography of the South Seas had long been a source of bewildered curiosity. The area became ever more tantalizing as Europe in the late eighteenth century focused increasingly on mapping the globe and shining the light of science on its dark corners. The growing pace of trade, along with increasing scientific self-confidence, drove European nations to more and more distant parts of the Earth, in the hope of commercial or strategic advantage along with the prestige that scientific knowledge brought with it in the Age of Enlightenment. The planning for the *Endeavour* voyage reflected this quest to map more fully the only sketchily known southern Pacific. Predictably, the instructions for the voyage issued by the Admiralty to Cook in July 1768 began with the most apparent and publicly proclaimed reason for the dispatch of the *Endeavour* from Britain in the following month. They recounted how the ship had been fitted out 'for receiving such Persons as the Royal Society should think fit to appoint to observe the Passage of the Planet Venus over the Disk of the Sun on the 3rd of June 1769' and that the appointed place for these observations was 'King Georges Island [Tahiti] lately discover'd by Captn Wallis' – though with contingency plans for some other site in the same area if that were needed (see number 10).

Secret additional instructions, however – which Cook was to open after the observations were completed – revealed plainly the wider purposes that the Admiralty hoped this scientifically prompted voyage might also serve. Its preamble referred to the growing European determination fully to map the globe, arguing at the outset that, 'Whereas the making Discoverys of Countries hitherto unknown, and the Attaining a Knowledge of distant Parts which though formerly discover'd have yet been but imperfectly explored, will redound greatly to the Honour of this Nation as a Maritime Power'. Particularly prompting the Admiralty's investment in the voyage was the hope that there was, indeed, as had been speculated about since the time of Ptolemy, a Great Southern Land to balance the great land masses of the Northern Hemisphere: 'Whereas there is reason to imagine that a Continent of great extent, may be found to the southward of the Tract lately made by Captn Wallis'. The possible whereabouts of this missing continent were exceedingly vague, however, and the instructions that Cook received were therefore expansive: 'You are to proceed to the southward in order to make discovery of the Continent above-mentioned until you arrive in the Latitude of 40° .... But not having discover'd it or any Evident signs of it in that Run, you are to proceed in search of it to the Westward between the Latitude before mentioned and the Latitude of 35° until you discover it, or fall in with the Eastern side of the Land discover'd by Tasman and now called New Zeland.' Should he discover this illusive land mass, he was to map it as fully as possible, an endeavour which extended to making as full a study of the flora, fauna and minerals, as circumstances permitted, with a view to their possible commercial value. A study of the human population was also important and, with the possiblity of colonization in mind, he was 'with the Consent of the Natives to take possession of Convenient Situations in the Country in the Name of the King'.[3]

The instructions issued to Cook for the *Endeavour* voyage mingled, then, a range of considerations – scientific, commercial and imperial. Attempts at drawing the South Seas into the European map of the globe had been made fitfully and partially since the time of Magellan's epochal circumnavigation of the Earth from 1519 to 1522. Some Spaniards had followed Magellan's lead with the result that Alvaro de Mendaña reached the Marquesas and the Solomons with his voyages of 1567–9 and 1595–7, and Pedro de Quirós, formerly Mendaña's pilot, landed on the island of Espritu Santo in Vanuatu in 1606 (claiming that it was the Great Southern Land) – though the location of these archipelagos had only been hazily determined. Once established in their East Indies empire at Batavia (Jakarta) in 1619 the Dutch later sent out Tasman to reconnoitre neighbouring territories, with the result that in 1642–3 he added Van Diemen's Land (Tasmania), New Zealand and the Tongan islands to the map and in 1644 provided further detail on the north of Australia. For the commercially minded Dutch, however, what Tasman found seemed to offer little hope of riches so his discoveries were not followed up. By Cook's time, however, the South Pacific was coming into much clearer focus in the overall world-view of the European imperial powers. Up to the end of the Seven Years' War (1756–63) the great preoccupation of the two major European super powers had been dominance in North America. The Treaty of Paris (1763) saw that struggle ending in British supremacy, transferring the rivalry to other parts of the globe. Britain began to be much more active in seeking out possibly advantageous sites in the South Pacific. John Byron's largely fruitless expedition across the Pacific in 1764 was followed by that of Samuel Wallis in 1766–8 with its epochal encounter with Tahiti.

Great though its achievements were to be, the *Endeavour* voyage built on the innovations of these two earlier voyages and particularly that of Wallis. The continuity between the voyages of Bryon and Wallis was underlined by the use of the same ship, the *Dolphin*, on which had served Philip Cartaret, captain of the *Swallow*, Wallis's companion vessel. On the *Endeavour* voyage there were a number of old hands from the *Dolphin*: Lieutenant John Gore had served under both Byron and Wallis, and Cook's master Robert Molyneux and two of his mates, Richard Pickersgill and Charles Clerke, had served under Wallis. A further point of continuity was the presence on board the *Endeavour* of a goat which had also circumnavigated the Earth on board Wallis's *Dolphin*. Many of Cook's methods for dealing with scurvy had already been pioneered on Samuel Wallis's 1766–8 voyage. The disease was one of the great scourges of Pacific voyages, on which it was difficult to renew supplies of fresh food – for, to put it in modern terms, the body cannot manufacture its own supply of Vitamin C. Wallis had returned without any loss of life from scurvy, a major achievement for a circumnavigatory voyage. As Cook was to do, he had used sauerkraut and, wherever possible, replenished the ship with fresh food. Like Cook, Wallis had tried to reduce exhaustion among his men by instituting three watches a day rather than the customary two; he had also ensured they were provided with warm clothing in high latitudes.[4]

The *Dolphin*, both under Samuel Wallis and on its previous voyage under John Byron, had in many regards been a site for experimentation for ways in which to make the vast distances of the Pacific more amenable to Europe incursion.[5] Dealing with scurvy was one such major issue, another was the use of distilled water, first on Byron's voyage and later on Cook's. An experiment which was tried on the *Dolphin* but not persisted with on Cook's voyage was the use of copper sheathing to prevent the ship being slowly eaten away by toredo worms in tropical waters. On Byron's expedition the *Dolphin* was only the second ship in the British navy to be fitted out with such sheathing. This, however, brought problems: the copper reacted with the iron fittings and it was not readily repaired in distant locations. The *Endeavour*, then, was coated not with copper but was 'sheathed and filled' – given an extra wooden skin over a layer of tar and matting intended to make the wood less attractive to the worms and then the outer surface virtually covered with wrought-iron nails with broad heads closely placed closely together.[6]

Most distinctively scientific was the increasing sophistication of the techniques used by the *Dolphin* (particularly under Wallis) for navigation. Greater accuracy in determining position at sea was made possible by the invention in 1757 of the sextant, which made calculating the angle between two objects much simpler and more accurate. This could be used to work out the angle between the two landmarks or, more routinely, the angle between the horizon and a celestial object, thus making the determination of latitude more exact. In skilled hands it could also be used to establish longitude by the laborious method of lunar differences – using the Moon as a reference point to determine its movement against a fixed backdrop of the stars.[7] With the appropriate tables it was then possible to calculate the difference from the Greenwich readings and thus determine the difference between local and Greenwich times – thus enabling an accurate measurement of longitude.

The accuracy and scope of such tables increased considerably over the course of the eighteenth century, helping to make farflung Pacific navigation and its accompanying mapping possible. On Wallis's *Dolphin* voyage the most up-to-date such tables were used, those produced by the astronomer royal Nevil Maskelyne in 1766, the very year in which the ship set off. Not only were Maskelyne's tables used, but also his method of lunar distance, which he had trialled on his voyage to St Helena to observe the 1761 transit of Venus. On the *Endeavour* voyage Cook, too, employed the method of lunar differences, though on the second and third voyages he also employed the newly invented chronometer (see number 11).

Determined not to be left behind, the French were also prompted by the voyages of Byron and Wallis to stake their claim in the Pacific. The most notable of their voyages before Cook was that of Louis Bougainville from 1766 to 1769, the highlight of which was a sojourn in Tahiti, which led to the French taking an even more idealized view of the attractions of that island than the British. This owed much to French Enlightenment speculation about 'the Noble Savage', prompting some of Bougainville's party (and especially his botanist, Philibert Commerson) to view Tahiti through rose- (or Rousseau-) tinted glasses. The rapid succession of Bougainville's and Cook's visits to Tahiti (in 1768 and 1769 respectively) underlines how such South Seas exploration represented a response to the end of Great Power rivalry in North America. Both powers had been active participants in the Seven Years' War, with Bougainville having been an aide-de-camp to General Montcalm, who died in the French defeat at Quebec in 1759. Cook, by contrast, had been an active participant in the English conquest of that French stronghold under General Wolfe, who also died in the battle. It was the Canadian theatre which made Cook's name and set him on the path to command an expedition to the South Seas.

Above all, what Canada had provided for Cook was rigorous training as a cartographer. These skills had been in large measure the result of an accidental meeting with the army engineer Samuel Holland after the British capture of the French citadel of Saint Louisbourg at the mouth of the Saint Lawrence River in 1758. Cook had received the customary basic training in nautical navigation, confirmed by his promotion in 1757 to the position of ship's master following an examination at Trinity House (the body charged with training pilots). What he learned from Holland, however, was that it was possible to apply to marine mapping some of the increasingly sophisticated techniques employed by land surveyors. Particularly important was the establishment of a reliable baseline and the use of triangulation and trigonometric readings from a series of landmarks to establish accurate coordinates. The use of such techniques was later to be further promoted by a meeting with another gifted land surveyor, James Des Barres, at Newfoundland in 1762. Meanwhile, the fruitful partnership of Cook and Holland had reached new heights of cartographical precision in 1759 with their joint 'New Chart of the River St Lawrence'. Its accuracy played a part in Wolfe's planning of the successful battle for Quebec, underlining how mapping could bring with it imperial advantage (see numbers 17, 18, 76, 104, 130 and 139).

With his reputation as a cartographer established, Cook played a part in another British victory with the mapping of the coast adjacent to St John's, Newfoundland, which had been recaptured by the French but was taken again by the British in 1762. Overall British victory in North America in 1763 still left Cook with plenty of mapping to do. From 1763 to 1767 Cook held the post of 'Surveyor of Newfoundland' as he accurately mapped the hitherto sketchily charted island, the better to incorporate it into the expanding British Empire. Given that bearings were often taken by reference to heavenly bodies, surveying and astronomy were closely linked. It was natural, then, that Cook observed closely an eclipse of the Sun from Newfoundland in 1766. This led to contact with the Royal Society, a body that would later loom large in his career. So impressed was the Society's foreign secretary John Bevis with the results recorded by Cook that he had them published in the *Philosophical Transactions of the Royal Society*, with the commendation that Cook was 'a good mathematician, and very expert in his Business'.[8]

When in 1768 the Royal Society considered who should command the voyage to observe the transit of Venus which it had set in motion, it was natural that Cook should be considered. He was not the first name put forward, however: that honour belonged to Alexander Dalrymple, a fellow of the Royal Society and an enthusiastic advocate of the view that there was a Great Southern Land to be discovered in the Southern Hemisphere. Dalrymple, however, demanded the command of the ship, which was too much for the Admiralty to contemplate, so, in April 1768, it went to one of the navy's own, James Cook. Assuming command of a Royal Naval vessel was a significant leap for this son of a Yorkshire agricultural labourer. Hitherto he had not yet passed the divide between officers and men, his rank being that of 'master', entrusted with sailing the ship but leaving the actual conduct of naval battle to the captain – a distinction which went back to the days when the chief form of naval combat was boarding a rival vessel. Being gazetted as an officer on 25 May (albeit only as a lowly lieutenant) was, then, a remarkable tribute to Cook's abilities and particularly those in the areas which were at the heart of this exceptional voyage, astronomy and cartography.

Appropriately, the ship that was to be under Cook's command was a Whitby collier, originally called the *Earl of Pembroke* but rechristened *Endeavour* by the navy. It was the sort of ship on which Cook had been apprenticed and on which he sailed before he joined the navy in 1755. The choice of ship and commander may, then, have been related – such a slow-sailing bulky vessel was, indeed, an unusual choice for the navy, though, as Cook showed, it had many advantages: it could carry copious amounts of stores, it had a low draft and so could come close in shore for mapping and it was readily beached for repairs, something which proved particularly important in June 1768, when *Endeavour* was nearly wrecked on the Great Barrier Reef, off what is now northern Queensland. So successful was the vessel that it was also the ship of choice on Cook's second and third great Pacific voyages in 1772–5 and 1776–80, the latter being brought home by Cook's subordinates after he died in Hawai'i in February 1779. The main vessel on both these voyages was the *Resolution*. A lesson that had been forcefully driven home after Cook's encounter with the Great Barrier Reef was that there should be a companion vessel, that on the second voyage being the *Adventure* and on the third the *Discovery* – both Whitby colliers (see numbers 7–9 and 114).

With the command of the *Endeavour*, then, Cook launched a career which formed a small but important niche in the great Royal Navy – that of a scientific explorer. Others were to follow in the same vein, looking up to Cook as the great exemplar, among them George Vancouver, who charted the Northwest Coast of America in 1791–4, and William Bligh, famous for the mutiny on the *Bounty* in 1789 but commander, too, of the *Providence* expedition (1791–3), which successfully brought the breadfruit plants from Tahiti to the West Indies – these having been unceremoniously thrown overboard by the mutineers on his previous expedition. Another naval officer who sought to make his name by scientific exploration was Matthew Flinders, who commanded the *Investigator* from 1801 to 1803 on its circumnavigation of Australia. There was something of an apostolic succession from Cook to these fellow scientifically minded naval personnel, since both Vancouver and Bligh served under Cook, and Flinders, in turn, served under Bligh. So great was Cook's reputation that when news reached England of Bligh's remarkable open-boat voyage to Timor in 1789 – a voyage which covered 3618 miles (5822 km) in six weeks, and during which Bligh and eighteen 'loyalists' were cast adrift by mutineers – a prominent politician of the day remarked to Lieutenant Burney (who had served under Cook): 'But what officers you are! you men of Captain Cook; you rise upon us in every trial!'[9]

Eventually, too, Cook's voyage was to serve as something of a template for the greatest of all scientific voyages, the voyage of the *Beagle* (1831–6) under Robert Fitzroy. The parallel with the *Endeavour* voyage was the closer since both carried gentlemen naturalists who financed their own voyages in return for the scientific riches the trip offered, Charles Darwin in the case of the *Beagle* and Joseph Banks on the *Endeavour*. Banks's presence on board was largely the work of the Royal Society, of which he was a member. For on 9 June 1768 the Society wrote to the Admiralty asking both that Tahiti should be the place chosen to observe the transit and that:

> Joseph Banks Esqr Fellow of this Society, a Gentleman of large fortune, who is well versed in natural history, being Desirous of undertaking the same voyage the Council very earnestly request their Lordships, that in regard to Mr Banks's great personal merit, and for the Advancement of useful knowledge, he also, together with his Suite, being seven persons more, that is, eight persons in all, together with their baggage, be received on board of the Ship, under the Command of Captain Cook.[10]

Banks's reputation, like Cook's, owed much to his service in Canada, though in a very different capacity. For Banks had earned his election to the Royal Society in acknowledgement of the work he had done as a naturalist on HMS *Niger* on its 1766 expedition around Newfoundland and Labrador. Though there is no evidence that Cook and Banks met at the time, they were both part of the consolidation of British control over the region after the British victory in the Seven Years' War. Banks's presence on board was, in the time-honoured tradition of the British establishment, due to his friendship with the commander of the voyage, his Eton chum Constantine Phipps. The voyage, with its combination of scientific and imperial goals, was something of a dress rehearsal for the *Endeavour* voyage, which in turn became a model for subsequent voyages of scientific exploration facilitated by the Royal Navy. Indeed, in 1781 Banks looked back proudly on the way in which he considered himself to have been 'the first man of Scientifick education who undertook a voyage of discovery & that voyage of discovery being the first which turned out satisfactorily to this enlightened age I was in some measure the first who gave that turn to such voyages, or rather to their Commander Capt Cook.'[11]

Banks's path to the role of naturalist was largely a self-made one. Born into a *nouveau* though very rich Lincolnshire landowning family, Banks had been given an education fitting his role as a gentleman – Harrow then Eton, followed by Christ Church, Oxford. He showed little enthusiasm for the classical education, then the mainstay of the curriculum, but embraced botany and natural history with enthusiasm, even while at school. There, he claimed in old age, he 'for want of more able tutors, submitted to be instructed by the women, employed in culling simples' (see number 2).[12] As a gentleman student at Oxford he was spared the tiresome necessity of sitting examinations or taking degrees, so again he followed his own interests in natural history. Since the Oxford professor of botany failed to give lectures, Banks imported at his own expense a private lecturer from Cambridge, Israel Lyons.

With the death of his father in 1761 and the inheritance of his Revesby estate, Banks was in a position to follow his own interests. Thus more and more he became part of the London world of learned clubs, making connections including a lifelong friendship with the Swedish Daniel Solander, who, from 1763, held a post at the recently founded British Museum, where he was engaged to catalogue their natural history collections. He no doubt owed this post to the high reputation of his great mentor Carl Linnaeus, who had taught Solander at the University of Uppsala. When Solander came to England in 1760 it was with the mission of promoting the Linnaean system of classification, and Banks was to be one of the most enthusiastic supporters of this form of ordering the diversity of nature. Indeed, Banks had planned a visit to the great Swedish naturalist but was diverted by the *Endeavour* voyage, on which he took with him Solander as a companion botanist (see number 4).

In some ways, then, the *Endeavour* voyage was another example of Linnaeus's wide global reach, as his students traversed the world endeavouring to bring order to the complex array of natural specimens, whether animal, vegetable or mineral, using the master's system of classification. Linnaeus, too, extended his investigations to the human world, with ethnological studies of the Lapps. His example was to be emulated by Banks, whose journal provides detailed anthropological discussions of the societies he encountered in the Pacific, much to the fascination of Europeans of the time. The close connection between Banks's presence on *Endeavour* and Linnaeus's work was remarked on by the naturalist and fellow of the Royal Society John Ellis in a letter to Linnaeus of August 1768 (the month in which *Endeavour* set off) acknowledging the wide reverberations of his influence: 'I must now inform you that Joseph Banks Esq., a gentleman of £6000 per annum estate, has prevailed on your pupil Dr Solander to accompany him on the ship that carries the British Astronomers to the newly discovered country in the South Sea [Tahiti]'. Ellis was particularly impressed by the lavish equipment that accompanied Banks and his party, which, along with Solander, consisted of another Swedish botanist, Herman Diedrich Spöring, two artists, Alexander Buchan and Sydney Parkinson, four servants and two dogs – a sizeable contingent in a ship with a complement of 94. 'No people,' wrote Ellis, 'ever went to sea better fitted out for the purpose of Natural History. They have got a fine library of Natural History; they have all sorts of machines for catching and preserving insects; all kinds of nets, trawls, drags and hooks for coral fishing, they have even a curious contrivance of a telescope, by which, put in the water you can see the bottom at a great depth, where it is clear … in short Solander assured me this expedition would cost Mr Banks £10,000.' And, concluded Ellis, 'All this is owing to you and your writings' (see numbers 15 and 16).[13]

The *Endeavour* voyage was, then, a manifestation of a cosmopolitan scientific impulse embracing natural history as well as astronomy. It was also part of the widening imperial reach of Britain as its commercial and strategic goals extended to the Pacific. With these mixed motives the South Pacific was brought into the European picture of the world, with consequences that endure today.

THE VOYAGE: AIMS AND ORGANIZATION

The Royal Society proposed a voyage to the South Seas as part of a collaborative international effort to observe the forthcoming transit of Venus on 3 June 1769. George III donated generously towards the expedition in response to a memorial from the Royal Society, and the Royal Navy provided a vessel and commander to carry an astronomer to the distant island of Tahiti, only recently visited for the first time by Europeans. This vessel was named *Endeavour*; its commander would be Lieutenant James Cook and the astronomer was an assistant at the Royal Observatory called Charles Green. Supplies and equipment were provided for a long voyage with scientific aims in astronomy, but a second mission priority was to search for the southern continent that many believed lay undiscovered somewhere in the southern oceans.

Instructions from the Admiralty therefore addressed not only observation of the transit but also the quest to find, chart and, if possible, claim such a geographical prize. It was a quest bound up both with the desire to explore the region's natural history and peoples and also with the imperial rivalries of Europe's leading maritime powers, for in these years the French also mounted their own expeditions to probe the vast Pacific. Joseph Banks used his contacts at the Royal Society and also within Admiralty circles to obtain a place on the mission as a supernumerary, and he took with him a team of illustrators and collectors as well as a great deal of equipment. Thus when HM Bark *Endeavour* set sail on 25 August 1768 it ventured into a Pacific world of astronomical, geographical and biological discovery.

6 'To the Kings most excellent Majesty The Memorial of the President, Council and Fellows of the Royal Society of London for improving Natural Knowledge', February 1768

The Royal Society of London. RS Misc. MSS V 39.
45.2 × 27.3 cm.

In the mid-eighteenth century the distances of planets and stars from the Earth were still unknown. To establish such measurements it was necessary to find the mean distance of the Earth from the Sun, a quantity now known as the Astronomical Unit (AU). It was understood that this value could be calculated from timings of the transit of the planet Venus across the face of the Sun. These observations had to be made from different points on the Earth's surface, which, for accuracy, had to be as far apart as possible. By combining the data from various transit observations conducted in these years, a distance equivalent to 153 million kilometres was obtained, which is within 2 per cent of the modern accepted value.

A memorial seeking royal support for a mission to the Pacific to observe the transit of Venus due on 3 June 1769 was drawn up by the Royal Society, to be presented to George III. It was signed at a council meeting on 15 February 1768. The president, James Douglas, Earl of Morton, was in the chair and fourteen other council members were present, among them Benjamin Franklin and the astronomer royal Nevil Maskelyne. Maskelyne had attempted in 1761 to observe the preceding transit of Venus in frustratingly poor weather conditions at St Helena, and it was he who prepared the astronomical instructions for the missions now being planned. George III gave £4,000 in response to the memorial, and the Royal Navy was ordered to supply a vessel and crew for a voyage to a suitable Pacific location.

Fortuitously, Samuel Wallis had just returned from a circumnavigation in HMS *Dolphin*, during which he was the first European to visit and describe what he called King George the Third's Island, afterwards known as Otaheite and now Tahiti, and his favourable reports regarding this island determined it as the destination for the forthcoming mission. James Cook, with his considerable experience as a navigator and surveyor, as well as his knowledge of astronomy, was appointed by the Admiralty to command the voyage to Tahiti. The Royal Society selected Charles Green of the Royal Observatory to be the mission astronomer. Cook would act as second observer of the transit. Other British expeditions were dispatched to Hudson's Bay and to Norway as part of an international effort to observe the transit, but the *Endeavour* mission is remembered as the most famous of those sent out due to the extensive discoveries it made of lands, peoples and natural history new to Europeans.

Earl Morton sent the voyagers his own advice on how to conduct themselves during the mission. Addressed from Chiswick on 10 August to Cook, Banks and Solander, his comments are modestly entitled 'Hints', but they reveal much about the aims and methods adopted by Europeans at this time in their quest for knowledge and territory. At the outset Morton stressed restraint in all dealings with local people, and respect for their rights as inhabitants of the lands to be visited. The voyagers should only use force in self-defence and, Morton argued, even if crewmen were killed by locals during any clash, attempts should afterwards be made to repair relations. Aware that there was considerable room for misunderstanding during first contacts, Morton recommended that signs of friendship be offered to build mutual confidence, and further that harmless demonstrations of the power of European weaponry be given in response to hostile threats in order to discourage actual fighting. Gifts, physical gestures, attempts to communicate through language and even music are all mentioned as means of achieving peaceful relations. Morton, himself an astronomer, was aware that poor weather might disrupt the transit observation, and so he suggested undertaking observations at more than one site, separated by a good distance, which Cook did. He speculated on the possibility of an inhabited continent unknown to Europeans lying in the South Seas, the search for one being the major secondary objective of this mission, and if it were found he felt that many new discoveries might be made there and, tellingly, that new commercial opportunities might open up.

Morton also outlined a series of observations of the natural landscape and its various inhabitants for the voyagers to address in a summary of what might broadly be regarded as a mode of scientific exploration that had, with the Royal Society's help, gradually emerged over the previous hundred years or so as an increasingly important influence in British travel and fieldwork. The description of new places, the collection of data from them and its sorting and interpretation in European centres were key aspects of such enquiries, with the resulting knowledge being deployed for a range of purposes intellectual and political. The present mission was to be no exception. Morton's suggestions, for he frames them only as such, indicate that to be of value any observations and collecting must be organized and systematic in approach. Consequently he lists in an ordered series some of the main things to look for when encountering new peoples, including their physical appearance, dress, tools and weapons, agriculture, economic systems, language, religion, government and social structure. Such a range is certainly reflected in Banks's journal of the *Endeavour* voyage as well as in the 'artificial' or man-made objects that were collected. Natural history is broken down by Morton into its main branches, comprising animals, plants and minerals. Less is said of plants since Banks and Solander were devotees of botany, but their commercial utility is highlighted, for example in the case of plants used in dyeing or medicine. With similar concerns in mind, Morton pointed out that precious metals and stones might be identified, but he stressed, too, the importance of obtaining as much information as possible about the immediate area in which any rocks were gathered and the local names for them. He felt that this might reveal more about their origin, including whether they resulted from volcanoes, and so deepen knowledge of natural history. Similar advice was offered in the case of animal specimens, to observe 'minutely' each in its habitat and seek its local name.

For all that his 'Hints' reflect a mix of priorities and assumptions underpinning the voyages of Pacific discovery launched with Cook, Morton's advice also incorporates a store of wisdom and insight that do him no little credit (see NLA MS 9/3/113–113h, and Chambers, ed., *Indian and Pacific Correspondence*, vol. 1, documents 2 and 12). For the inherent difficulties with which Banks and Cook grappled when trying accurately to report the voyage and its discoveries, see also Markman Ellis on 'Tails of Wonder', in Lincoln, *Science and Exploration in the Pacific*, esp. pp. 179–82. The unofficial advisory role here adopted by Morton as Royal Society president was later developed substantially by Banks when producing plans and instructions for numerous explorers and plant collectors.

That by neglecting to take the necessary precautions in due time, the passage of the Planet in the year 1761 was not observed in some places from whence the greatest advantages might have redounded to the improvement of Astronomy.

The Memorialists are humbly of opinion, that Spitzbergen, or the North Cape, in the higher northern latitudes; Fort Churchill in Hudson's Bay, and any place not exceeding 30 degrees of Southern latitude, and between the 140th and 180th degrees of longitude, West of Your Majesty's Royal Observatory in Greenwich park, would be proper stations for observing the ensuing Transit; to each of which places two Observers ought to be sent.

That a correct Set of Observations made in a Southern latitude would be of greater importance than many of those made in the Northern: But it would be necessary that the Observers who are to pass the Line should take their departure from England early in this spring; because it might be sometime before they could fix upon a proper place for making the Observation within the limits required.

That the Expence of having the Observations properly made in the Places above specified, including a reasonable gratification to the persons employed, and furnishing them with such Instruments as are still wanting, would amount to about 4,000 pounds, exclusive of the Expence of the Ships which must convey and return the Observers that are to be sent to the Southwards of the Equinoctial Line, and to the North Cape.

That the Royal Society are in no condition to defray this Expence; their Annual Income being scarcely sufficient to carry on the necessary business of the Society.

The

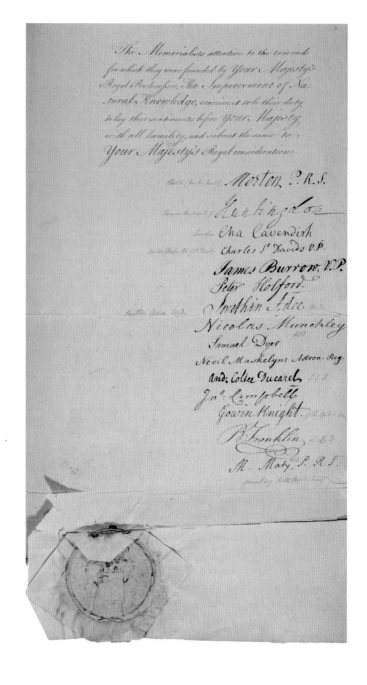

The Memorialists attentive to the true ends for which they were founded by Your Majesty's Royal Predecessor, The Improvement of Natural Knowledge, conceive it to be their duty to lay their sentiments before Your Majesty, with all humility, and submit the same to Your Majesty's Royal consideration.

Morton, P.R.S.
Willoughby
Cha Cavendish
Charles St Davids V.P.
James Burrow, V.P.
Peter Holford.
Inothin Idee
Nicolas Munckley
Samuel Dyer
Nevil Maskelyne Astron. Reg.
And: Coltee Ducarel
Jno. Campbell
Gowin Knight
B Franklin
M. Maty, S.R.S.

7 HM Bark *Endeavour* model

By A.J. Berry-Robinson, 2000.

Private collection. Length 82, height 72 and width 34 cm. Scale 1:55. Wood, cloth and brass or copper.

Built at Whitby in June 1764 by Thomas Fishburn, the *Earl of Pembroke* was a collier operating in the coal trade along the north-east coast of England to London. It measured just 32 metres in length and 8.9 metres in width. This vessel was purchased by the Admiralty in March 1768 and renamed *Endeavour*. Stable and with good stowage, she had a shallow draught making her suitable for survey and other work close inshore as well as easy to beach for repair, as happened on the east coast of Australia (then New Holland; see number 114). Alterations according to James Cook's specifications were made at Deptford dockyard in order to prepare her for a voyage to the Pacific.

The present model is constructed in the Navy Board style, fully rigged and under sail. Navy Board style means that the hull is not entirely planked so their lordships could see the volume of the hold. The model hull is made of Brazilian lemon wood as are the blocks. The masts and spars are made of English lime. The fittings are brass or copper. The flag represents the Queen Anne Union design.

8 Ship plan showing the decks of HM Bark *Endeavour*

By Adam Hayes.
The National Maritime Museum, Greenwich, London. ZAZ6593. 40.6 × 47.7 cm. Scale 1:96.

In July 1768 Adam Hayes, the master shipwright at Deptford dockyard, oversaw the alterations to *Endeavour* in order to prepare the vessel for its voyage to the Pacific. Originally a Whitby-built collier, this vessel was sturdy and had a large cargo capacity, both important qualities for a long mission into the Pacific requiring much equipment and supplies. A third deck was squeezed into the ship to house extra cabins and storerooms. Note the bluff bow and broad hull. Note, too, the modifications to accommodate Banks and his party. They would share the great cabin with Cook, using it for sorting, illustrating and preserving natural history specimens, and for writing their notes and journals. Cook learned to sail on the tough North Sea and Baltic coal trade routes before he joined the Royal Navy in 1755, and so was very familiar with vessels of this design, the best, he thought, for exploration. Accordingly, this would be the vessel type chosen for his next two Pacific expeditions.

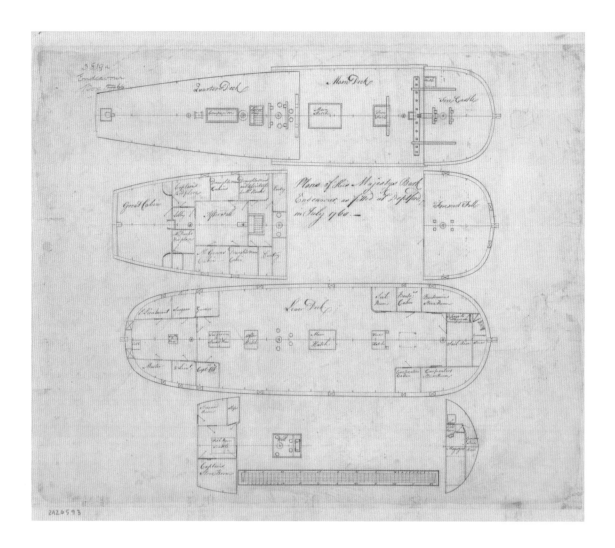

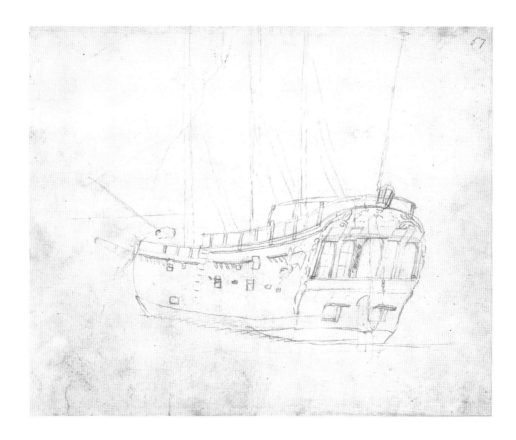

9 Sketch of HM Bark *Endeavour*, port side view

By Sydney Parkinson.

The British Library, London. BL Add. MS 9345, f. 57. 18.4 × 23.5 cm.

This unfinished pencil sketch by Sydney Parkinson of *Endeavour*, possibly drawn at Endeavour River in present-day Queensland, is one of three Parkinson sketches thought to have been drawn at about the same time, the other two being at BL Add. MSS 9345, ff. 15 verso and 50. A pencil note on folio 50 indicates that this sheet had an earlier illustration on it that was made at Huahine in the Society Islands, and afterwards erased to sketch *Endeavour*. Since the expedition was at Huahine in mid-July 1769, the folio 50 sketch must have been made after that date.

It was during an enforced stay at Endeavour River after almost being wrecked on a reef in June 1770 that *Endeavour* was repaired (see number 114). She was careened to perform this task, so the trio of sketches mentioned above may not show her there at all. Moreover, in the present sketch the anchors are visible and these would have been landed before the vessel was beached. Nevertheless, she appears to sit high out of the water, as would have been the case when all her stores were ashore prior to or just after being careened. It may therefore be that the sketches were made at Endeavour River, but not while the ship was on land. If so, they may be preparatory drafts for a lost Parkinson illustration of the ship on the river bank being repaired, which after the mission was engraved for publication in John Hawkesworth's edition of the official voyage account, vol. 3, plate 19.

Original depictions of *Endeavour* are rare. A further Parkinson sketch of the ship in heavy seas, perhaps in Le Maire Strait, exists at BL Add. MS 9345, f. 16 verso, although the strait was encountered much earlier in the mission (see number 27). The most famous image of the vessel, leaving Whitby Harbour in 1768 while still named the *Earl of Pembroke*, is that by Thomas Luny, in oils on canvas, now held at the National Library of Australia, Canberra. In 1768 Luny was just a boy, and since his first paintings date from 1777 this painting cannot be of the vessel actually sailing from Whitby. The painting in fact dates to about 1790.

By &c.ª

Whereas we have, in obedience to the King's Commands, caused His Majʸ Bark the Endeavour, whereof you are Commander, to be fitted out in a proper manner for receiving such Persons as the Royal Society should think fit to appoint to observe the passage of the Planet Venus over the Disk of the Sun on the 3ᵈ of June 1769, and for conveying them to such Place to the Southward of the Equinoctial Line as should be judged proper for observing that Phænomenon; And whereas the Council of the Royal Society have acquainted us that they have appointed Mr Charles Green, together with yourself, to be their Observers of the said Phænomenon, and have desired that the Observation may be made at Port Royal Harbour in King George's Island lately discovered by Capᵗ Wallis in His Majesty's Ship the Dolphin, the Place thereof being not only better ascertained than any other within the Limits proper for the Observation, but also better situated, & in every other respect the most advantageous; You are hereby required & directed to receive the said Mr Charles Green, with his Servant, Instruments & Baggage, on board the said Bark, & proceed in her, according to the following Istructions.

You are to make the best of your way to Plymouth Sound, where

we

10 Instructions for the *Endeavour* voyage from the Admiralty, London, 30 July 1768, signed by Lord Edward Hawke, Sir Peircy Brett and Lord Charles Spencer

National Archives, Kew. Adm. 2/1332, ff. 160–9. 41 × 26.5 cm.

Cook's Admiralty instructions for the *Endeavour* mission came in two parts, both marked secret in the National Library of Australia manuscript version, NLA MS 2 ff. 54–9, but so marked only on the second part of the National Archives version. Here is the first folio side of the first part of the latter version. The first part of Cook's instructions referred to his primary task of observing the transit of Venus at Tahiti. The second comprised sealed additional instructions to be followed once the transit of Venus had been observed and Cook had departed the Society Islands (see *Cook's Journals*, vol. 1, pp. cclxxix–cclxxxiv). The first part opens as follows: 'Whereas we have, in obedience to the King's Commands, caused His Maj$^{ys.}$ Bark the Endeavour, whereof you are Commander, to be fitted out in a proper manner for receiving such Persons as the Royal Society should think fit to appoint to observe the passage of the Planet Venus over the Disk of the Sun on the 3$^{d.}$ of June 1769, and for conveying them to such Place to the Southward of the Equinoctial Line as should be judged proper for observing that Phænomenon; And whereas the Council of the Royal Society have acquainted us that they have appointed M$^{r.}$ Charles Green, together with yourself, to be their Observers of the said Phænomenon, and have desired that the Observation may be made at Port Royal Harbour in King George's Island lately discovered by Cap$^{n.}$ Wallis in His Majesty's Ship the Dolphin, the Place thereof being not only better ascertained than any other within the Limits proper for the observation, but also better situated, & in every other respect the most advantageous; You are hereby required & directed to receive the said M$^{r.}$ Charles Green, with his Servant, Instruments & Baggage, on board the said Bark, & proceed in her, according to the following Instructions.' The remainder of this first part relates to the course to be taken and the surveying work that Cook was to undertake during the voyage.

The sealed additional instructions required Cook to look for a supposed unknown southern continent, *Terra Australis Incognita*. Such a discovery would, it was hoped, yield riches and fame to the nation that achieved it. Cook was ordered to survey any such land that he found, to observe its coast, natural resources and inhabitants, and to establish friendly relations with any local people with a view to securing an alliance and their consent to claim for the Crown what are termed 'convenient Situations in the Country'. If uninhabited, any continent was to be claimed by Cook. In pursuit of these objectives, Cook was instructed to proceed south from the Society Islands to latitude 40°, and from there westward between latitudes 40° and 35° until the east coast of New Zealand was reached. He did so, discovering no new land on the way, and thereby proved that no large landmass lay in those waters. Cook went on to produce a brilliant chart of the coastal outline of New Zealand's two main islands, also proving that they were not part of a continent as some had previously thought. However, to disprove finally the theory of such a landmass elsewhere in the southern oceans he would embark on a second Pacific voyage to circumnavigate the globe at the highest latitudes then possible.

During this second voyage, one of epic endurance, Cook three times penetrated the Antarctic Circle, but encountering only a barrier of ice he was forced to turn back. It would not be until the next century that Antarctica was sighted, although Cook had certainly sailed deep enough south to have encountered land had he been at the right longitude. In his journal Cook memorably observed: 'I who had Ambition not only to go farther than any one had done before, but as far as it was possible for man to go, was not sorry at meeting with this interruption as it in some measure relieved us, at least shortned the dangers and hardships inseparable with the Navigation of the Southern Polar Regions; Since therefore, we could not proceed one Inch farther to the South, no other reason need be assigned for my Tacking and Standing back to the north, being at this time in the Latitude of 71° 10´ S, Longitude 106° 54´ W' (*Cook's Journals*, vol. 2, p. 323). Whatever lay hidden to the south, it was not a fertile continent, rich with potential for trade and human contact.

11 The Rev. Nevil Maskelyne, *The Nautical Almanac and Astronomical Ephemeris for the Year 1768. Published by Order of the Commissioners of Longitude* (London: W. Richardson and S. Clark, 1767), with the ensuing edition for 1769 (London: W. Richardson and S. Clark, 1768)

The National Maritime Museum, Greenwich, London. PJN0037 and PJN0038.

1. PJN0037, light brown full leather binding, with gilt inlaid House of Commons Library stamp on front board. Title and year for tables, 1768, gilt lettering on black label on spine. 21.3 × 13.8 cm (closed).

2. PJN0038, light brown full leather binding. Title and year for tables, 1769, gilt lettering on red label on spine, gold lattice tooling above and below. 22.3 × 14.3 cm (closed).

Nevil Maskelyne commenced publication of the *Nautical Almanac* in 1766 to facilitate calculation of longitude, which was necessary for the purposes of accurate navigation. The *Nautical Almanac* simplified the complex calculations required to determine longitude using the relative positions of the Moon and stars. It was published annually, with each edition providing data for the next year. The 1767 and 1768 editions were ready when James Cook sailed in command of *Endeavour* in August 1768, but those published in 1769 and 1770 were not. The lack of these ensuing editions considerably increased the labour of lunar-distance calculations later in the voyage (from Cape Maria van Diemen on the North Island of New Zealand at the end of 1769), something about which Cook complained in his journal.

Prior to his death in the Indian Ocean on the way home due to fever contracted in the Dutch East Indies, the astronomer Charles Green amply proved his worth by assisting Cook and his officers in the

11.1

Apparent Times of the external and internal Contacts of VENUS with the Sun's Limb at several Places, as they may be expected to happen on June the 3d 1769.						
	1st ext. cont.	1st int. cont.	2d int. cont.	2d ext. cont.	Latitude.	Supposed Longitude from Greenw
	h. m.	h. m.	h. m.	h. m.		h. m. f.
Greenwich	7 6	7 25	—	—	51 29 N	0 0
Edinburg	6 53	7 12	—	—	55 58 N	0 13 13 W
Dublin	6 41	7 0	—	—	53 20 N	0 24 54 W
Tornea	8 43	9 2	14 56	15 15	65 51 N	1 36 48 E
Kittis	8 43	9 2	14 56	15 15	66 48 N	1 36 48 E
Attengaard	8 39	8 58	14 51	15 10	69 59 N	1 32 27 E
Wardhus	9 12	9 31	15 24	15 43	70 35 N	2 5 36 E
North Cape	8 51	9 10	15 3	15 22	71 23 N	1 44 48 E
Bear Island	8 12	8 31	14 24	14 43	74 32 N	1 5 36 E
Spitzbergen, Bell-Sound	7 57	8 16	14 8	14 27	77 15 N	0 51 2 E
Petersberg			15 21	15 40	59 56 N	2 1 20 E
Tobolski			17 53	18 12	58 12 N	4 32 51 E
S. John's, Newfoundland	3 40	3 59	—	—	47 32 N	3 31 13 W
Quebec	2 29	2 48	—	—	46 55 N	4 39 36 W
Hudson's Bay	0 49	1 8	6 54	7 13	58 56 N	6 19 40 W
Boston	2 26	2 45	—	—	42 25 N	4 42 29 W
Williamsberg	2 3	2 22	—	—	37 20 N	5 6 20 W
Jamaica, Port Royal	2 4	2 23	—	—	18 0 N	5 7 2 W
Mexico	0 19	0 38	6 14	6 33	20 0 N	6 54 40 W
Cape Corientes	23*48	0 6	5 44	6 3	20 50 N	7 25 40 W
Cape St. Lucar	23*31	23*50	5 28	5 47	23 15 N	7 41 40 W
Cape Conception	22*34	22*53	4 33	4 52	35 30 N	8 37 40 W
Fernambuca, Brazil	4 50	5 9	—	—	8 13 S	2 20 8 W
Conception, Chili	2 24	2 43	—	—	36 43 S	4 50 44 W
Bombay			18 9	18 28	19 18 N	4 47 48 W
Madras			18 41	19 0	13 13 N	5 20 9 E
Calcutta			19 14	19 33	22 30 N	5 53 43 E

N. B. The Contacts marked with Asterisks belong to the 2d Day of June 1769, according to Astronomical Time.

11.2

demanding task of navigation using the methods then available. In particular, from the end of 1769 onwards, the precision of Green's astronomical observations were critical to the running surveys of the coasts encountered, not least the east coast of what became known as Australia. Cook recorded this debt in his journal in August 1770 when praising Green for his services.

The first volume included here, that published in 1767, is open at pages 94 and 95. These show a lunar distance table for August 1768, such tables being used to determine longitude. In this one angular distances from the centre of the Moon to the Sun and bright stars are tabulated against Greenwich time at three hourly intervals. By measuring these angles accurately with a sextant and comparing them with those in the tables it was possible to work out Greenwich time at the time of the observation. The difference between this and local time could then be used to calculate longitude, since an hour of difference corresponds to 15 degrees of longitude. The expected accuracy of this method when Cook sailed was about ¼ of a degree in longitude, which corresponds to about 15 nautical miles at the equator. This was the best method of determining longitude at sea until chronometers that kept accurate time over long voyages were generally available. Cook would carry such chronometers on his next Pacific mission.

The second volume, that published in 1768, is open at its last printed page, showing a table entitled 'Apparent Times of the external and the internal Contacts of VENUS with the Sun's Limb at the several Places, as they may be expected to happen on June the 3^d 1769'. This table displays the approximate predicted times of the transit of Venus at different places around the world. Times are given in relation to local noon for the place concerned, so that 7 h 6 m is six minutes past seven in the evening at Greenwich. The table gives the time when Venus first appears to contact the edge of the Sun, and then when it is first completely inside the Sun. Similar times are then given for the planet's exit. The blanks correspond to where the Sun is no longer visible, so for Greenwich the Sun has set before the transit is complete.

12 Portable Gregorian reflector telescope

Made by James Short, Edinburgh, 1734.

National Maritime Museum, Greenwich, London. AST0958. Length 62.5 cm; barrel 56.5 × 42 cm; aperture diameter 7.6 cm; focal length ca. 45.7 cm. Brass, glass and metal.

An early example of a telescope made by James Short similar to telescopes used by James Cook and the *Endeavour* mission astronomer Charles Green to observe the transit of Venus. The telescope is adjusted using the semi-circle at the top of the stand and the arc at its base, and in the barrel itself are primary and secondary metal specula, the latter of which is used to focus the instrument by adjusting the threaded rod fixed to the barrel.

13 Regulator clock

Made by John Shelton, 1756.

National Museum of Scotland, Edinburgh. T.1978.1. Height 16.25, width 4.5 and depth 3 cm. Wood, glass, silvered brass, brass and steel.

This much-travelled clock by John Shelton has been used as the master regulator in astronomical and gravitational experiments all over the world. Not all of its history is clear, but it seems certain that it was purchased by the Royal Society of London in 1760 and thereafter taken on HM Bark *Endeavour* to Tahiti for the observation of the 1769 transit of Venus. In 1774 it was used by Nevil Maskelyne in his gravitational measurements at Schiehallion. Extensive experiments with compound pendulums were made in the 1820s and 1850s, with this clock as one of the principal reference standards. It latterly became the master clock at the observatories at Kew, Ben Nevis and Eskdalemuir.

For the 1769 transit event it was set to display the correct local time using astronomical observations, such as those made with the portable twelve-inch quadrant shown in number 14 below. Local time when compared with the time at Greenwich could be used to determine the longitude of the mission observatory, since each hour of difference corresponds to 15 degrees of longitude. An accurate clock was also essential to measure the duration of the transit itself.

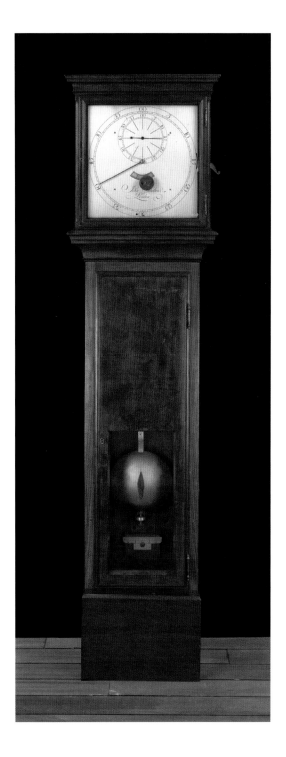

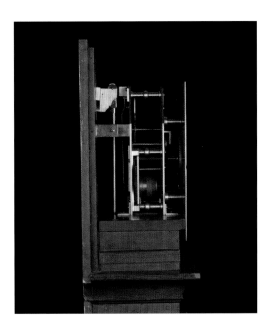

14 Twelve-inch astronomical quadrant

Made by John Bird, 1760–9.

The Science Museum, London. 1876-572/1. 64 × 45 cm. Brass and glass.

The astronomical quadrant is used to measure the angular altitudes of the Sun and stars. These readings can be used to determine time and so calibrate an accurate regulator clock. The astronomical quadrant was essential to the observation of the transit of Venus because it was necessary to record the exact length of time that Venus took to pass across the disc of the Sun. The altitudes of fixed stars were also needed to determine the latitude of the mission observatory.

When the quadrant was stolen by a Tahitian man the day after first being brought ashore from *Endeavour*, the transit observation was in jeopardy. The theft was noticed at about 10 in the morning of 2 May 1769. Banks, Charles Green, a midshipman and a local chief called Tupura'a i Tamaita (Tubourai to Banks, or, in a more light-hearted vein, Lycurgus, after the Spartan lawgiver, because of the help he gave in retrieving a stolen snuffbox and opera glass) immediately went in pursuit of the thief and retrieved this vital instrument. James Cook followed with a party of marines, having sealed off Matavai Bay to all exiting canoes, a tactic that he employed a number of times during his Pacific voyages in order to recover stolen property, ultimately with fatal consequences to himself.

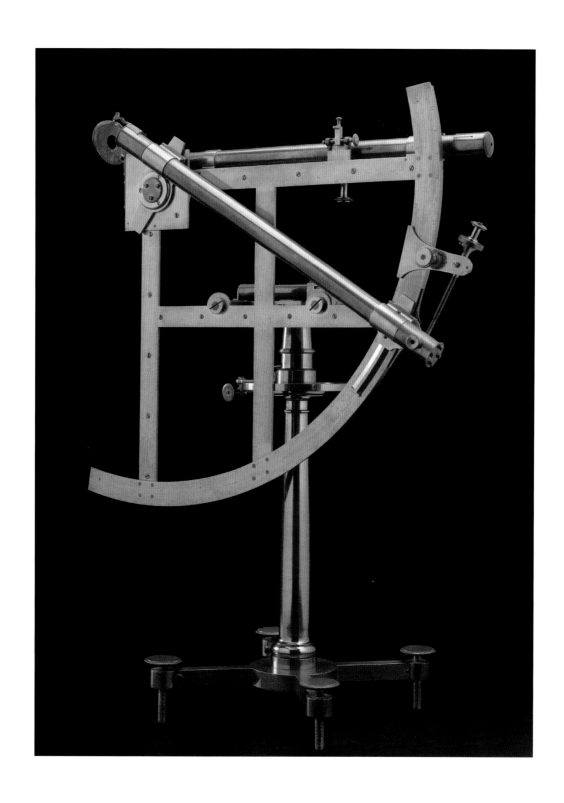

15 Carl Linnaeus, *Systema Naturæ per Regna Tria Naturæ, secundum Classes, Ordines, Genera, Species, cum Characteribus, Differentiis, Synonymis, Locis.* 2 tom., Editio Decima (Holmiæ: Laurentius Salvius, 1758, 1759)

The British Library, London. 956.e.6,7.

1. Vol. I, rebound on guards, green cloth. Title, place and date of publication, gilt lettering on spine. Author and volume given. 21.6 × 17 cm (closed).

2. Vol. II, rebound on guards. Half bound in green leather and green cloth boards. Title, place and date of publication, gilt lettering on spine. Author and volume given. 21.2 × 16.4 cm (closed).

Joseph Banks carried a select reference library with him on *Endeavour* and this was used to help sort and identify the specimens that were gathered during the voyage. Key works by Linnaeus were obvious choices to accompany him and his party into the Pacific. The Swedish naturalist Carl Linnaeus laid the foundations of modern plant and animal taxonomy and nomenclature. He classified plants into groups based on the number and arrangement of their sexual organs – stamens and pistils – and, using this system, first published in his *Systema Naturæ* (Leiden, 1735), plants were grouped into 24 classes each of which was further divided into orders, genera and species. This artificial system produced many unnatural groups, and Linnaeus freely admitted his classification to be an artificial one. It was, however, precise and simple, and since he used it in his major publications *Fundamenta Botanica* (Amsterdam, 1736) and *Classes Plantarum* (Leiden, 1738), which other naturalists found indispensable, it became widely accepted.

Whereas in the first edition of *Systema Naturæ* the system is laid out in only eleven pages, by the time of this, the tenth edition, there were a total of 1,384 pages in two volumes, volume I for animals and volume II for plants. Botanical and zoological nomenclature begins with Linnaeus, the names of plants accepted as valid today starting with those published in his *Species Plantarum* (Stockholm, 1753; see number 16 below), and of animals with those in the first volume of the tenth edition of his *Systema Naturæ*. This copy of the work is from Banks's own library and it probably travelled with him on *Endeavour*. Volume I is open at the page showing Banks's library stamp, matching the form of his usual signature, underneath Linnaeus's Latin paraphrase of Psalm 103, verse 24 of the Latin vulgate Bible, which translates as 'O God, How manifold are your works! How wisely have you made them! How full is the Earth with your riches!' Linnaeus once wittily observed, 'Deus creavit, Linnaeus disposuit' – God created, Linnaeus organized – a phrase that became his personal motto. Volume II is open at page 837, showing Linnaeus's classification of plants based on the number of stamens and pistils in the flower.

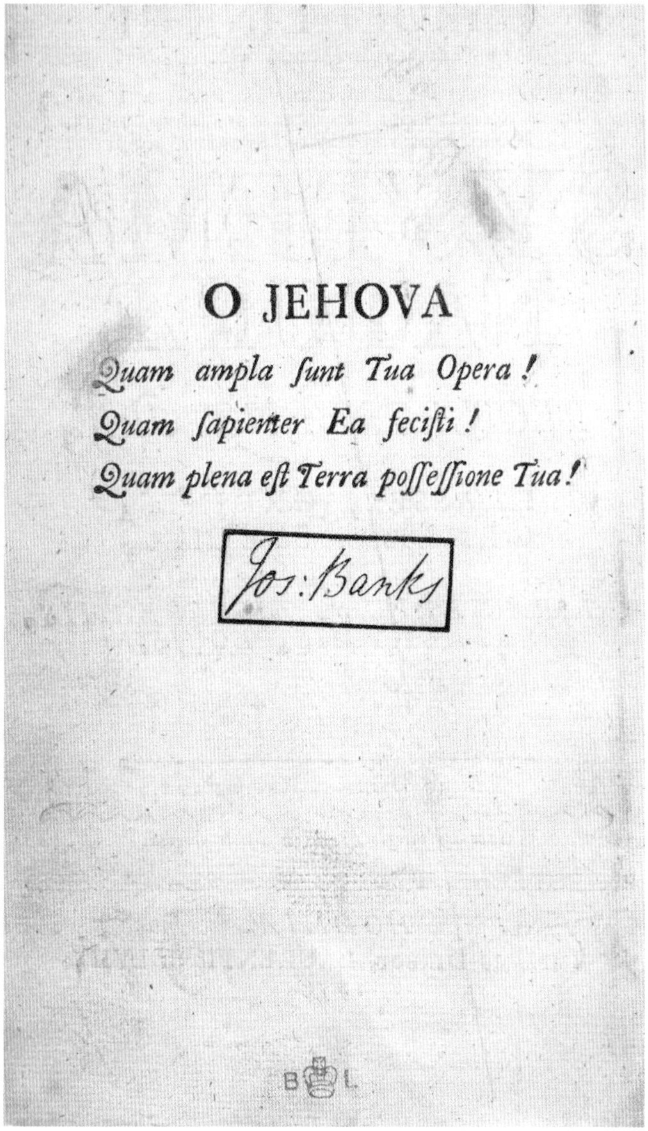

O JEHOVA

Quam ampla sunt Tua Opera!
Quam sapienter Ea fecisti!
Quam plena est Terra possessione Tua!

Jos: Banks

15.1

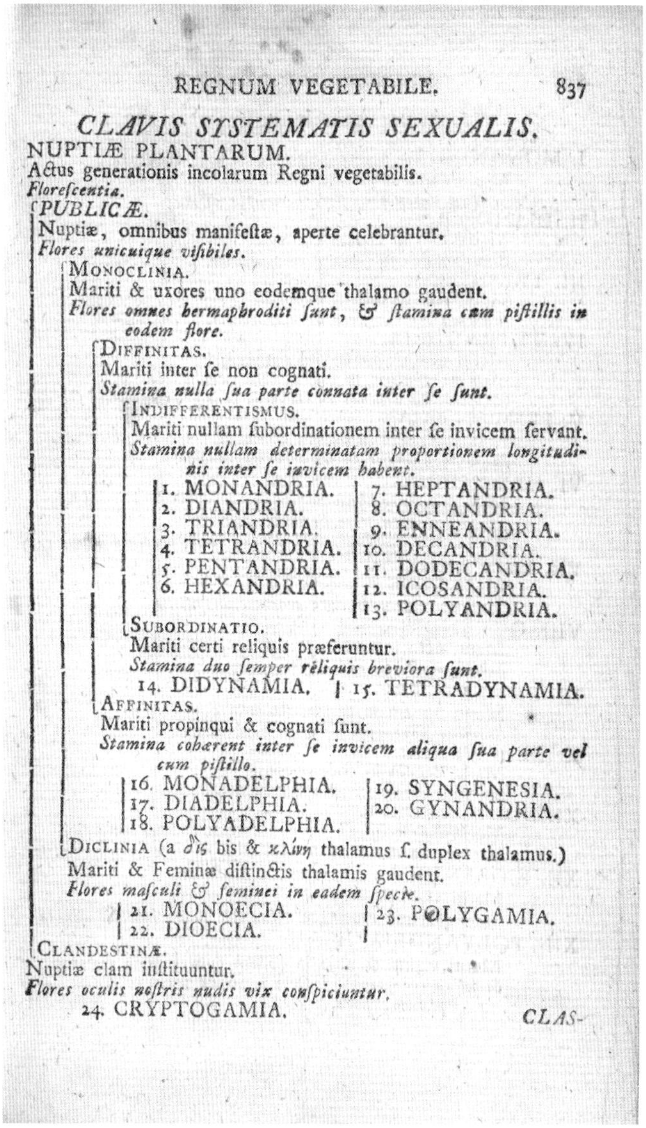

REGNUM VEGETABILE. 837
CLAVIS SYSTEMATIS SEXUALIS.
NUPTIÆ PLANTARUM.
Actus generationis incolarum Regni vegetabilis.
Florescentia.
PUBLICÆ.
Nuptiæ, omnibus manifestæ, aperte celebrantur.
Flores unicuique visibiles.
 MONOCLINIA.
 Mariti & uxores uno eodemque thalamo gaudent.
 Flores omnes hermaphroditi sunt, & stamina cum pistillis in eodem flore.
  DIFFINITAS.
  Mariti inter se non cognati.
  Stamina nulla sua parte connata inter se sunt.
   INDIFFERENTISMUS.
   Mariti nullam subordinationem inter se invicem servant.
   Stamina nullam determinatam proportionem longitudinis inter se invicem habent.
1. MONANDRIA.	7. HEPTANDRIA.
2. DIANDRIA.	8. OCTANDRIA.
3. TRIANDRIA.	9. ENNEANDRIA.
4. TETRANDRIA.	10. DECANDRIA.
5. PENTANDRIA.	11. DODECANDRIA.
6. HEXANDRIA.	12. ICOSANDRIA.
	13. POLYANDRIA.
   SUBORDINATIO.
   Mariti certi reliquis præferuntur.
   Stamina duo semper reliquis breviora sunt.
   | 14. DIDYNAMIA. | 15. TETRADYNAMIA. |
  AFFINITAS.
  Mariti propinqui & cognati sunt.
  Stamina cohærent inter se invicem aliqua sua parte vel cum pistillo.
16. MONADELPHIA.	19. SYNGENESIA.
17. DIADELPHIA.	20. GYNANDRIA.
18. POLYADELPHIA.	
 DICLINIA (a δὶς bis & κλίνη thalamus s. duplex thalamus.)
 Mariti & Feminæ distinctis thalamis gaudent.
 Flores masculi & feminei in eadem specie.
  | 21. MONOECIA. | 23. POLYGAMIA. |
  | 22. DIOECIA. | |
CLANDESTINÆ.
Nuptiæ clam instituuntur.
Flores oculis nostris nudis vix conspiciuntur.
 24. CRYPTOGAMIA.

CLAS-

15.2

16.1

16 Carl Linnaeus, *Species Plantarum, Exhibentes Plantas Rite Cognitas, ad Genera Relatas, cum Differentiis Specificis, Nominibus Trivialibus, Synonymis Selectis, Locis Natalibus, secundum Systema Sexuale Digestas.* 2 tom., Editio Secunda (Holmiæ: Laurentius Salvius, 1762, 1763)

The British Library, London. 439.b.17,18. Rebound in half brown leather and brown cloth. Title, place and date of publication, gilt lettering on spine. Author and volume given.
1. Vol. I, 20.7 × 14 cm.
2. Vol. II, 20.7 × 14 cm (both closed).

This is Joseph Banks's two-volume library copy of Linnaeus's second-edition *Species Plantarum*. In the first edition, of 1753, Linnaeus classified every species of plant known at the time into genera, and his was the first work consistently to apply binomial names, giving a one-word name to a genus and a two-word name to an individual species within the genus. This system superseded longer variable descriptive names for plants and animals, and rapidly became the standard method for naming species during the second half of the eighteenth century. Linnaeus was not the first to use binomials, but he was the first to use them consistently. This work is therefore the starting point for the modern naming of plants. By the time of the publication of this, the second edition of *Species Plantarum*, the work had grown from the 1,200 pages of the first edition to 1,684 pages, reflecting the increasing number of plant species being described for the first time.

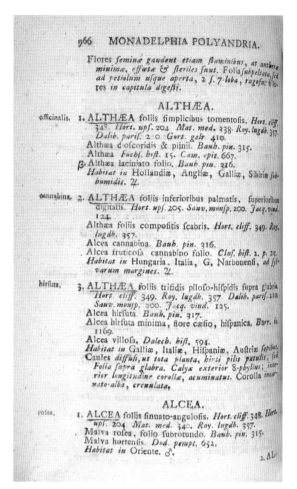

16.2

Interestingly, the present copy from Banks's library is annotated on one of the blank pages at the beginning of volume 1 by Jonas Dryander, Banks's Soho Square librarian from 1777 until his death in 1810: 'This is the handwriting of John Earl of Bute'. There follows a handwritten note by John Stuart, 3rd earl of Bute: 'In these volumes all the Plants are markd of the Gronovian Hortus Siccus purchasd by me in 1778'. Then there is a final comment by Dryander: 'and by Sir Jos. Banks, in 1794'. Lord Bute was tutor to George III in his youth, and thereafter one of the king's closest advisors, going on to become prime minister. Bute had a lifelong passion for botany and he developed the Royal Gardens at Kew into the botanical centre that Banks would later supervise and enlarge. Jan Frederik Gronovius was a Dutch botanist and patron of Linnaeus, and his herbarium contained large numbers of specimens sent to him for identification by John Clayton, one of the early collectors of plants in Virginia. Many of these were studied by Linnaeus and were among the earliest North American specimens that he had seen. In *Species Plantarum* Linnaeus's knowledge of North American species was based heavily on Clayton's specimens. Since Clayton published almost nothing himself, many of his specimens are types of Linnaean names and therefore of considerable nomenclatural importance, and would have been included in the Gronovius specimens later purchased by Bute in 1778 and then acquired after his death by Banks in 1794. Hence the annotation at the front of this particular copy of Linnaeus's work.

During the *Endeavour* voyage some 30,400 individual plant specimens were collected representing more than 3,600 described species. Of the latter total it is estimated that 1,400 species were then new to Western science, as against Linnaeus's total of nearly 6,000 names in the first edition of his *Species Plantarum*, and the more than 9,000 valid plant names that Linnaeus published in his lifetime. The specimens of these species are thus type specimens, that is the specimens on which all subsequent nomenclature of the species is based. More than 1,000 species of animals were collected, mainly comprising birds, fishes, arthropods and molluscs, but only five mammals were among them. Few animal specimens from this voyage now survive, but a number of arthropods and molluscs may be found at the Natural History Museum, London. Banks's herbarium specimens were complemented by the drawings of Sydney Parkinson, made from living specimens collected by Banks and Solander during the voyage, and by a very large collection of manuscript lists by Banks, Solander and Herman Diedrich Spöring. After the expedition's return Banks intended to prepare the botanical results for publication, but for a variety of reasons the engraved plates that he commissioned were not published until the 1980s (see numbers 122–6). Banks's failure to publish the new *Endeavour* species in his planned *Florilegium* has meant that his name is not as prominent in taxonomic literature for the Pacific as it might otherwise have been, and the plants gathered during the voyage remained to be described by other naturalists.

Here volume 1 is open at the frontispiece, showing an engraved portrait of Linnaeus as well as the title page. The engraving is thought to be by Carl Bergqvist. Volume 2 is open at page 966, showing Linnaeus's classification of the Marsh Mallow under the binomial *Althaea officinalis* (see number 2).

17 'A CHART OF THE GREAT SOUTH SEA OR PACIFICK OCEAN SHEWING THE TRACK AND DISCOVERIES MADE BY THE ENDEAVOUR BARK IN 1769 AND 1770.'

By James Cook.

The British Library, London. BL Add. MS 21593.D. 45.4 × 114.6 cm. Scale approx. 1:18,970,000.

This chart, a manuscript original, shows the track of HM Bark *Endeavour* southwards along the east coast of South America, round Cape Horn and then across and finally out of the Pacific, November 1768–January 1771. It is inscribed in ink with the title given above, and is also annotated in ink 'Commodore Roggewins Track According to the Supposition of M. Pengre.' and 'NB The Lands discovered by the Endeavour are Shaded. Those Without Shade are Copied from Charts. The Prick'd Line shews her Track.' The chart is contained in a portfolio of charts by various Pacific explorers formerly in the possession of Sir Joseph Banks, which passed to the British Museum with his library and herbarium collections in 1827. It may well be that a similar manuscript chart by James Cook and Isaac Smith also showing *Endeavour*'s track was based on this one, BL Add. MS 7085, f. 1. This latter chart is bound with other charts, plans, coastal views and drawings thought to have been selected by Cook as the official graphic record of his voyage to match the fair-copy journal that he submitted to the Admiralty as a full account of the mission, perhaps with future publication in mind. Cook produced various other manuscript drafts of his mission journal. BL Add. MS 7085, f. 1, appears to be the version that was engraved for the *Endeavour* account edited and published by Dr John Hawkesworth in 1773 (see number 18 immediately below). A number of items in BL Add. MS 7085 were included by Hawkesworth in that work.

Isaac Smith was Cook's wife's nephew and he sailed on *Endeavour* as an able seaman. He made a significant contribution to the cartographic results of the voyage. Smith had already served on the *Grenville* under Cook while charting the coasts of Newfoundland, proving himself to be a capable sailor and apprentice hydrographic surveyor. During the *Endeavour* voyage Cook continued the education of his young kinsman in surveying, promoting him along the way from able seaman to midshipman and then to master's mate. After the voyage Cook commended Smith to the Admiralty for his good conduct, particularly in helping with the mission's survey work. Smith appears to have been the draughtsman who produced a number of fair-copy charts that now survive in *Endeavour* map collections, although the actual surveys on which these were based would certainly have been led by Cook. Most famously, Cook allowed Smith to go ashore at Botany Bay before anyone else from *Endeavour*, thus ensuring Smith's place in history as the first Briton to set foot on that coast. Smith sailed on Cook's second Pacific voyage in HMS *Resolution*, during which he continued to excel at drawing charts, much to Cook's satisfaction. Smith went on to become a rear admiral. In his retirement he lived for a while with his widowed aunt, the aged Elizabeth Cook.

Jacob Roggeveen was a Dutch navigator. In 1721 he was sent by the Dutch West India Company in search of *Terra Australis Incognita*, and his voyage – also included on the map – effectively marked an end to Dutch attempts to find a supposed southern continent. After sailing to the Falkland Islands, Roggeveen passed through the Le Maire Strait, and then entered the Pacific Ocean. He visited the Juan Fernández Islands, off Chile, and named Easter Island, at which he arrived on Easter Sunday 1722, before sailing to Batavia (Jakarta) by way of the Tuamotu Archipelago, the Society Islands and Samoa. He eventually reached the Netherlands in 1723 following the seizure of his ships by the Dutch East India Company, but was compensated along with his crew after a court case.

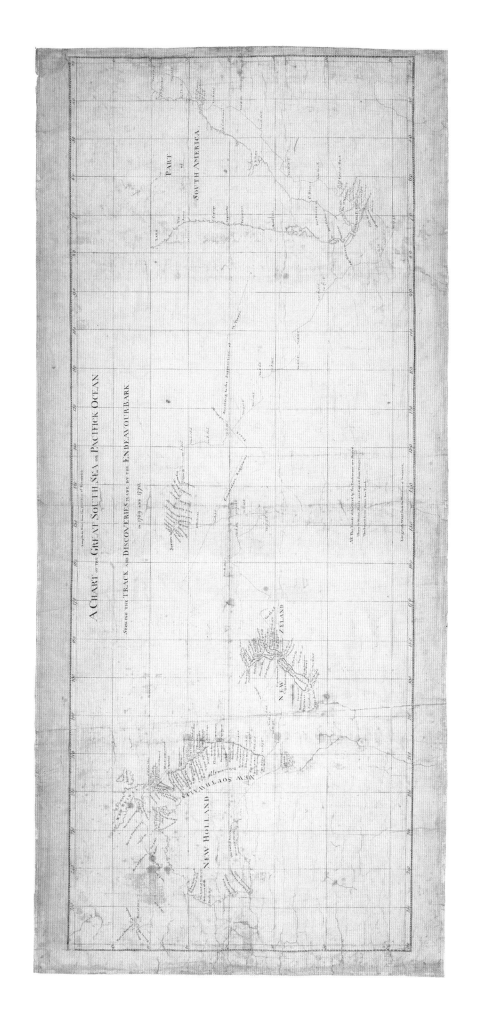

18 'CHART of part of the SOUTH SEA, Shewing *the Tracks & Discoveries made by His MAJESTYS Ships* Dolphin, *Commodore Byron, &* Tamer, *Capt*^n. *Mouat*, 1765. Dolphin, *Capt*^n. *Wallis, &* Swallow, *Capt*^n. *Carteret*, 1767. *and* Endeavour, *Lieutenant Cooke*, 1769.'

John Hawkesworth, ed., An account of the voyages undertaken by the order of His Present Majesty for making discoveries in the southern hemisphere, and successively performed by Commodore Byron, Captain Wallis, Captain Carteret, and Captain Cook, in the Dolphin, the Swallow and the Endeavour: drawn up from the journals which were kept by the several commanders, and from the papers of Joseph Banks, Esq, 3 vols (London: Printed for W. Strahan and T. Cadell, 1773). Vol. I, frontispiece. By J. Cook and I. Smith. Engraved by W. Whitchurch.

The British Library, London. 455.a.21. Rebound in half blue leather and blue cloth. Title, place and date of publication, gilt lettering on spine. Editor and volume given. 29 × 24.2 cm (closed). Plate 37.2 × 65 cm. Scale approx. 1:35,000,000.

This is the engraved general chart of the *Endeavour* voyage that was printed for the official mission account as edited and published in 1773 in three volumes by Dr John Hawkesworth. An incomplete proof copy exists at BL MAPS C.2.a 7.(3). A manuscript chart at BL Add. MS 7085, f. 1, appears to be the original on which this printed version was based. The published chart seen here was given prominence as the frontispiece to Hawkesworth's work, which rapidly sold out and was soon reprinted. The Pacific tracks of John Byron and Frederic J. Mouat, Samuel Wallis, and Philip Carteret are included in the chart since Hawkesworth edited the voyage accounts of these earlier missions along with that of James Cook. Here, then, we see the impressive *Endeavour* track, reaching deeper into the South Pacific than those of the previous voyagers, and showing Cook's major discoveries of New Zealand as two main islands, not a larger landmass, and his historic run along the east coast of Australia (New Holland at the time). *Endeavour*'s track is clearly indicated by a dotted line.

The primary task of the *Endeavour* mission was to observe the transit of Venus from Tahiti, but a subsidiary objective was to look for an unknown southern continent that some, the hydrographer Alexander Dalrymple foremost among them, supposed lay hidden somewhere in the southern oceans. Dalrymple, who had worked for the East India Company in India, and who had undertaken his own expeditionary voyages to the Philippines, Borneo and Sulu, argued strongly for the existence of such a landmass, adhering to the view that it would counterbalance those found in the northern hemisphere – the old 'counterpoise' theory. He was even put forward by the Royal Society to lead the *Endeavour* mission, an idea rejected outright by the Admiralty, much to Dalrymple's disappointment. Consequently he played no part in the voyage.

In the decade prior to the launch of *Endeavour* the British had dispatched successive Royal Navy vessels into the Pacific with the aim of establishing what territories and trade existed there. Commodore John Byron captained HMS *Dolphin*, accompanied by the sloop *Tamar*, during a circumnavigation of the world lasting from June 1764 to May 1766. His mission was to explore the South Atlantic for any significant landmass, to claim the Falkland Islands and then to enter the Pacific and seek the Northwest Passage. In 1765 he claimed the Falkland Islands for Britain, having shortly beforehand been preceded there by the French explorer Louis Antoine de Bougainville, who established a small colony on East Island. During his voyage Byron visited Patagonia, islands in the Tuamotu group (where he could not anchor), the King George Islands and the Gilbert Islands, as well as Tinian in the Northern Mariana Islands. However, he narrowly missed sighting Tahiti. More significantly, given his mission objectives, he did not explore northwards in the Pacific, and his voyage, lasting only two years, was considered a failure. The flamboyant poet Lord Byron (see number 1) was Commodore Byron's grandson.

Following Byron's return, Samuel Wallis was given command of HMS *Dolphin* and sent to explore for the fabled southern continent in August 1766. Wallis sailed in the company of Philip Carteret, who commanded HMS *Swallow*, but the two vessels were permanently separated after a gruelling passage through the Strait of Magellan. Due to contrary winds and scurvy among his crew, Wallis was unable to explore as far southward in the Pacific as instructed, although he did discover a number of uncharted islands. He is primarily remembered as the first European navigator to visit and describe Tahiti. From 18 June to 27 July 1767 he was at this famous island where, despite initial clashes with the Tahitians, he and his officers were able to establish friendly relations and to observe the people and their customs. Wallis's health was poor, and he did not investigate reports of land sighted to the south of the Society Islands, preferring instead to sail for home. This was reached in May 1768, before *Endeavour* departed and in time to bring news of King George the Third's Island, as Wallis called it, otherwise Otaheite or now Tahiti. Located at a

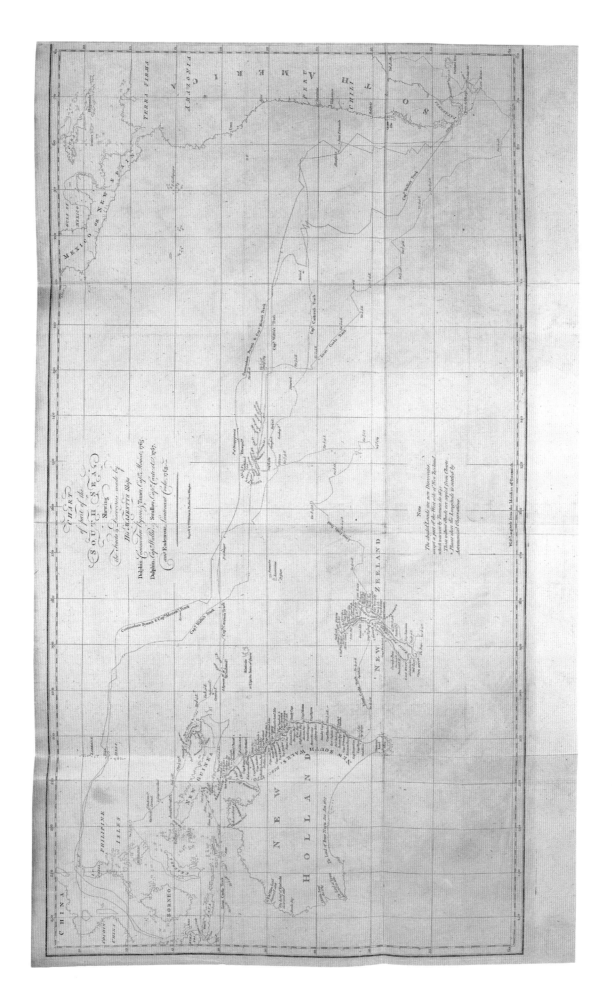

suitable latitude to observe the transit of Venus, this island could also provide essential supplies, and was therefore the one to which James Cook would sail in *Endeavour*.

Philip Carteret had previously sailed in the *Tamar* and also the *Dolphin* under Byron. Despite the poor condition of *Swallow*, Carteret made a number of discoveries, among them Pitcairn Island, and he named various islands, some already known, in the Tuamotu Archipelago, the Santa Cruz Islands and the Solomons. He discovered the islands of New Britain, New Ireland and New Hanover as well as useful straits lying between them. He also named the Admiralty Islands, which Abel Tasman had previously explored, and along with New Britain, New Ireland and New Hanover these today form part of the Bismarck Archipelago. Carteret arrived home a year after Wallis, and after *Endeavour* had sailed from Plymouth in August 1768, carrying among her crew a small number of experienced veterans from Wallis's circumnavigation (one, Third-Lieutenant John Gore, had served in the *Dolphin* under both Byron and Wallis and would later sail with Cook on his final Pacific mission).

The missions described here helped prepare the way for that of James Cook in *Endeavour*. They suggested Tahiti as a base for astronomical observation, but at the same time failed to dispel the notion that an undiscovered continent lay hidden somewhere in the southern oceans. Moreover, the Northwest Passage remained a tantalizing prize for any navigator who might discover it. These would now be tasks for Cook to undertake, necessitating his three great Pacific expeditions.

THE ATLANTIC STAGE

HM Bark *Endeavour* sailed from Plymouth on 25 August 1768 and made landfall at Madeira from 12 to 18 September, there taking on additional supplies before departing for South America. Stops such as this allowed Banks and his party to go ashore to collect plants and animals and to observe the local people. Once brought aboard, their collections could be sorted, recorded, illustrated and stored, so periods spent at sea after leaving land were also a busy time for the naturalists. Moreover, Banks travelled with a considerable amount of equipment for collecting from the oceans, and birds were also gathered if they came near or landed on the ship.

South America offered the first opportunity to explore territory largely new to Europe's naturalists, and this part of the voyage was eagerly anticipated by Banks and Daniel Solander. Landings on the continent were few, however, before *Endeavour* passed through the Strait Le Maire and rounded Cape Horn to enter the Pacific. Thereafter it would be some one and three-quarter years before she again encountered European civilization. Of the stops in South America, that at Rio de Janeiro from 13 November to 7 December was a frustrating disappointment due to the viceroy confining most of the expedition on ship, and that at the Bay of Good Success from 15 to 21 January 1769 was a near disaster for Banks and his team after being caught overnight in a blizzard while ashore.

Banks's party on the voyage consisted of Daniel Solander (naturalist), Sydney Parkinson (natural history artist), Alexander Buchan (landscape artist), Herman Diedrich Spöring (clerk and draughtsman), and the servants and field assistants James Roberts, Peter Briscoe, Thomas Richmond and George Dorlton.

19 *Heberdenia bahamensis* (Gaertner) Sprague, *J. Bot., Lond.* 61: 241 (1923)

Engraving by Daniel Mackenzie after Sydney Parkinson.

The Natural History Museum, London. Diment et al., Catalogue, Part 2 Botany (1987), M 9, col. engraving 1985 BF: pl. 395. Plate height 46 × width 29.5 cm; plant ill. height 26 cm.

HM Bark *Endeavour* anchored at Madeira from 12 to 18 September and there took on supplies. The stop gave Banks and his party the opportunity to explore and collect ashore at what was already a location well known to naturalists, but nevertheless one with a rich island flora including many endemic species. While at Madeira, Banks made the acquaintance of Thomas Heberden, a physician and a fellow of the Royal Society. Heberden aided Banks in the collection of 329 plant species, and Banks and Daniel Solander named the tree *Heberdenia excelsa* (later *H. bahamensis*) after him. The plant was subsequently named *Ardisia excelsa* Aiton, but the taxonomy and nomenclature of this species currently remain unresolved. It appears here in a coloured print from a copper plate engraved by Daniel Mackenzie, based on a finished 1768 watercolour drawing by Sydney Parkinson. There are also two Banks herbarium specimen sheets for this plant at the Natural History Museum, London. The drawings and plate were intended for Banks's planned but unpublished *Endeavour Florilegium* (see numbers 122–6).

20 *Polybius henslowii* Leach, 1820

By Alexander Buchan.

The Natural History Museum, London. Wheeler, Catalogue, Part 3 Zoology (1986), 221.(3:8). 29.1 × 23.5 cm.

A swimming crab caught off the coast of Spain is here depicted in a finished watercolour by Alexander Buchan, annotated in ink on the recto 'Cancer Depurator Sept. 4. 1768' and on the verso 'off the coast of Spain. Septʳ· 4ᵗʰ· 1768'. During the *Endeavour* expedition Banks and his party were kept busy at sea sorting and recording plants that had been collected on land, but they also gathered marine life and this too was recorded (see Carter *et al*., 'The Banksian Natural History Collections of the *Endeavour* Voyage and their Relevance to Modern Taxonomy', pp. 62–4). Banks was well equipped with nets, fishing tackle, preservatives and containers for collecting from the ocean. The *Endeavour* voyage has therefore been called the 'first extensive survey of marine life to be carried out systematically during a voyage round the world' (Lysaght, 'Banks's Artists and his *Endeavour* Collections', p. 37). According to Banks some 500 fish preserved in alcohol were brought back from the voyage, although few of these specimens now survive (see Kaeppler *et al*., *James Cook and the Exploration of the Pacific*, p. 228).

21 *Anthias anthias* (Linnaeus, 1758)

By Sydney Parkinson.

The Natural History Museum, London. Wheeler, Catalogue, Part 3 Zoology (1986), 169.(2:79a). 26.3 × 32.9 cm.

This finished watercolour illustration by Sydney Parkinson is annotated in ink on the recto 'Perca Imperator. Sydney Parkinson pinx^t 1768. S. Parkinson' and on the verso 'Madeira'. Daniel Solander's suggested name, 'Perca Imperator', does not seem to have been employed by later naturalists. It derives from his note of the Madeiran vernacular 'Emperador' (Wheeler, *Catalogue*, Part 3 Zoology, p. 101). Counted among the earliest of Parkinson's fish paintings during the voyage, this excellent finished piece was preceded by a detailed preliminary colour sketch, which is also held at the Natural History Museum, London. Interestingly, a comment by Parkinson on the verso of this sketch indicates that Banks thought the colours were too pale, showing that Banks was guiding his artist in depicting specimens, including those of animals. There is at the museum a specimen of this fish in spirit that was almost certainly gathered during the *Endeavour* voyage.

The fact that Parkinson had time to draft and then produce a finished version of this painting is indicative of his generally lighter workload during the initial part of the *Endeavour* voyage. He managed to produce considerably more finished work in the Atlantic, and to focus on a wider range of natural history specimens than just plants, but in the virgin Pacific, where plant collecting increased greatly, botanical illustration prevailed and annotated sketches for later completion preponderate. In his zoological drawings Parkinson appears to have concentrated on sea creatures, in particular fish. Delicate marine invertebrates are also depicted in significant numbers, as are to a lesser extent birds, especially sea birds. Mammals, reptiles, molluscs, insects and other arthropods are, however, far fewer in number. Possibly this was the result of a conscious policy on the part of the artist and his employer since fish colours fade more rapidly than those of other animals, for which skins and shells may be kept, and because the forms of soft-bodied animals are generally difficult to preserve. It would seem, too, that plant illustration sometimes abated long enough at sea to allow work on animals of the ocean, whereas once ashore the numerous plants that were found had to be quickly recorded while still fresh, leaving little or no time for illustrating land animals.

Parkinson's animal illustrations from the voyage held at the Natural History Museum are the largest collection of surviving zoological work by him from the *Endeavour* voyage, and have been of considerable importance to naturalists seeking to describe the various species gathered by Banks and his party given that actual specimens were widely dispersed and mostly lost in ensuing years (see Wheeler, *Catalogue*, Part 3 Zoology, pp. 23–30; see also number 5). For this latter reason, some of the animal illustrations now have status as types for taxonomic purposes.

The present species may be found in the Mediterranean and the North Atlantic from southern Biscay to the Moroccan coast. It lives at around 30 metres and schools around the entrances to caves or near rocky overhangs and, since it is common around Madeira, doubtless such habits made it a relatively easy specimen to obtain. Like others of the genus, a school of this species is all female save for one large male. Should that male be removed, the leading female becomes a functional male. Typically these fish grow to 24 cm.

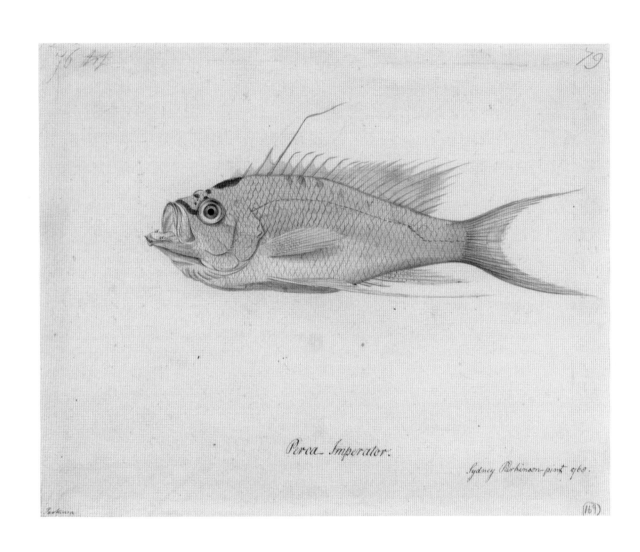

22 'A VIEW of the Town of RIO JANEIRO from the Anchoring-Place.'

By Alexander Buchan.

The British Library, London. BL Add. MS 23920, ff. 7–9. Each section 26.7 × 50.8 cm.

These three pencil drawings of Rio de Janeiro by Alexander Buchan comprise a panorama showing the harbour and town. On folio 8 the set is entitled as given above. Each section is additionally entitled in ink above the drawing itself, 'Rio de Janeiro'. Folio 7 is unsigned. Folios 8 and 9 are signed, 'A. Buchan Delint. Nov. 1768' and 'A. Buchan Delint 1768' respectively. Herman Diedrich Spöring labelled two of the section drawings in ink with a key. Folio 7 key reads, 'A. Ilha dos Cobros with the Sugar-loaf B appearing behind it. C.C. Fort S^t· Sebastian. D. Careening-place. E. Way the Boats went to the Town.' Folio 9 key reads, 'F. The Guard-Boat. G. The Old Ambuscade. H. Convent of Benedictines. I. The Bishops Palace. K. A decay'd Fort. L. Fishermens Houses.'

Having crossed the Atlantic the expedition arrived at Rio de Janeiro on 13 November and stayed there until 7 December. This would be the last chance for the vessel to resupply at a major port before entering the Pacific, and the first chance for Banks to visit a country then largely unexplored by naturalists. To his intense frustration, however, the Portuguese viceroy Don Antonio Rolim de Moura, Conde d'Azambuja, could not grasp the astronomical purposes of the mission and suspected that the British were spies or engaged in some form of illicit trade. Between them Cook and Banks wrote several memorials to the viceroy in an attempt to explain their real intentions, but to no avail. Consequently, only Cook and a few of his men were allowed off the vessel to obtain supplies, although Banks and Solander risked imprisonment by slipping ashore to gather a limited number of specimens, Solander in the guise of the ship's surgeon. The naturalists also inspected plants brought aboard to feed the crew and animals in hopes of finding something new, and Banks apparently paid some of the sailors who were allowed ashore to gather a few specimens.

Note the guard boat that kept watch over *Endeavour*, folio 9 (22.2). When the ship departed Banks recorded that: 'the pilot desired to be dischargd and with him our enemy the guard boat went off, so we were left our own masters and immediately resolved to go ashore on one of the Islands in the mouth of the harbour ...' (*Banks's Journal*, vol. 1, pp. 194–5). Having left the port, Banks and his party botanized on nearby Ilha Rasa, a small compensation for the vexation of their previous confinement.

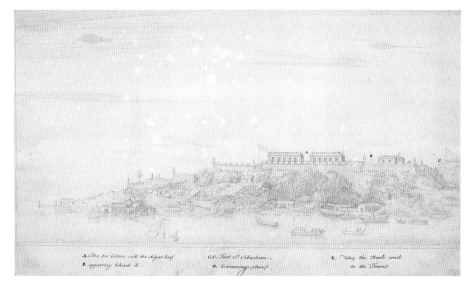

22.1

22.2

22.3

23 To James Douglas, 14th Earl of Morton, President of the Royal Society

From HM Bark Endeavour, *Rio de Janeiro, 1 December 1768.*

The British Library, London. BL Add. MS 34744, ff. 38–44. 18.4 × 15 cm.

This is the first folio side of a copy letter in which Banks reported on voyage progress to the president of the Royal Society in London (see number 6). Banks gives an account of events leading to the arrival of *Endeavour* at Rio de Janeiro, including his own seasickness for about a week during the journey to Madeira. He notes that his collecting at sea produced new species of great interest and that Thomas Heberden at Madeira was helpful both to him and his party (see number 19). Banks then vents his considerable frustration at the behaviour of the Portuguese viceroy at Rio de Janeiro, who insisted that most of the voyagers remain confined aboard ship (see number 22). Banks closed his letter to Lord Morton with the comment: 'the Viceroy upon being told that the Ship was fitted out to observe the transit of Venus Gravely askd whether that was the Passing of the North Star to the South pole this alone will I think Sufficiently Shew your lordship the State of Learning in this place'. Although Banks could not have known it, Lord Morton was dead by this time (12 October 1768), but such a remark was hardly calculated to raise the viceroy in the estimation of a man whose chief interest in science had been astronomy (see Chambers, ed., *Indian and Pacific Correspondence*, vol. 1, letter 19).

24 To William Perrin

From HM Bark Endeavour, *Rio de Janeiro, 1 December 1768.*

Derbyshire Record Office. DRO D239 M/F 15882. 23.5 × 18.5 cm (one folded and two separate sheets).

Perrin was educated at Eton and at Christ Church, Oxford, as was Banks. He was also a fellow of the Royal Society like Banks. It was probably with Perrin that Banks went plant collecting in the Weald of Kent in the summer of 1765. Banks's friendship with Perrin was at this time a close one. Accordingly his letter to Perrin about the frustrations of being confined aboard ship in Rio de Janeiro harbour is informal and frank. It is written in his own hand: 'O perrin you have heard of Tantalus in hell you have heard of the French man laying swaddled in linnen between two of his Mistresses both naked using Every possible means to excite desire but you never heard of a tantalizd wretch who has born his situation with less patience than I have done mine I have cursd swore ravd stampd & wrote memorials to no purpose in the world they only Laugh at me & exult in their own penetrations to have defeated so deep laid a scheme as they suppose ours to have been'. One shore trip was undertaken despite the ban on such forays: 'I have venturd ashore once Evading a boat load of Soldiers who took after us & found such things as well repaid my risk tho the next Morn the Viceroy had intelligence that such a thing had been done so I dare not venture any more'.

In the heat and annoyance of his unexpected incarceration, Banks noted that the harbour at Rio was a fine but poorly defended one, and a ripe prize for any fleet that might sail in to take it, although he ruefully added that British policy then ran counter to such a measure. With this unhappy thought, and since the *Endeavour* was now being pressed by the Portuguese authorities to depart, Banks concluded his letter to Perrin thus: 'it is probably the last you will receive or any of my Freinds till we have met adieu success attend your Endeavours & undertakings Shall be my warmest wish even at the antipodes Your Sincere & affectionate Friend Jos: Banks'.

James Cook also noted the weakness of the fortifications at Rio de Janeiro. For more on this subject, see Banks's comments in *Banks's Journal*, vol. 1, pp. 203–4; see also Chambers, ed., *Indian and Pacific Correspondence*, vol. 1, letter 20. Consider in this context, too, Buchan's detailed views of the harbour its buildings, defences and layout (number 22).

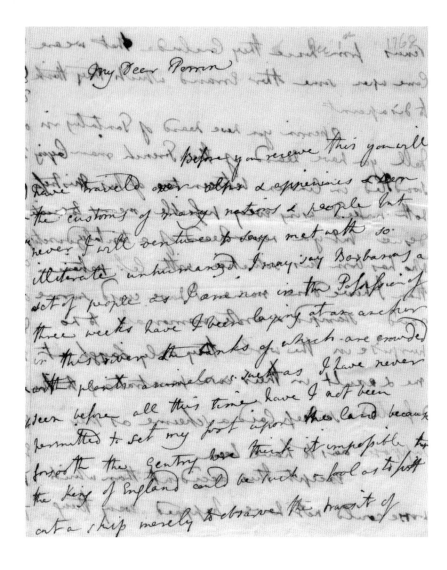

59

nenew from hence they conclude that we are come upon some other errand which they think to disapoint

Opinion you never heard of Tantalus I dell you have heard of the french man being swaddled in linnen between two of his Mistresses both naked using every possible means to excite desire but you never heard of a tantalus destroyed who has born his situation with less patience than I have done mine I have curd swore raved stamped & wrote memorials to no purpose in the world they only laugh at me & exult in their own penetrating to have defeated so deep laid a scheme as they suppose ours to have been

Except this accident than which worse could not have happened every thing

has been favourable the winds & seas have combined to make our passages pleasant 18 days brought us to Madeira where tho we staid only 5 we collected above 300 species of plants 200 of insects &c &c & this late in their autumn the worst time of the year for vegetables two months more brought us here seas mild & calm as they you know always are between the tropicks so that great part of our time has been spent in fishing for fish mollusca &c. & not without great success as new genera in the lost of those species are as common as new species in every other part of nat Hist. drawing has gone forward every day so that I have I do assure you time more to my satisfaction than those two months beyond any with the hopes of

what we should find here hopes not without foundation which nothing but the unexampled barbarity of these rascaly Portugese could have disapointed.

You know that I am a man of adventure & as I scaped hanging in England have now taken it for granted that I am born to be different in pursuance of this opinion I have ventured ashore once evading a boat load of soldiers who look after us & found such things as well repaid my risk tho the next morn the Viceroy had intelligence that such a thing had been done so I dare not venture any more

Since we come here a Spanish packet boat arrivd here from Buenos Ayres you as well as myself know that the spaniards are the natural enemies of the Portugese yet they received with uncommon marks of politeness this I must confess almost drove me mad but I was without remedy

The people here live in a state of Slavery hardly to be equald I believe in the world I hope some time or other to be able in person to tell you what little I have learnt of them which appears realy more like fable than reality their town is as large as I believe any Trading town in England Bristol or Liverpool not at all defended as is possible scarce even an appearence of fortification & what there is mounted with rusty guns falln off from their carriages so much I have seen from my Glass for one day Just under Iloa de Cobras on which is the cheif fortification they offerd to the town as for the Cruz which Guards the Entrance of the river so easy it is good which it is not the sea breeze that blows

here every day from 12 till 6 or 7 would carry a fleet of ships by it without danger as if they chuse their time well they may go in at the rate of 6 or 7 knots an hour the harbour we are in is certainly a very fine one capable of containing any number of ships which may heave down close under the fort in four or five fathom water scarcely they have if the Portugese continue to treat us as they have for some time done a fleet of ships sent here would be a medecine very easily administered & very likely to make a compleat cure But this is a measure so contrary to the present Politicks of our countrey that much as I should from revenge wish it I cannot hope to see it.

My time here is now unexpectedly shortend I have have so broad a hint given us as to think it necessary to Leave the

port tomorrow morn so you must excuse my shortning my letter more than I should have wishd as it is probably the last you will receive or any of my freinds till we have met adeiu success attend your endeavours & undertakings shall be my warmest wish even at the antipodes

Your sincere & affectionate
Freind Jos: Banks

Rio de Janeiro December ye 1st 1768

P.S. Dr Solander who is included in the we so often repeated in this letter desires his best Complts. & joins in my wishes

25 Hummingbird nest

The Natural History Museum, Tring. 1901.1.30.15. 6 × 5 cm (nest) and 16 cm (branch).

This hummingbird nest was probably collected by Banks or Daniel Solander at Rio de Janeiro, judging from label notes in the Natural History Museum, London, where it is held. For a long time its provenance was forgotten among some 2,000 bird nests in the museum collections, but in 2003 it was rediscovered and described. It is thought to be the work of a Glittering-bellied Emerald hummingbird, *Chlorostilbon aureoventris*, and sits in the fork between two branches of the original shrub on which it was found, mainly being attached to the vertical twig. The nest itself was made from fine plant fibres and downy seeds woven together with cobweb. The outside is camouflaged with different species of lichen. Banks and Solander were mostly frustrated in their attempts to explore at Rio de Janeiro, only venturing ashore in secret due to the Portuguese authorities forbidding most of *Endeavour*'s men from landing, except when necessary to resupply. Few natural history specimens were therefore obtained at this location. *Chlorostilbon aureoventris* breeds about a metre above the ground and that probably made the nest easy for Banks or Solander (or a helpful crew member) to collect. This species is still fairly common around Rio de Janeiro.

26 *Globivenus rigida* (Dillwyn) as *Ventricola rigida* (Dillwyn), in G.L. Wilkins, *Bulletin of the British Museum (Natural History)*, vol. 1 (3) 1955, p. 92, and plate 19, figs. 23 and 24

The Natural History Museum, London. Height 4.9, width 2 and length 5.3 cm and height 4.9, width 2 and length 5.3 cm, for two valves.

These shells are part of a collection now held at the Natural History Museum, London, that originally belonged to Banks and which, according to Guy Wilkins, passed from him to the Linnean Society in 1815. It was subsequently received by the British Museum in 1863. The collection remained untouched until Wilkins published his descriptive catalogue in 1955 in which he identified the specimens and correlated them with Daniel Solander's original manuscript labels and descriptions. Wilkins noted that when the Banks shell collection came to the British Museum it was clearly incomplete, and suggested that it may have been inadvertently divided in 1863, the missing portion perhaps having been disposed of at an auction sale. It is likely that the present specimen, originally described by Solander as *Venus rigida*, and labelled 'Rio de Janeiro' probably by Banks, was gathered from Ilha Rasa, where Banks and his party were able to collect freely having departed Rio de Janeiro and the unwelcome attentions of the Portuguese viceroy's guards. This specimen is of taxonomic importance, having been designated by Wilkins as the lectotype (the single specimen from those collected which he designated the type) of the name *Venus rigida* (see numbers 35 and 115).

Note that the transfer to the Linnean Society by Banks of the cabinet containing his shells and insects took place in 1805 according to the naturalist Edward Donovan, and not at a date around 1815, which both Wilkins and later Peter Whitehead thought to be the case, for which see Donovan's *An Epitome of the Natural History of the Insects of New Holland, New Zealand, New Guinea, Otaheite, and other islands in the Indian, Southern, and Pacific Oceans...* (London, 1805), p. iv. Banks's library copy of this work, in the British Library, is at pressmark 435.f.16.

27 *Endeavour*, off the coast of Tierra de Fuego, 12 January 1769

By Alexander Buchan.

The British Library, London. BL Add. MS 15507, f. 6(b). 13.3 × 37.5 cm.

The distance between Rio de Janeiro and Cape Horn is over 2,500 miles, and as *Endeavour* sailed southwards the winds and seas rose, and the weather became less clement. On 6 January Banks described a rough night and its consequences for his belongings: 'In the Evening blew strong, at night a hard gale, ship brought too under a mainsail; during the course of this my Bureau was overset and most of the books were about the Cabbin floor, so that with the noise of the ship working, the books &c. running about, and the strokes our cotts or swinging beds gave against the top and sides of the Cabbin we spent a very disagreable night'. Evidently the pounding did less harm to the ship than to Banks's books, for the next day he commented: 'The ship has been observd to go much better since her shaking in the last gale of wind, the seamen say that it is a general observation that ships go better for being what they call Loosen in their Joints, so much so that in chase it is often customary to knock down Stantions &c. and make the ship as loose as possible'.

This gouache coastal scene, painted by Alexander Buchan, is annotated in Banks's hand on the verso 'Jan 12. 1769 Lat.'. Tierra del Fuego was sighted to the south-west the day before, appearing cold and mountainous inland but, Banks noted in his journal, with what looked to be more fertile, wooded land nearer the coast. The sombre mood of the painting reinforces the sense of *Endeavour*'s isolation when set against the rugged backdrop formed by the coast at this point in the voyage. Smoke from fires was seen ashore, and appropriately so for what Ferdinand Magellan had in 1520 named the Land of Smoke, later to become the Land of Fire in recognition of the many fires maintained by the local people to stay warm and prepare food. A harbour was sought and the next day Banks wrote that 'After dinner a small breeze sprung up and to our great Joy we discoverd an opening into the land and stood in for it in great hopes of finding a harbour; however after having ran within a mile of the shore were obliged to stand off again as there was no appearance of shelter and the wind was on shore'.

At this point in the voyage *Endeavour* was near the Le Maire Strait, into which at the fourth attempt she would pass. During that stormy passage, amid heavy gales and powerful currents, *Endeavour* pitched and rolled violently, but proved herself well able to withstand the notorious fury of these waters. Whitby's shipbuilders had done their job well (see *Banks's Journal*, vol. 1, pp. 212–15).

28 'A VIEW of the ENDEAVOUR'S Watering-place in the Bay of GOOD SUCCESS'

By Alexander Buchan.

The British Library, London. BL Add. MS 23920, f. 11(b). 24.8 × 33.7 cm.

This is a painting by Alexander Buchan signed 'A. Buchan delin.ᵗ', and annotated on the recto probably by Herman Diedrich Spöring as given above. In gouache on vellum, it shows the Bay of Good Success, where *Endeavour* anchored on 15 January 1769. Buchan depicts the men of *Endeavour* ashore gathering water and wood, and in contact with the first non-European people encountered during the mission, the nomadic Ona of Patagonia. He made a preparatory pencil sketch for this painting, perhaps on 20 January following a disastrous inland excursion with Banks and after two days of poor weather when landing proved impossible, BL Add. MS 23920, f. 11(a). He may then have completed the final painting once *Endeavour* was back at sea on its way to Tahiti. The painting shows his limitations as a draughtsman, especially with regard to figure work. Possessing little or no formal training in landscapes, Buchan does not invest the scene with any of the picturesque elements then becoming so fashionable among British artists. Instead, his simplicity of style brings with it a directness of observation that must have suited Banks's requirement for a faithful record of the places and peoples visited during the voyage.

The day before arriving at the Bay of Good Success, Banks and Solander went ashore for four hours to collect plants at Thetis Bay (to them Vincent's Bay), and there they obtained some hundred new plants. This must have fired Banks's enthusiasm following the earlier disappointment encountered at Rio de Janeiro, where most of the voyagers were confined aboard ship by the Portuguese viceroy. Banks and Solander led the way in establishing friendly relations with the local people at the Bay of Good Success, but the name of this bay belies harsh lessons soon to be learned about the dangerously unpredictable conditions sometimes encountered during missions of this kind. The day after landing Banks led a party inland, including some crewmen, Alexander Buchan, Daniel Solander, the astronomer Charles Green as well as Banks's Revesby Abbey servant Peter Briscoe, and also two black servants Thomas Richmond and George Dorlton. To begin with the mid-summer weather was fine, but later it unexpectedly deteriorated into a blizzard. Heavy undergrowth and Buchan fainting (he suffered from epilepsy) caused considerable difficulties while making for some hilltops to gather alpine plants. Banks recorded the trek towards the high ground thus: 'Soon after we saw the plains we arrivd at them, but found to our great disappointment that what we took for swathe was no better than low bushes of birch about reaching a mans middle; these were so stubborn that they could not be bent out of the way, but at every step the leg must be lifted over them and on being plac'd again on the ground it was almost sure to sink above the anckles in bog' (*Banks's Journal*, vol. 1, p. 219). The problematic plant was almost certainly a species of the southern beech *Nothofagus*, probably *N. betuloides*, a species with leaves resembling a birch (see number 29).

Banks and some others, Solander among them, reached the hilltops while the rest huddled around a fire, but Solander fell exhausted on the return journey, as did Richmond. Snow was descending and it became clear that the party would not make it back to *Endeavour* and, worse, Richmond and Dorlton could not be moved to an overnight camp, having collapsed from the cold and the effects of rum (while alone they drank the party's entire supply). Both men were left for the night under a thick covering of branches but tragically neither survived. The next day Banks and his party reached the ship again. Unfamiliar terrain, the changeable weather and, perhaps too, over-enthusiasm following the earlier frustration experienced at Rio de Janeiro led to a near disaster for the expedition in the form of the loss of its two naturalists and designated astronomer. Banks, however, was soon netting fish, collecting shells and plants and visiting nearby settlements.

The expedition departed on 21 January. From the Le Maire Strait, *Endeavour* went on to round Cape Horn in mild weather conditions with only some light fog. Astronomical readings were possible, and using them the position of Cape Horn was calculated to Cook's considerable satisfaction. It was usual after entering the Pacific by this route for vessels to hug the coastline of South America, sailing northwards in order to avoid the prevailing ocean and wind currents from the west. Once in the tropics they availed themselves of favourable ocean currents running from the east to cross the Pacific in warmer latitudes. However, Cook's orders enjoined him to pursue a course north-west from the Cape into the mid-ocean in search of a supposed southern continent, heading eventually for Tahiti. This he did after sailing to latitude 60° 4′ South. Day by day the ship proceeded into the central Pacific, turning more and more of the imagined land before it into salt water. The next safe anchorage would not be until 13 April, when the ship finally sailed into Matavai Bay at Tahiti.

In the 1784 reissue of Sydney Parkinson's voyage account, John Fothergill noted an obscure detail relating to the voyage aftermath. According to Fothergill, Banks provided for the wife of one of his deceased black servants. Fothergill does not say which one, nor whether the other black servant had a wife, but it seems unlikely since none is mentioned. At this time, Banks also paid Parkinson's family the outstanding £151 of salary owing to the artist, who was yet another mission casualty, further enhancing this sum to £500 in recognition of Parkinson's outstanding service and as a final settlement with his family for his papers and object collections. Fothergill explained in his 'Explanatory Remarks' to the *Journal of a Voyage*, p. 6, that 'J. Banks very readily fell in with the proposal, and settled at the same time a pension upon a black woman, the wife of a faithful black servant who went out with him, and perished by the cold of Terra del Fuego'. The death rate on the *Endeavour* mission was much higher than that on either of the two ensuing Cook voyages into the Pacific, a fact in no way due to neglect of the crew's health on the part of their commander, but rather to diseases contracted at ports in the Dutch East Indies as the ship made its way home (see numbers 131 and 133).

29 *Nothofagus betuloides* (Mirbel) Oersted, *K. dansk. Vidensk. Selsk. Skr.*, ser. 5, 9: 354 (1871)

Engraving by Gerald Sibelius after Sydney Parkinson.

The Natural History Museum, London. Diment et al., Catalogue, Part 2 Botany (1987), TF 66, col. engraving 1988 BF: pl. 727. Plate height 46 × width 29.5 cm; plant ill. height 29 cm.

This is a coloured print from a copper plate engraved by Gerald Sibelius, based on a finished 1769 watercolour drawing by Sydney Parkinson. A specimen of the plant, gathered at Tierra del Fuego in January 1769, the original Parkinson watercolour and the Sibelius plate are all held at the Natural History Museum, London.

While at Tierra del Fuego, Banks lost two of his servants, who died in a blizzard on an inland trip, and the unpredictable conditions almost claimed Daniel Solander and Banks's landscape artist Alexander Buchan too (see number 28 above). They were making for hills to collect alpine plants. The way ahead looked to be grass-covered, but turned out to be brush-covered bog, which proved heavy going and contributed to the problems. A blizzard set in and Banks and his companions were forced to camp overnight in the bitter open air. Banks later complained in his journal about what he called birch as being the problematic shrub. This was, in fact, almost certainly a species of the southern beech *Nothofagus*, either *N. antarctica* or *N. betuloides*, a species with leaves resembling a birch, with which tree it was confused by both Banks and subsequently Charles Darwin. *N. betuloides* was named *Betula antarctica* by Solander, and later *Fagus betuloides* by Charles-François Mirbel, who correctly classified it as a beech and not a birch. Beeches growing in the southern hemisphere were transferred to the genus *Nothofagus* by Carl Ludwig Blume in 1850. The southern beeches are large forest trees, but at altitude in wet soils their growth can become stunted and shrubby, and it was probably dense stands of such plants that caused serious difficulties for Banks's party.

Later on, Charles Darwin would encounter similar problems when exploring Tierra del Fuego. Referring to the Bay of Good Success during the voyage of the *Beagle*, Darwin wrote on 20 December 1832: 'One side of the harbour is formed by a hill about 1500 feet high, which Captain FitzRoy has called after Sir J. Banks, in commemoration of his disastrous excursion, which proved fatal to two of his party, and nearly so to Dr. Solander.... I was anxious to reach the summit of this mountain to collect alpine plants; for flowers of any kind, in the lower part, were few in number'. Darwin described having to 'crawl blindly among the trees. These, from the effects of the elevation, and of the impetuous winds, were low, thick, and crooked. At length we reached that which from a distance appeared like a carpet of fine green turf, but which, to our vexation, turned out to be a compact mass of little beech-trees about four or five feet high. These were as thick together as box in the border of a flower-garden, and we were obliged to struggle over the flat but treacherous surface' (Darwin, *Beagle Narrative*, vol. 3, p. 232).

30 'INHABITANTS of the Island of TERRA DEL FUEGO in their Hut.'

By Alexander Buchan.

The British Library, London. BL Add. MS 23920, f. 14(a). 26 × 36.2 cm.

Signed by Alexander Buchan 'A. Buchan Delin^{t}', and annotated in ink on the recto probably by Herman Diedrich Spöring as given above, this gouache painting must date to the visit of *Endeavour* to Tierra del Fuego in January 1769.

Of Scottish descent (probably from Berwickshire), Alexander Buchan was employed by Banks on the *Endeavour* mission to depict landscapes and figures, which he did in a simple and unaffected manner. It is not clear how Banks came across Buchan or what training as an artist, if any, he might have had before departing on the voyage. There are a small number of finished invertebrate and fish watercolours by Buchan during the Atlantic stage, probably completed under Parkinson's instruction when no land was in sight. There are no plant illustrations by him. Buchan drew some good coastal profiles in the Atlantic but, like his landscapes, these tailed off after Tierra del Fuego. It may be that some of Buchan's work has been lost, or that his declining health in the Pacific prevented him producing many drawings there, but his untimely death at Tahiti curtailed what was in any case a modest overall contribution to the mission artwork. His main effort

was devoted to landscape and figure work, in which he appears free from the picturesque influence sometimes apparent in Sydney Parkinson's scenes, but otherwise lacking in Parkinson's delicacy and skill of execution. Buchan suffered from epilepsy, dying of a fit on 17 April 1769, much to Banks's dismay. Thereafter his duties were performed by Parkinson and by Banks's clerk Herman Diedrich Spöring. Spöring was a useful support to Parkinson in this regard because the quantity of natural history illustration undertaken by Parkinson grew enormously as the voyage progressed.

In the current painting, Buchan records a group of Fuegians huddling in their hut before a warming fire, perhaps while readying a meal of shellfish. There is no background. Attention is firmly fixed on the hut and its occupants in a depiction that appears severely factual. Not untypically Buchan appears to have struggled with the figures, but he fared rather better with the hut. This scene was witnessed during Banks's last shore visit at this location on 20 January, the day before *Endeavour* departed, and of it he wrote: 'We returnd on board to dinner and afterwards went into the Countrey about two miles to see an Indian town which some of our people had given us intelligence of; we arrivd at it in about an hour walking through a path which I suppose was their common road tho it was sometimes up to our knees in mud. The town itself was situate upon a dry Knowl among the trees, which were not at all cleard away, it consisted of not more than twelve or fourteen huts or wigwams of the most unartificial construction imaginable, indeed no thing bearing the name of a hut could possibly be built with less trouble. They consisted of a few poles set up and meeting together at the top in a conical figure, these were coverd on the weather side with a few boughs and a little grass, on the lee side about one eighth part of the circle was left open and against this opening was a fire made. Furniture I may justly say they had none: a little, very little, dry grass laid round the edges of the circle furnishd both beds and chairs, and for dressing their shell Fish (the only provision I saw them make use of) they had no one contrivance but broiling them upon the Coals. For drinking indeed I saw in a corner of one of their hutts a bladder of some beast full of water: in one side of this near the top was a hole through which they drank by elevating a little the bottom which made the water spring up into their mouths.

In these few hutts and with this small share or rather none at all of what we call the necessaries and conveniences of life livd about 50 men women and children, to all appearance contented with what they had nor wishing for any thing we could give them except beads; of these they were very fond preferring ornamental things to those which might be of real use and giving more in exchange for a string of Beids than they would for a knife or a hatchet' (*Banks's Journal*, vol. 1, p. 224).

Banks alludes to the apparent contentment of the Fuegians, as though they might almost be 'noble savages' living at one with their bleak environment and desiring little from Europeans except perhaps a few beads, but his physical description is nevertheless close in detail and tone to Buchan's uncompromisingly documentary view. In the case of the Fuegians, at least, Banks appears to have found it relatively easy to resist the temptation to make romantic comparisons and throughout his voyage journal, despite the use of assorted classical and other allusions, he generally attempts to observe peoples and their customs as he found them. See, also, Buchan's pen and wash illustration of the Fuegian village situated among some trees backed by a hill, BL Add. MS 23920, f. 12.

Unmentioned by Banks, note what appear to be drying frames, one on the hut roof to the viewer's left, and a second leaning against the hut side to the viewer's right. Such frames were used for drying skins, a process that took two to three weeks, with the skins being stretched on the frame, which was set upright and exposed to air currents. Rods were positioned lengthwise and crosswise, and fixed through holes made in the edge of the skin, thereby holding the skin taut. The frames appear to be located on the skin's fur-side. Buchan's placing of two examples to show the back and front of the frames is evidently a device to make their construction and use clear (see Joppien and Smith, *The Art of Captain Cook's Voyages*, vol. 1, pp. 82–3).

31 'Natives of Terra del Fuego with their Hut'

By Sydney Parkinson.

The British Library, London. BL Add. MS 23920, f. 13.
29.8 × 48 cm.

This pen and wash drawing by Sydney Parkinson is annotated in ink on the folio as given above, with a partly erased inscription possibly by Banks reading 'Natives of Terra del Fuego with their Hut'. A note is inscribed in pencil on the verso in a later unknown hand, 'inhabitants of Terra del Fuego with their Hut'. A preparatory sketch for one of the native faces is at BL Add. MS 23920, f. 18(e). Assuming it was completed at or shortly after leaving Tierra del Fuego, this is the earliest voyage drawing by Parkinson of a landscape with figures. It is different in various ways to the same scene as drawn by Alexander Buchan in number 30 immediately above.

Parkinson's Fuegian hut is, for example, shown against a background, whereas Buchan omits any context, but Parkinson's setting is not a sheltered woodland glade behind a hill as was actually the case. Instead his hut appears amid pleasing shoreline glades with an extensive background vista. He also includes more figures than Buchan chose to record, and these are decorously arranged before the hut. There are token references to ethnographic details such as the water bladders, a bow and the hut itself, although the latter lacks any drying frames. Parkinson avoids the nakedness of the male Fuegian. Alterations such as these were in keeping with a picturesque sense of design apparent in much of Parkinson's landscape work on the voyage, one that seems here to pervade what was in any case a preliminary foray into the scenic field. The picturesque as a term defined and applied by the writer and artist William Gilpin had, of course, yet to enter general usage but the ideas on which it was based were already present in English landscape art.

Parkinson's ethnographic and landscape drawings form a richly evocative record of the voyage that includes full landscapes, landscapes with figures and objects, detailed studies of individuals and their dress and possessions, draft sketches for finished illustrations and many sketches of buildings and, in particular, canoes and other vessel types. Some three-quarters of the surviving ethnographic and scenic drawings made during the voyage are attributable to Parkinson. His varied work is generally reliable in its detail and convincing in terms of approach. His landscapes and scenes really commence in earnest once he reached Tahiti and Buchan had died. These often display picturesque elements of organization and style within which new information about the terrain, its vegetation and the manners and customs of local people is framed. The near and middle distance of a view might be staffed with various typical figures or feature particular activities or plants that are intended to be plausible examples of their kind. Sketches and specific drawings of ethnographic subjects such as man-made structures or the implements and activities of daily life are otherwise carefully rendered. For these Parkinson's approach tended to be more obviously documentary, just as it was to a greater degree with natural history specimens, and for that matter with the coastal profiles that he and his colleagues undertook for Cook (though Andrew David thought he detected some imaginative touches in one or two of Parkinson's profiles, *The Charts & Coastal Views of Captain Cook's Voyages*, vol. 1, p. xliii).

Rarely did he make personal portraits. Instead Parkinson tried in the main to capture characteristic differences between island peoples. Like his fellow voyage artists, Parkinson struggled with figures, but he nevertheless attempted to convey a clear visual impression of the peoples encountered, and largely avoided classical or other artificial poses. When the latter do appear in illustrations attributed to him, it is usually in engravings produced by others in London after the voyage, Parkinson by then being dead and his original drawings for such engravings now being lost. It is not therefore clear whether he drew these particular illustrations in a stylized way, but that seems unlikely given the approach adopted in most of his surviving ethnographic work. Parkinson's figures gradually improved as the voyage progressed so that his existing individual and group studies from New Zealand demonstrate a more confident hand. Close studies of heads clearly show facial contortions, as seen in dances at Tahiti or in the Māori *haka*. They might incorporate, too, ornamental items and clothing as well as skin markings, notably the iconic *moko* observed in New Zealand. See numbers 78 and 84. Studies of whole figures or groups incorporate such aspects too, and they also reference activities such as dancing, rowing, sailing, spear throwing or making barkcloth. A number of sketches by Parkinson show the care with which he observed

these and other details, including particular locations and man-made structures, and how using such sketches in draft form he might prepare for larger, finished works. See numbers 46 and 63. Parkinson was never, however, as starkly factual in his rendering of people as was Buchan in his version of the present scene.

In the Pacific, Parkinson also started to draw inventive composite scenes showing elements of local life and action in a single set-piece illustration, sometimes making preparatory sketches of some of the details to be included as explained above. Groups of figures, their dress and other possessions as well as their activities, are featured together in these works. Such illustrations were a means of presenting a lively scene in which selected aspects of indigenous life appear against a visually striking backdrop. The ethnographic details may be realistic, but the composition itself is to a great extent imagined. In such illustrations the background is intended to be artistically pleasing and to suggest the exotic locations that were visited. Palms and other foliage, bay outlines or clearings, buildings and distant topography help to invoke these places. While in the foreground what is aimed at is a sense of people and their habits and surroundings, with these displayed more faithfully and in greater detail for closer scrutiny. At the very least, such an approach had the practical advantage of enabling Parkinson to capture more of the expedition in a single illustration, thereby somewhat reducing his workload, a not unimportant consideration as the artistic burden on him increased greatly during the voyage. A good example of this sort of illustration may be seen at number 71.

On other occasions Parkinson's landscapes seem almost to be a picturesque outlet from the rigours of his main scientific task of accurately recording the plants and animals gathered on the voyage. At times like this Parkinson might embellish his scenery, making it more dramatic or romantic than actually was the case, heightening hills to mountains, focusing on or even adding decorative features such as cascades or trees, giving the tropical sky and weather a prominent place and rearranging topography to suit his artistic taste. Ethnographic and natural history content might be less prominent in such illustrations, or be altogether absent from them, and the landscape as a subject dominates. This brings with it greater creative freedom, but it introduces too European conventions of style and presentation. A number of landscape illustrations like this find Parkinson at his most inventive during the voyage and are fine examples of his personal response to events and places seen on it (see numbers 94 and 95).

32 'A View of the Indians of Terra de Fuego in their hut.'

John Hawkesworth, ed., An account of the voyages undertaken by the order of His Present Majesty for making discoveries in the southern hemisphere, and successively performed by Commodore Byron, Captain Wallis, Captain Carteret, and Captain Cook, in the Dolphin, the Swallow and the Endeavour: drawn up from the journals which were kept by the several commanders, and from the papers of Joseph Banks, Esq, 3 vols (London: Printed for W. Strahan and T. Cadell, 1773). Vol. II, plate 1. Drawn by J.B. Cipriani. Engraved by F. Bartolozzi.

The British Library, London. 455.a.22. Rebound in half blue leather and blue cloth boards. Title, place and date of publication, gilt lettering on spine. Editor and volume given. 29 × 24.2 cm (closed). Plate 22.6 × 30 cm.

This is an engraved print from the official account of the *Endeavour* voyage, edited in three volumes by Dr John Hawkesworth and published in 1773. With Alexander Buchan and Sydney Parkinson dead, preparation of this scene for publication was given to artists and engravers in London, who used Buchan's and Parkinson's work, but made significant changes to elements of its content and style. These London illustrators had never been to the Pacific. Lacking any direct experience of the scenes to be completed for the account, they tended to render those that they prepared according to the stylistic conventions with which they were most familiar. Moreover, they worked under an editor who had only been selected to oversee the publication on the recommendation of Charles Burney because the first lord of the Admiralty, Lord Sandwich, could not think of anyone better to do it. Ponderous to read for some, Hawkesworth was a former schoolmaster who had worked successfully with Samuel Johnson on the *Gentleman's Magazine* and then the *Adventurer* as well as penning various moral essays, criticism, translations and oriental tales, and might have seemed a fair choice except that, like his illustrators, he had no real experience of the sea.

Regardless, Sandwich gave Cook's and Banks's journals to Hawkesworth, who was required to blend them so that they read as one, and further to edit and publish the journals of John Byron, Samuel Wallis and Philip Carteret in the same work. Influenced by concepts of the noble savage, Hawkesworth tended to see the indigenous peoples that were encountered during these voyages, including the Fuegians eking out a sparse life in their often hostile surroundings, as exemplars of primitive man living in natural harmony with his environment, seemingly free from the material attachments and other trappings of European civilization. Peoples inhabiting areas of natural abundance such as the Society Islands might likewise be viewed as living in 'a state of nature', except that they apparently enjoyed an existence of simple innocence in a tropical paradise of plenty. Even disturbing aspects, such as the wars and sometimes cruel religious practices witnessed in these islands, might be explicable in terms of the natural condition and primitive innocence of their inhabitants (which is more than could be said of many Europeans then or since). In what was, to him, a familiar moralizing mode Hawkesworth added to the voyage narratives various comments along these lines while at the same time retaining the first person, thereby giving the impression that he was putting words in the mouths of the explorers. Unfortunately for Hawkesworth this, the unseaman-like classical allusions, various technical errors and the frankness with which he treated the sexual freedoms of the South Sea islanders all drew withering criticism following publication. Cook dismissed the book, a reaction perhaps exacerbated by the exceptional fee of £6,000 that Hawkesworth received for the rights from the publishers Strahan and Cadell. The unhappy doctor never recovered from the shock of it all and died of a fever a few months after publication. Nevetheless the first edition rapidly sold out and further editions were soon being issued by publishers eager to cash in on the Europe-wide popularity of 'Cook's voyage account'.

It was against this background that at least two versions of the Fuegians in their hut were produced for the official account, both probably completed in 1772. The little-known John James Barralet produced a watercolour and ink version of Parkinson's hut scene, but it was not used. Instead, Hawkesworth appears to have preferred a wash and watercolour version by the fashionable Italian neo-classical artist Giovanni Battista Cipriani, based on Parkinson's and Buchan's illustrations of the same scene. Respectively the Barralet and Cipriani versions are now held in the Knatchbull Collection, Brabourne Estate, Kent, and in the Dixson Library, Sydney, DL PXX2, 43. Cipriani's work was predictably enough engraved by his lifelong friend and fellow-Florentine Francesco Bartolozzi, a leading engraver in the capital in these years. Both men had trained together in their home city, and during their time in England a large number of Cipriani's designs were engraved by Bartolozzi. Both men were founding members of the Royal Academy in 1768.

Cipriani transformed the reality of life on Tierra del Fuego as recorded by Buchan into something resembling a classical idyll. His treatment of the hut and the figures crouching in it contrasts with Buchan's austerely rendered view, with the addition of three extra figures situated in front of the

dwelling evidently being suggested by Parkinson's drawing, although their dress and posture are much altered from those in the Parkinson original. Cipriani's hut is framed by a tree on one side and a crag on the other. Joppien and Smith convincingly suggest that some of the figures placed before the hut reference earlier work by Cipriani at Osterley Park, where he was employed by Robert Adam to produce an over-mantel for the fireplace in the dining room, a work entitled *An Offering to Ceres* in celebration of the harvest and rural life. Adams revamped Osterley in the 1760s for the wealthy banking dynasty the Childs, and so this room would have been the setting for family meals and entertaining. Such a link, if correct, is ironic to say the least given the actual situation of Buchan's Fuegians. Where once a group of Fuegians sheltered from the elements, attired in traditional skins and furs, Cipriani now gives us 'comely youths and maidens, and wise old men enjoying the delights of nature's simple plan' (Joppien and Smith, *The Art of Captain Cook's Voyages*, vol. I, pp. 16–19). Some basic ethnographic facts are retained more or less intact, most notably the hut itself, as well as the bladders holding water and one of the drying frames, but these are inevitably out of place in such a depiction. Even allowing for the fashionable tastes of the day, as also for theories of the noble savage then current among London's literary classes, if not all of its circumnavigating explorers, the result of Cipriani's efforts was in many respects a world away from life in the Fuegian village visited in January 1769.

33 'A MAN of the Island of TERRA DEL FUEGO'

By Alexander Buchan.

The British Library, London. BL Add. MS 23920, f. 16. 36.8 × 26.7 cm.

Signed by Alexander Buchan 'A. Buchan', this gouache painting is annotated in ink on the recto probably by Herman Diedrich Spöring as given above. The date of painting is likely to be of January 1769. In his journal Banks recorded in summary various details about the landscape, natural history and people in the Bay of Good Success area: 'As this is to be the last time of our going ashore on this Island I take this opportunity to give an account of such things [as] the shortness of my stay allowd me to observe'. This was typical of his approach to journal writing during the voyage. A general account of a location was normally written upon departure from it, encompassing a range of phenomena including the topography, plants, animals, inhabitants and their way of life. These reflections reveal much about the places visited and the contacts that took place, and they served to enrich the official account of the voyage that was eventually published.

Of the men and women of the Bay of Good Success, Banks wrote: 'The inhabitants we saw here seemd to be one small tribe of Indians consisting of not more than 50 of all ages and sexes. They are of a reddish Colour nearly resembling that of rusty iron mixd with oil: the men large built but very clumsey, their hight from 5 f' 8 to 5 f' 10 nearly and all very much of the same size, the women much less seldom exceeding 5 f'. Their Cloaths are no more than a kind of cloak of Guanicoe [Guanaco, a South American lama] or seal skin thrown loose over their shoulders and reaching down nearly to their knees; under this they have nothing at all nor any thing to cover their feet, except a few of them had shoes of raw seal hide drawn loosely round their instep like a purse. In this dress there is no distinction between men and women, except that the latter have their cloak tied round their middle with a kind of belt or thong and a small flap of leather hanging like Eve's fig leaf over those parts which nature teaches them to hide; which precept tho she has taught to them she seems intirely to have omitted with the men, for they continualy expose those parts to the view of strangers with a carelessness which thoroughly proves them to have no regard to that kind of decency' (*Banks's Journal*, vol. 1, p. 227).

Banks took care to record the height of the men and women that he met at this location, the myth having spread in Europe that Patagonia was peopled by giants. Note that Buchan also produced a close study of a male headband of the kind seen in the present illustration, BL Add. MS 23920, f. 21(a), as well as of some Fuegian necklaces and a bracelet, BL Add. MS 23920, f. 20(a,b) and BL Add. MS 15508, f. 1. At the Museum of Archaeology and Anthropology, Cambridge, there is a shell necklace from Tierra del Fuego that was gathered on the *Endeavour* voyage, but it does not precisely match any of the Buchan studies just cited, D.1914.77. For Ona interest in gifts of beads, doubtless for use in decorative items of this kind which were popular among them, see also number 30 above and number 34.

34 'A WOMAN of the Island of TERRA DEL FUEGO'

By Alexander Buchan.

The British Library, London. BL Add. MS 23920, f. 17.
36.8 × 26.7 cm.

Completed in the same period as the preceding male figure, this gouache painting by Alexander Buchan is annotated in ink on the recto probably by Herman Diedrich Spöring as given above. Sydney Parkinson produced a pencil sketch apparently of the same female attired and positioned in similar fashion, BL Add. MS 23920, f. 18(b). Upon encountering indigenous peoples during this and subsequent Cook voyages into the Pacific, mission artists would often illustrate certain individuals to show their physical appearance, dress, ornaments, hair (how it was cut and adorned) and skin markings if they had any. These illustrations were frequently divided into one of a man and one of a woman, featuring head and shoulders or the whole figure as here, but thereby creating a pair. Such illustrations convey, among other things, an impression of a people through close studies of selected individuals who display representative details of physical appearance and local attire. Joppien and Smith noticed the tendency in the *Endeavour* artwork to portray typical rather than known persons, presumably the better to demonstrate broad cultural and ethnic traits, and that accordingly only two print engravings in Parkinson's posthumously published voyage account are of named individuals (these must be Taiato and Te Kuukuu, numbers 48 and 84 below; see Joppien and Smith, *The Art of Captain Cook's Voyages*, vol. 1, pp. 21–4).

In his journal Banks recorded: 'Their food at least what we saw them make use of was either Seals or shell fish. How they took the former we never saw but the latter were collected by the women, whose business seemd to be to attend at low water with a basket in one hand, a stick with a point and barb in the other, and a satchel on their backs which they filld with shell fish, loosning the limpits with the stick and putting them into the basket which when full was empty'd into the satchel' (*Banks's Journal*, vol. 1, p. 228). It seems likely, given the close agreement between Banks's description of Ona life and dress and Buchan's paintings of them, that Banks supervised the work of his artist at Tierra del Fuego.

On 21 January the *Endeavour* departed the Bay of Good Success, supplies having been gathered and the bay itself surveyed by Cook.

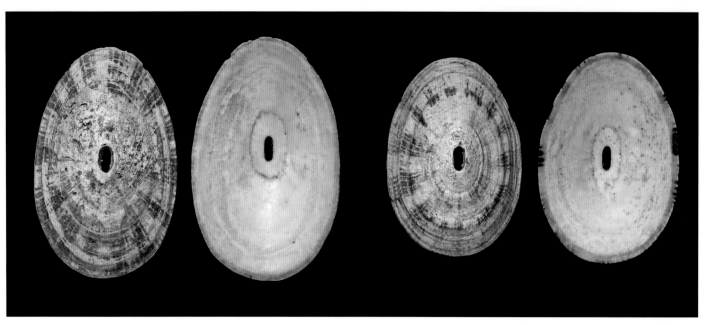

35.1

35 *Fissurella picta* Gmelin as *Fissurella picta* Lamarck and *Acanthina monodon* form *imbricata* as *Acanthina calcar* (Martyn) in G.L. Wilkins, *Bulletin of the British Museum (Natural History)*, vol. 1 (3) 1955, pp. 93–4

The Natural History Museum, London.

1. Height 8.1, width 2.7 and length 5.2 cm and height 7, width 2.3 and length 4.8 cm, for two specimens.

2. Height 5.5, width 3 and length 4 cm and height 3.5, width 2 and length 2.5 cm, for two specimens.

These specimens were gathered on 20 January 1769 by Banks and Daniel Solander from the beaches of Tierra del Fuego and were mentioned by Banks in his voyage journal (see *Banks's Journal*, vol. 1, pp. 223–9). Banks's comments demonstrate that in the short time that was available to him, he collected from the range of shell material on the beaches for later study, rather than simply collecting the largest or most attractive specimens. Both of the present species were new to him and he particularly mentions them in his journal. For the species now known as *Fissurella picta*, Banks noticed the difference between the shell of this species, which has an apical opening, and others lacking this feature. That clearly identifies the present specimen as the one he collected. Among the other shells collected on this day that he also mentions is a whelk with a long tooth, a character which identifies it as the second specimen shown here, originally classified by Solander as *Buccinum monodon*. These species almost certainly formed part of the diet of the local people, see numbers 30–2 and 34 above.

35.2

36 *Phaethon rubricauda melanorhynchos* Gmelin, 1789

By Sydney Parkinson.

The Natural History Museum, London. Wheeler, Catalogue, Part 3 Zoology (1986), 31.(1:31). 29 × 31.5 cm.

This is a signed watercolour by Sydney Parkinson, annotated on the recto in ink and pencil 'Phaëton erubescens. Sydney Parkinson pinx^t 1769. Tawai'. This subspecies is now confined as a breeding bird to the Society Islands and Palmerston Island. It was seen by Banks and noted in his journal during March 1769 as *Endeavour* crossed the Pacific towards the Tuamotus and thereafter the Society Islands. On 21 March Banks wrote: 'Calm this morn: went out in the boat and shot Tropick bird *Phaeton erubescens*, and *Procellaria atrata*, *velox* and *sordida*' (*Banks's Journal*, vol. 1, p. 240). Banks brought back a specimen that was used, possibly with the present Parkinson illustration, by the naturalist John Latham to describe the species in his *General Synopsis of Birds* (1785) as the black-billed tropic bird. J.F. Gmelin later used that description when he named the species *Phaethon melanorhynchos* in his *Systema Naturae* (1789), so the painting has some systematic importance.

As an adult the red-tailed tropic bird has long slender red streamers in its tail and flies with only periodic flaps of its wings. It often follows ships for hours or even days, circling, diving for fish and sometimes landing on the water. Its distinctive feathers were used for decorative purposes in Polynesia. The white feathers fringing the frontal shields of Society Islands *fau* are probably from the white-tailed tropic bird *Phaethon lepturus*, a related species which is also widespread in tropical waters (see number 54). Interestingly, this illustration is the only painting by Parkinson during the *Endeavour* voyage of a bird in flight, possibly because he had a chance to observe one following the ship on its course across the Pacific to Tahiti. It is considered to be one of Parkinson's more convincing bird paintings and evidence of his growing stylistic maturity in this particular branch. However, during the Pacific stage of the voyage plant illustrations dominated Parkinson's output, and he was increasingly unable to do anything more than produce annotated sketches of botanical specimens since these were being gathered in ever greater numbers. Of the 54 species of bird described by Daniel Solander all but five were oceanic species, and three of the five land birds actually flew aboard the *Endeavour* while at sea.

Phaëton erubescens.

Sydney Parkinson pinx.t 1769.

Dressing Up, Taking Over, and Passing On:
Joseph Banks and Artificial Curiosities from the Endeavour Voyage

JEREMY COOTE

Pitt Rivers Museum, University of Oxford

WHAT POLYNESIAN ARTEFACTS – 'artificial curiosities' in eighteenth-century parlance – did Joseph Banks acquire on the *Endeavour* voyage, what did he do with them and where are they now? We are a long way from being able to answer these superficially simple questions, but in attempting to do so here I can perhaps say something interesting about Banks's engagement with Polynesian art and culture more generally. Banks had precious little to say in his voyage journal about collecting artefacts, as opposed to natural history specimens, and no inventory survives. There are, however, a range of records of various sorts that we can draw on in order to get at least some idea of the things Banks had in his collection, what he did with them and where they are – or might be – now.

In November 1772, the newly appointed keeper of the Ashmolean Museum at Oxford, William Sheffield, visited Joseph Banks at New Burlington Street. On 2 December he provided his friend Gilbert White with 'an imperfect sketch' of his visit:

> It would be absurd to attempt a particular description of what I saw here; it would be attempting to describe within the compass of a letter what can only be done in several folio volumes. His house is a perfect museum; every room contains an inestimable treasure. I passed almost a whole day here in the utmost astonishment, could scarce credit my senses. Had I not been an eye-witness of this immense magazine of curiosities, I could not have thought it possible for him to have made a twentieth part of the collection. I have excited your curiosity; I wish to gratify it; but the field is so vast and my knowledge so superficial that I dare not attempt particulars. I will endeavour to give you a general catalogue of the furniture of three large rooms. First the Armoury; this room contains all the warlike instruments, mechanical instruments and utensils of every kind, made use of by the Indians in the South Seas from Terra del Fuego to the Indian Ocean – such as bows and arrows, darts, spears of various sorts and lengths, some pointed with fish, some with human bones, pointed very finely and very sharp, scull-crackers of various forms and sizes, from 1 to 9 or 10 feet long, stone hatchets, chisels made of human bones, canoes, paddles, &c. It may be observed here that the Indians in the South Seas were entire strangers to the use of iron before our countrymen and Monsieur Bougainville arrived amongst them; of course these instruments of all sorts are made of wood, stone, and some few of bone …. The second room contains the different habits and ornaments of the several Indian nations they discovered, together with the raw materials of which they are manufactured. All the garments of the Otaheite Indians and the adjacent islands are made of the inner bark of the *Morus papyrifera* [*Broussonetia papyrifera*] and of the bread tree *Chitodon altile* [*Artocarpus altilis*]; this cloth, if it may be so called, is very light and elegant and has much the appearance of writing paper, but is more soft and pliant; it seems excellently adapted to these climates. Indeed most of these tropical islands, if we can credit our friend's description of them, are terrestial Paradises. The New-Zealanders, who live in much higher southern latitudes, are clad in very different

manner. In the winter they wear a kind of mats made of a particular species of Cyperus grass. In the summer they generally go naked, except a broad belt about their loins made of the outer fibres of the cocoa nut [*sic*], very neatly plaited; of these materials they make their fishing lines, both here and in the tropical isles. When they go upon an expedition or pay or receive visits of compliment, the chieftains appear in handsome cloaks ornamented with tufts of white dog's hair; the materials of which these cloaks are made are produced from a species of *Hexandria* plant [*Phormium tenax*] very common in New Zealand, something resembling our hemp, but of a finer harl and much stronger, and when wrought into garments is as soft as silk: if the seeds of this plant thrive with us, as probably they will, this will be perhaps the most useful discovery they made in the whole voyage.[1]

As A.M. Lysaght has pointed out,[2] what Sheffield saw in November 1772 would have included material Banks had collected in Newfoundland and Labrador in 1766 as well as what had been collected on the *Endeavour*, but it is nevertheless worth quoting his account at length as it is the best description we have of what was in Banks's possession after the *Endeavour* voyage and before the return of the *Resolution* and *Adventure* from Cook's second voyage of 1772–5. Of course, Banks may already have given some things away, but as Sheffield's description is so general this hardly matters for present purposes.

Given that Sheffield's description lacks 'particulars', what can we do to flesh it out? First, of course, there is the famous portrait, painted by Benjamin West in 1771–2, the frontispiece to the present volume (see also number 143). Here we see Banks dressed in a Māori cloak with an intricate *taniko* border, with to his right a Tahitian *fau* (or headdress), a Māori staff and paddle, and to his left – on the floor – a Tahitian hafted adze and barkcloth beater and what appears to be a Māori greenstone patu or club. We can be confident that these items were in Banks's collection, though – unfortunately – we do not know where any of them is now. Similarities between the design of the border of the cloak Banks is wearing and that of a cloak in the collections of the British Museum have been pointed out by Adrienne L. Kaeppler,[3] while I have suggested that the cloak may be identified with one that survives in the Banks collection at the Pitt Rivers Museum (see number 81),[4] but doubts remain. Similarly, the *fau* was previously thought to be the one surviving at the British Museum,[5] but Karen Stevenson and Steven Hooper have recently shown that it is not (see number 54). The other objects also remain unlocated.

The next body of evidence comprises drawings of objects by John Frederick Miller that survive in the British Library. Unfortunately, although it is known that Miller was employed by Banks as a draughtsman, it seems likely that he also drew objects in the 'official' collection that passed from Cook to Lord Sandwich and thence via the Lords of the Admiralty to the British Museum;[7] certainly a number of objects now in the British Museum can be identified with objects drawn by Miller – moreover, many of the objects featured in the engravings published in Hawkesworth's *Voyages* are based on Miller's drawings.[8] Thus, while the drawings provide good evidence of the sorts of objects that Banks is likely to have had in his collection, they are not necessarily drawings of the actual objects Banks had. A probable – and remarkable – exception to this generalization is the Māori *poupou* or house panel from Pourewa Island in Uawa (Tolaga Bay) that was drawn by John Frederick Miller and identified recently in the collections of the University of Tübingen, Germany.[9] This must have been acquired on 28 October 1769, when Banks was one of party that visited Pourewa from the *Endeavour*. Banks remarks in his journal on the elaborately carved canoe and house he saw there, commenting 'all the side posts were carved in a masterly stile of their whimsical taste which seems confined to the making of spirals and distorted human faces'.[10] In fact, the *poupou* is the only known carving from the Te Rawheoro school and its survival is of great significance to the local Te Aitanga-a-Hauiti people (see number 87).[11]

Another body of evidence is provided by the references in contemporary letters, catalogues and other written sources – both published and unpublished – to objects collected by Banks.[12] With such references, however, it is rarely clear whether the objects referred to, listed or discussed can be identified as having been collected on the *Endeavour* voyage; for Banks is known to have received large numbers of objects that were collected on Cook's second and third Pacific voyages and it is often difficult if not impossible to be confident as to which voyage particular surviving objects are from. Thus, Banks is a probable source of at least some of the material in the collections of the British Museum. However, no surviving object can be definitively identified as having been collected on the *Endeavour* voyage and given to the British Museum by Banks.[13] Indeed, we also know that some of what Banks gave the British Museum was later withdrawn and taken to the Continent by Danish entomologist Johann Christian Fabricius.[14]

Similarly, there is good evidence to suppose that the Pennant–Denbigh Collection in the University of Cambridge's Museum of Archaeology and Anthropology contains objects given to Thomas Pennant by Banks, but it is unclear how much – if any – of this collection came from the *Endeavour* voyage.[15] Banks is certainly, via the Alströmer Museum, the source of the Cook-voyage collection in the Etnografiska Museet in Stockholm,[16] though again it is impossible to know whether any of the objects in the collection came from the *Endeavour* voyage. Through his friend Charles Greville, a son of the Earl of Warwick, Banks is also likely to have been the source of what are thought to be some Cook-voyage objects, including perhaps some from the *Endeavour* voyage, in the collection at Warwick Castle, sold in 1969 and now dispersed to a number of private collections.[17] Similar difficulties surround the objects Banks may have given Sir William Hamilton,[18] and those he gave the Duchess of Portland.[19]

Among collections said to be from *Endeavour* it would be wonderful to trace the present whereabouts of what was in the possession of Richard Greene of Lichfield. In the first edition of the catalogue of his 'Rarities', published in 1773, Greene refers to articles 'brought by Mr. *Banks* and Dr. *Solander* from *George Island* [Tahiti], South Seas': namely 'A Fish[h]ook, made of the Bone of a Fish tied to a piece of Wood', 'A Fish[h]ook of Sea Shell', and 'An Adze of Flint, curiously bound to an handle of wood'.[20] Neither Banks nor Solander is included in 'A List of Benefactors to the Museum' appended to the *Catalogue*, but it does includes the names of Matthew Boulton, Thomas Pennant and others known to have been associated with Banks.

Another collection it would be wonderful to trace is that apparently given by Banks to the bibliophile John Lloyd of Wygfair, near St Asaph in Denbighshire. We only know of this collection because it was included in the sale of Lloyd's library in January 1816: 'Curiosities brought over by Sir Joseph Banks, Knt. collected during his Voyage with Captain Cook, and presented by him to the late Mr. Lloyd'.[21] These curiosities included: '2314 A Patoo, a weapon of War, from *New Zealand* … 2315 An Adze, made of Stone, from *Otaheite* … 2316 A curious Comb for the Hair, from *Otaheite* … 2318 Fish Hooks and lines, from *Otaheite*, very great curiosities, beautiful workmanship… 2319 Battle-Axe, made of hard wood, from *New Zealand*. 2320 A curious Belt, formed of Beads, &c. from ditto. 2321 Lot of Wooden Gods, curiously wrought, decorated with *Pearl* Eyes, &c.', and '2323 Portrait of Dr. Solander, with a Lock of his Hair'.

Until little more than a decade ago, the above would have been a fair summary of all we knew about the fate of Banks's collection of artificial curiosities from *Endeavour*. In 2002, however, I had the great fortune to discover at the University of Oxford's Pitt Rivers Museum a previously unrecognized collection of some thirty objects from the Society Islands and New Zealand that Banks had sent to Oxford by 16 January 1773 – and which thus must have been collected on the *Endeavour*. I have discussed the complicated history of this collection in detail elsewhere,[22] so suffice it to say here that the objects now held at the Pitt Rivers were previously held at Christ Church, which had been Banks's college when he was an undergraduate at the university. As yet it is not known whether Banks sent the collection directly to his old college, or perhaps to his undergraduate friend John Parsons, who in 1767 had returned to Christ Church to take up the post of Lee's Reader as well as a university lectureship in anatomy.

According to Banks's biographer Harold B. Carter, Banks visited Parsons at Christ Church in January 1768, and the two friends are known to have corresponded at least until this time (unfortunately, the present whereabouts of any surviving correspondence between them from after 1768 is not known).[23] No doubt they met again after the voyage, when Banks and Solander were given honorary degrees by the University on 21 November 1771. Both Parsons and the Anatomy School would have been well established at Christ Church by then, and it may thus have been as a result of a visit to his old college that Banks made the donation. What we do know for sure is that he had sent the collection by 16 January 1773, for on that day Banks's friend Thomas Falconer, classical scholar and Recorder of Chester, wrote to him: 'I was highly entertained at Oxford with a sight of some curiosities you sent from Otaheita & new Zealand'.[24]

The surviving collection comprises fourteen objects from Tahiti and thirteen from New Zealand, though given the lack of any lists or inventories it is impossible to know if the collection was previously more extensive. The Tahitian material comprises four pieces of barkcloth, an adze blade, a shark-hook, a canoe bailer, a head-rest, a barkcloth beater, a chisel or gouge, a food pounder, a breast ornament or *taumi*, a noseflute and a piece of matting. The Māori material comprises three belts, two cloaks, a fish-hook, a canoe-baler, five cleavers and a weaving peg.[25] These are all typical of the sorts of objects collected on the first voyage, and many of them resemble closely the objects drawn by Miller (as discussed above) and illustrated in Hawkesworth's *Voyages* (see numbers 38, 44, 45, 47, 54, 55, 66, 81, 82, 83, 92 and 99).

Listed like this, the collection may appear rather random. On closer inspection and reflection, however, it may be seen as systematic. There are two fish-hooks – one Tahitian, one Māori, two canoe balers – one Tahitian and one Māori, and two textile implements – a Tahitian barkcloth beater and a Māori weaving peg. Moreover, the three Māori belts are of contrasting styles, and the five Māori cleavers illustrate some of the range of materials and forms that so impressed Banks that he was led to commission 40 Māori-style cleavers in brass – an example of which is exhibited – to take on what we now know as Cook's second voyage (on which, in the end, Banks did not go).[26] Moreover, the four pieces of Tahitian barkcloth are of varying quality, thickness and colour: a medium thick yellowish piece; a thin, cross-ribbed piece, dyed brown; a thin white, finely ribbed piece; and a thick, felted piece, lightly stained red on one side. Even more strikingly, the two Māori cloaks illustrate perfectly Banks's own account in his voyage journal:

> they have several kinds of Cloth which is smooth and ingeniously enough workd: they are cheifly of two sorts, one coarse as our coarsest canvass and ten times stronger but much like it in the lying of the threads, the other is formd by many threads running lengthwise and a few only crossing them which tie them together. This last sort is sometimes stripd and always very pretty, for the threads that compose it are prepard so as to shine almost as much as silk; to both these they work borders of different colours in fine stitches something like Carpeting or girls Samplers in various patterns with an ingenuity truly surprizing to any one who will reflect that they are without needles but the great pride of their dress seems to consist in dogs fur, which they use so sparingly that to avoid waste they cut it into long strips and sew them at a distance from each other upon their Cloth, varying often the coulours prettily enough.[27]

It is, of course, immensely frustrating that no correspondence or list survives relating to the curiosities Banks sent to Oxford. It would be wonderful to have in Banks's own words an account of what was sent and why. It is, however, perhaps not too fanciful to see the collection as an attempt by Banks to give Parsons and his Oxford colleagues an idea of the variety of materials and techniques he had observed during the voyage.

Unfortunately, we do not know where and when the individual artefacts in the collection were obtained. Banks regularly recorded in his journal the taking of zoological specimens and, less frequently, the collecting of botanical specimens. The few references to the acquisition of 'artificial curiosities', however, are tantalizingly brief. During the stay in Tahiti, from 13 April to 13 July 1769, for example, Banks records being presented with barkcloth on four or five occasions. The day after *Endeavour* arrived, Banks and Cook were escorted to a house where they met a Tahitian chief named Tutaha: 'a peice of Cloth was presented to each of us perfumd after their manner .... My piece of Cloth was 11 yards long and 2 wide: for this I made return by presenting him with a large lacd silk neckcloth I had on and a linen pocket handkerchief'.[28] There is no cloth of that size in the Banks collection at Oxford, though it is not inconceivable that such a large piece was divided up and thus could be the source of one of the pieces in the collection.

We know even less about where and when Banks acquired curiosities during the circumnavigation of New Zealand. The voyagers spent little time ashore and, as a consequence, most of the trading took place on board *Endeavour* or between ship and canoe. One tantalizingly brief reference occurs in Banks's journal for 22 November 1769, when *Endeavour* was at anchor in the Wairou River (Cook named it the 'Thames') on the east coast of the North Island. Cook and Solander had set off in the pinnace, but Banks had decided not to join them:

'there were many Canoes about the ship with which I traded for their clothes, arms &c, of which I had got few so I stayd on board'.[29] It thus could well have been in this place and on this day that at least some of the Māori artefacts in the collection were acquired.

According to Banks, 'they sold cheifly for paper'. Indeed, paper and Tahitian barkcloth were the items most favoured by Māori when trading with the voyagers on *Endeavour*. While from a twenty-first century perspective, the exchange of paper for fine cloaks and other *taonga*, the Māori word loosely translatable as 'treasures', may look distinctly exploitative, it does seem that both parties felt they were getting a good deal. Such questions are the subject of much current debate and no doubt new understandings of the exchanges made on the voyage will emerge in due course. In the meantime, from the point of view of a museum curator, what is important is that the *taonga* have survived in museum collections to be, amongst many other things, a focus for such debates.

A brief account of Banks's collecting activities leads to a discussion of the work carried out under his auspices by those who accompanied him. As is well known, when *Endeavour* set sail from Plymouth on Thursday 25 August 1768 Banks had with him in his entourage – and at his own expense: Daniel Solander, as his companion and co-scholar; Herman Didrich Spöring, as secretary; Sydney Parkinson, as botanical and natural history draughtsman; Alexander Buchan, as landscape and figure artist; and four assistants – Peter Briscoe, James Roberts, Thomas Richmond and George Dorlton.[30] Richmond and Dorlton died at Tierra del Fuego before *Endeavour* reached the Pacific, Buchan died at Tahiti and Spöring and Parkinson died on the voyage home, but each in their own way contributed to the collecting and recording that constituted so much of the success of the voyage. As the wealthy sponsor of all this activity, it was Banks who was lionized when *Endeavour* returned to England, and he has continued to receive the credit ever since. While not wishing to diminish his importance – if it were not for him and his wealth, after all, Solander, Parkinson and the rest would not have participated in the voyage, nor would Cook have allowed Tupaia to join it – a focus on Banks should not obscure the achievements of his co-voyagers.

Solander's role, for example, has been almost completely overshadowed in the literature by the attention given to Banks. If he had left us a voyage journal, perhaps, he would be better known. Moreover, it was only in 1997 that Tupaia, the priestly navigator from Ra'iatea who joined *Endeavour* in Tahiti under Banks's care, was identified as the person who had painted the dozen watercolours formerly assigned to the anonymous 'Artist of the Chief Mourner'.[31] More recently, Anne Salmond has suggested that the *poupou* from Pourewa may have been given by local Māori to Tupaia,[32] and Paul Tapsell has discussed the possibility that at least the two cloaks, if not other *taonga*, in the Banks collection at Oxford might have been presented to Tupaia and appropriated by Banks after Tupaia's death in Batavia.[33] Pending the discovery of some hitherto unknown documentation, it seems unlikely that this suggestion will remain other than speculative. However, it seems certain that Tupaia, who Māori apparently regarded as the leader of the voyagers, would have been presented with *taonga* and have brought personal possessions – and perhaps objects to trade – with him on the voyage. No surviving object has otherwise been associated with him, but it seems likely that his possessions would have been appropriated by Banks after his death, so some of the surviving 'artificial curiosities' were surely once his – though which ones we may never know.

As well as possibly appropriating Tupaia's paintings and possessions after his death, it would seem likely that Banks also appropriated the possessions of those members of his entourage who died on the voyage. We know nothing of what Richmond, Dorlton, Buchan and Spöring acquired before their deaths, but we do know that there was a heated argument between Banks and Parkinson's brother and executor Stanfield about Parkinson's collections and manuscripts. Whatever the legalities of that dispute, it seems clear that Banks thought himself entitled to Parkinson's collections (see number 5).[34] Altogether, it is difficult to avoid the conclusion that although many objects collected on the *Endeavour* voyage were at one time or another in Banks's possession it is difficult to say with any certainty of any particular object that Banks collected it. No doubt he was given things in both Tahiti and New Zealand and obtained other things through trade; what we are never likely to know for sure is exactly what he obtained, where and when.

In the eighteenth century, the artefacts made by Pacific Islanders and brought back by Banks and the other voyagers were known generally as 'artificial curiosities'. Our knowledge of how they were regarded by Banks and his contemporaries is limited. We know that illustrations of them were published in the accounts of the voyages and we also know that selections were made from larger collections to be presented to Oxford and Cambridge colleges, individual collectors and museums. Some of these given collections appear to be systematic. They seem to have been organized to provide a representative sample of the material culture of individual societies, or at least of those types of objects that had been brought back. So, as we have seen, collections did not contain duplicates but a range of types: – of Māori weapons, of types of barkcloth and so on. We know little about how they were displayed, to either private or public visitors, or how they were interpreted, if they were interpreted at all. In many cases, as in West's portrait, it seems that they were regarded more as signs of the adventures and escapades of the voyagers than as evidence of previously unknown ways of life.

More than two hundred years later, the artefacts collected on Cook's voyages are of greater interest and significance than ever before. In recent years, anthropologists and historians have paid increasing attention to the material world and to the evidence contained in museum specimens. Because the voyage literature is so rich, the collections brought back by Banks, Cook and their co-voyagers do not stand alone, but can be interpreted alongside the accounts that the collectors themselves have left us of their experiences. Taken together, the material artefacts and the literary texts comprise a rich resource for studying the history and culture of the Pacific in the late eighteenth century. In addition, the collections provide hard evidence of the relationships between islanders and voyagers. Again, taken together with the voyagers' accounts and local histories preserved in oral traditions, the collections provide key materials for reassessing the events of the 1760s and 1770s.

More importantly, as the peoples of the Pacific regain control of their cultural presents and futures, they turn with increasing interest and determination to the interpretation of their cultural pasts. The Cook-voyage collections held in museums, mostly though not exclusively in Europe, preserve the cultural histories of the peoples of the Pacific. Those artefacts collected on the first voyage, including those that have concerned us here, are of especial importance as they preserve examples of forms that were to be transformed by the changes that followed missionization and colonization.

THE SOCIETY ISLANDS

In the decade prior to the launch of *Endeavour*, the British dispatched successive Royal Navy vessels into the Pacific with the aim of establishing what territories and trade existed there. Commodore John Byron sailed in HMS *Dolphin*, accompanied by the sloop *Tamar*, on a circumnavigation lasting from June 1764 to May 1766. He was followed in August 1766 by Samuel Wallis in the same vessel, accompanied by Philip Carteret in command of the sloop HMS *Swallow*, until the ships were separated after a stormy passage through the Strait of Magellan. Wallis was the first European navigator to visit and describe Tahiti. He stayed at the island from 18 June to 27 July 1767 and, despite an initial clash with the Tahitians, was able to establish friendly relations and to observe the people and their customs. Wallis reached England in May 1768, before *Endeavour* departed and in time to bring news of King George the Third's Island, as he called it, more usually Otaheite at the time or today Tahiti. Located at a suitable latitude to observe the transit of Venus, the island would also be able to provide essential supplies for the *Endeavour* expedition. The British were not the only nation exploring the Pacific at the time, for the French also launched a mission – under Louis Antoine de Bougainville, carrying the naturalist Philibert Commerson – which visited Tahiti in April 1768 during the first ever French circumnavigation.

On his arrival at Tahiti in April 1769, Cook selected Matavai Bay (previously named Port Royal by Wallis) as the location for his observation of the transit of Venus, and Tutaha, the chief of the local district, allowed him to make a base on the headland that became known as Point Venus. A brisk trade commenced with the islanders for water, food, 'natural and artificial curiosities' (as natural and man-made products were then known) and sex, in return for which metallic objects such as iron nails were eagerly sought by the Tahitians, since in Polynesia the making of metal was unknown. On all his Pacific voyages Cook struggled to regulate the exchange of the latter two commodities, which led to the spread of disease on the one hand and on the other caused the stripping by sailor and Polynesian alike of metal from his vessels. The Quaker influence in his background may partly explain Cook's own abstinence from the usual excesses of navy life, but he was also genuinely concerned about the welfare of local peoples as also the health and therefore efficiency of his men.

Artefacts were mainly obtained through gift or exchange, with each side deciding on the relative value that it attached to items offered or desired. With the later arrival on a permanent basis of Europeans, Polynesian customs, beliefs, modes of life and manufacture were changed forever. The surviving collections from *Endeavour* are therefore uniquely important for preserving material and illustrated evidence of a culture on the eve of irreversible change.

37 Two hafted adzes *to'i*

1. The Museum of Archaeology and Anthropology, University of Cambridge. D.1914.20 (front). Handle length 56.5 and blade length 16 cm. Stone (basalt) blade; wood handle; coconut fibre binding.

2. The Museum of Archaeology and Anthropology, University of Cambridge. D.1914.21 (rear). Handle length 56 and blade length 22.2 cm. Stone (basalt) blade; wood handle; coconut fibre binding; banana leaf padding.

Adzes were important tools in the Society Islands and were handled with great skill by their owners. They served for felling trees, working timber into canoes, dwellings and various household implements as well as for making figure carvings. A number of adzes gathered from the Society Islands during James Cook's voyages are held in modern collections. They fall into two main types, one with an ordinary hafted blade head and the other with a removable mounted blade head used as a swivel-blade adze. The present adzes are of the former sort. Both were gifted by Lord Sandwich to Trinity College, Cambridge, in October 1771.

As first lord of the Admiralty, Sandwich admired and patronized Cook, who after the *Endeavour* voyage gave him a collection of 'artificial curiosities' as such objects were then known. Having studied at Trinity, Sandwich decided to pass some, but not all, of this material to his old college, first having much of it sorted and listed broadly by type and area of origin. The resulting 'delivery' list, numbering 90 entries but with unexplained gaps in the sequence, appears to be of categories of objects from Cook, although the actual number of objects in each category is not stated. Nevertheless, the list gives an impression of the size and content of Cook's original collection, even though it is unlikely that it encompasses everything passed on by the explorer. It may even be that this list was based on another now lost version, still larger and more comprehensive in its coverage. The types and number of items received by Trinity are indicated on the list's left-hand side by Thomas Green, the college librarian, and a note by him appears at the foot of the list. This note refers to objects from three categories that were not present in the material sent to Cambridge and it also refers to two gaps for items actually received but not given a numbered category. These Green evidently filled in himself as they are inserted at the appropriate places, suggesting that he at least had sight of another list or perhaps of labels showing what the categories for these objects were. However, any labels that might have come to Trinity with the Sandwich gift no longer survive. The material actually received by the college is detailed in a later, geographically arranged list, apparently written in Green's hand. With the aid of such documentation, still held in the college archives, scholars have been able to identify more than 100 surviving objects donated to Trinity by Sandwich, who may well have followed these with ensuing gifts of objects from Cook's second voyage (see Gathercole, in Lincoln, ed., *Science and Exploration in the Pacific*, pp. 103–9; see also MS Add. a. 106, ff. 108r–9v and MS Add. a. 106, ff. 211r–212v). In 1914 and 1924 the Sandwich collection, as it is here called, was deposited at the university's Museum of Archaeology and Anthropology. It would appear too that Sandwich retained some Cook-voyage material, including Polynesian musical instruments in which he would have been interested not only as collector in his own right but also as a keen musician, and as a result in 1922 two Māori trumpets formerly belonging to him were passed to the museum by the 9th Earl.

In October 1771, the same month that Sandwich made his donation to Trinity, he also gave part of Cook's collection to the British Museum, and this set an important precedent, for in the years that followed the museum increasingly acted as a repository for material from discovery voyages. At the time, however, museum registration was primitive and individual items were rarely recorded. Most incoming material was simply added to existing collections and it is difficult or impossible now to be sure which surviving objects originally came from Cook's voyages. Yet modern research has revealed some clues leading to possible candidates for the high distinction of first-voyage status. A small number of unattached labels that still accompany objects plausibly of *Endeavour* origin have numbers and descriptions on them. These numbers and descriptions appear to agree with those given for categories in Sandwich's delivery list at Trinity, implying that their accompanying objects were part of the collections passed on by Sandwich in October 1771. Peter Gathercole noted this possibility in 1998 in *Science and Exploration in the Pacific*, pp. 108–9, and latterly five museum objects with such labels have been put forward as probable candidates for having been given by Cook to Sandwich after the first Pacific expedition (see paper by Salmond, *Artefacts of Encounter*, 2011 version). A sixth candidate might be a nephrite adze blade of clear Māori origin with a label that precisely corresponds to category number 25 in the Sandwich delivery list, 'Edge Tools of green Stone from [New Zeland]', NZ.153. A few of the mysterious gaps in this list may also be resolvable using some of the British Museum labels, which seem to fit the numerical sequence and evidently indicate the missing object categories at those points. Perhaps gaps were left in the delivery list for those categories not intended for Trinity but instead sent to the British Museum. Whatever the case, it is increasingly apparent that Sandwich distributed the collections that he received from Cook with some thought and organizational care.

An early voyage ordinary adze from the Society Islands is held at the British Museum, Oc,TAH.88, although whether it was collected during the *Endeavour* voyage remains open to question as no documentary evidence to that effect exists. It may have been gathered on this or a later Cook voyage, or even afterwards. A series of donations of so-called artificial curiosities were made to the museum following Cook's first Pacific voyage – by Sandwich, Cook, Banks and others. Subsequent voyage donations added to the influx of material arriving there as European penetration of the Indo-Pacific region increased, a trend that Banks actively promoted as a museum trustee from 1778 until his death in 1820 by channelling many objects from later voyages to the museum and by the bequest of his library and herbarium. Since record keeping at the museum was limited in this period, there are just fourteen entries relating to donations from Cook's voyages in the Book of Presents and these all lack specific details. Of the fourteen entries, seven mention artefacts alone, four mention natural history specimens alone and the remaining three mention both. Only at the end of the nineteenth century were ethnographic objects of possible Cook-voyage origin assigned the designation 'Cook collection', by the Pacific collector James Edge-Partington in his registration slips compiled for material accessioned before 1860. The evidence for so allocating them remains largely circumstantial. Many museum objects from the Cook voyages had by then been lost, destroyed or exchanged, and those that remained were mixed with other similar material so that their history was obscured. Some object labels, contemporary illustrations and other evidence survives to suggest possible links to Cook and even to Banks, but, useful though these are, they are in most cases by no means conclusive. Charles Hercules Read and Ormond Maddock Dalton, successive keepers of the Department of British and Mediaeval Antiquities and Ethnography at the turn of the twentieth century, supplemented Edge-Partington's slips with comments and slips of their own.

Designated a 'Cook collection' object by Edge-Partington, Oc,TAH.88 has been likened to an adze depicted for Joseph Banks by one of his post-voyage artists J.F. Miller, BL Add. MS 15508, f. 31 (see Kaeppler, *Artificial Curiosities*, p. 154). The resemblance is not close since folio 31 features a swivel-bladed adze, not one with an ordinary blade head like that of Oc,TAH.88. Perhaps closer is the Miller illustration BL Add. MS 15508, f. 30, featuring the correct adze type and a length measurement of 1 foot 11 inches that approximates to the length of the museum adze at $23^{14}/_{16}$ inches. Another Miller drawing of the same adze type is at BL Add. MS 23921, f. 55(a). The match is, however, not exact in either case and, as with many Miller illustrations of *Endeavour*-voyage objects, definite evidence to support visual links of this kind, written or otherwise, does not exist. Neither of the two Cambridge adzes shown here appear to have been illustrated by Miller.

At the Pitt Rivers Museum, Oxford, there is an unhafted adze blade of black basalt, 1887.1.10. This was donated by Banks to Christ Church, Oxford, at an unknown date before mid-January 1773 as part of a collection of Pacific objects acquired directly by him or by other expedition members during the *Endeavour* voyage. Banks had been an undergraduate at Christ Church. It is possible that his original donation was to a college fellow and friend, the anatomist John Parsons, but a personal gift seems less likely than an institutional one, and in any case the collection remained in the care of the college until it passed in part to the Ashmolean Museum in 1860, and then was deposited in its entirety at the Pitt Rivers Museum in 1886. Like Sandwich before him, then, Banks favoured his old college with what is now an extremely valuable first-voyage collection. The Cambridge and Oxford collections seem to have been intended to provide representative samples of the material culture of the individual societies encountered during the voyage, or at least of the sorts of objects that had been gathered from them. Each provides evidence of objects connected with at least some key human activities, such as fishing, weapons, working tools and items of dress, and in consequence displays the material from which these were made, how they were designed and sometimes even a range of types (see, for example, numbers 43, 65, 66, 82, 92, 99, 100, 101, 102 and 107).

First-voyage material, man-made and natural, survives in various collections around the world today. Research to establish what artefacts still exist from this and other missions, and where it now resides, is ongoing. Documentary evidence to link particular artefacts to a voyage or collector is often lacking, so attributions are frequently at best circumstantial. Illustrations can help, but connections between contemporary drawings and objects in present-day collections are not now in most cases certain or possible. John Frederick Miller's illustrations of artefacts from the *Endeavour* voyage are nevertheless an important record of objects brought back from that mission. Miller was one of the team of illustrators assembled by Banks after the voyage to work on his private collections, and as such Miller depicted both artefacts and natural history specimens with documentary accuracy. Although not themselves voyagers, these illustrators are nonetheless mission collection artists in their own right, taking the place of and building on the work of their deceased counterparts from *Endeavour*. Note that an adze with an ordinary blade head appears to the viewer's bottom right in Benjamin West's portrait of Joseph Banks surrounded by material gathered during the *Endeavour* voyage.

38 Chisel *tohi*

The Pitt Rivers Museum, University of Oxford. 1887.1.390. Length 21 and diameter 4 cm. Bone, wood and coconut fibre.

This is a chisel from the Society Islands, probably Tahiti, that was given by Banks to Christ Church, Oxford, which he attended as an undergraduate. The chisel was part of a collection of Pacific objects donated by Banks some time before mid-January 1773, and later passed to the Pitt Rivers Museum. This collection is extremely valuable for being a known first-voyage group of objects, but was only identified as such in recent years by Jeremy Coote of the Pitt Rivers Museum. The chisel itself is a piece of bone, very possibly human, secured to a wooden handle with coconut fibre bindings. Such objects were used in carpentry and were an important working tool. Another example from the *Endeavour*, donated in October 1771 by Lord Sandwich to his former college Trinity, Cambridge, is now deposited at the university Museum of Archaeology and Anthropology, D.1914.22. After the *Endeavour* voyage Banks's London artist J.F. Miller produced illustrations of such chisels alongside a selection of other Society Island artefacts, BL Add. MS 15508, f. 31, and BL Add. MS 23921, f. 54(a). For the latter illustration, see number 39 below (see also Kaeppler, *Artificial Curiosities*, p. 154, and Coote, *Curiosities from the Endeavour*, pp. 14–15). Although similar to the chisels mentioned here, Miller's illustrations are not provably of the same objects. Surviving chisels of this kind from Cook's three Pacific voyages are few in number, and so those at Oxford and Cambridge show the importance attached to these working tools by collectors on the first voyage.

Note that another likely Cook-voyage *tohi* chisel is held at the Museum of Archaeology and Anthropology, in its Pennant collection, 1925.423. This collection belonged to Banks's contemporary, the Welsh naturalist Thomas Pennant. It may be that Banks gave the chisel concerned to Pennant, although if he did the particular voyage from which it came remains uncertain since its source is undocumented and Banks had access to material from successive voyages by Cook and others. For Pennant, see number 128 and the essay by N. Chambers in this volume.

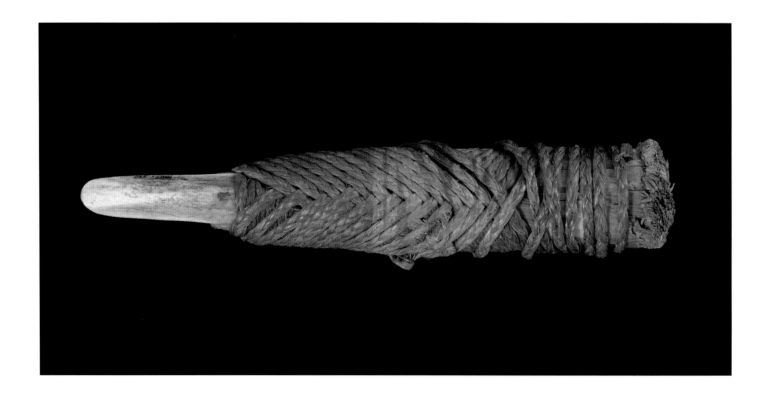

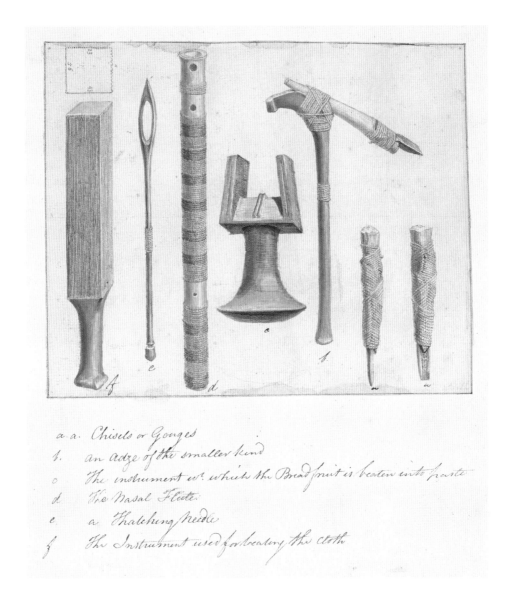

39 Tools and instruments from the Society Islands

By John Frederick Miller.

The British Library, London. BL Add. MS. 23921, f. 54(a). 20.3 × 16.5 cm.

This pen and wash illustration by J.F. Miller, probably completed in 1771, shows a barkcloth beater, needle, noseflute, food pounder, swivel-blade adze and two chisels. A key appears on the folio, 'a.a. Chisels or Gouges b. An Adze of the smaller kind c The instrument w⁺ʰ which the Breadfruit is beaten into paste d The Nasal Flute. e. a Thatching needle f The Instrument used for beating the Cloth'. The barkcloth beater is headed by a plan view of its beating end showing the number of grooves on each face, 11, 23, 43 and 56. It was used to beat bark into thin pieces for making cloth. Beating would start with the face possessing the least and hence largest grooves and, as the barkcloth increased in fineness, move through to the face with the most and so narrowest grooves. The present illustration appeared in the 1773 Hawkesworth edition of the *Endeavour* voyage, vol. 2, plate 9, engraved by Record. A similar set of objects was depicted by Miller at BL Add. MS. 15508, f. 31.

40 *Artocarpus altilis* (Parkinson) Fosberg, *J. Wash. Acad. Sci.* 31:95 (1941)

By Sydney Parkinson.

The Natural History Museum, London. Diment et al., Catalogue, *Part 2* Botany *(1987), SI 2/37. Plate height 46 × width 29 cm; plant ill. height 35.5 cm.*

The breadfruit, or *uru*, and its cultivation and use by local people, was observed with great interest by Banks and Solander while they were in the Society Islands. They described the plant itself, noted the ease with which it appeared to be grown by the islanders and gathered some specimens for Banks's herbarium. This illustration is a finished watercolour by Sydney Parkinson, annotated in ink on the recto 'Sitodium altile. Sydney Parkinson pinx:ᵗ 1769.' There are pencil notes 'Artocarpus incisa, L. fil Otaheite'. Some part sketches of breadfruit by Parkinson also exist as well as a finished watercolour by J.F. Miller. These are all held at the Natural History Museum, London, along with two Banks herbarium specimen sheets, one of which is marked for Tahiti, and the other for Tahiti, Huahine and Ra'iatea.

Breadfruit was probably introduced to the Society Islands by the early Maohis when they arrived there and, along with existing coconut palms, came to occupy extensive areas of the coastal plains and adjacent lower hill slopes. Perhaps unsurprisingly for such a widely used and valuable food, vegetative propagation and selection over a long period have resulted in a large number of separate cultivars, and DNA fingerprinting has revealed that some of these may be the result of hybridization and selection from other species of *Artocarpus*. Breadfruit trees provided timber for the construction of dwellings and, although not ideal for the purpose, could be used to make canoe hulls. The inner bark of the smaller branches was made into barkcloth (see number 44) and the tree also produced a resin that was useful for tasks such as sealing canoe seams. The broad leaves of the high, shady canopy were utilized to wrap food and to cover pit ovens. To increase the number of breadfruit trees it was only necessary to plant cuttings of roots or young shoots, and then ensure their safety until the plant reached fruit-bearing age at about five or six years. A fully grown breadfruit may weigh up to ten pounds and possesses a fibrous core of starchy, white pulp covered by a rough rind. Breadfruit was a staple food source throughout the islands and was harvested at different times of the year depending on where it grew and the local climate. Usually a boy or young man would scale a tree and then twist loose the fruit.

Once gathered the breadfruit could be prepared in a number of different ways. The fruit was commonly picked at its first stage of ripening, when the pulp was firm but sweet, and then baked in pits either whole in its rind or with rind and core removed and cut into large chunks. A wooden splitter was employed to cut the pulp (see number 42). Stones heated by fire were used to bake or steam the fruit, with the contents of the pit being covered while cooking took place inside. Breadfruit could be baked and rebaked, and during the breadfruit season pits of up to ten feet in diameter were dug and filled with scores of fruit. Large pits of breadfruit might require three or four days to cook. Prepared in this way, a breadfruit might keep for some weeks. At later stages in its ripening, the pulp of the breadfruit becomes softer and sweeter. Another method of preserving it was to remove the rind of a number of breadfruit, and to ferment their pulp in a heap for three or four days. Fermentation was hastened by adding some overripe fruit or by leaving in the cores. Once stored in a leaf-lined pit, and covered by earth and stone, the breadfruit pulp continued to soften and ferment for some months. When needed, the resulting paste or *mahi* was taken out and baked by itself or with other ingredients. Soups could also be made using the pounded pulp of breadfruit (see numbers 41 and 43).

Breadfruit was not available all year round, and there might be a scarcity of it at times of war or after heavy storms or other natural disasters. When this happened, the people resorted to other foods such as sweet potatoes and pandanus or they might search the mountains and valleys for plantains, wild fruits and vegetables. There were too, of course, the coconut palm and the sea from which to gather supplies. The preparation and eating of food was governed by various and strict social codes. In the Society Islands food intended for men and that for women was generally cooked separately and the sexes did not eat together. Men were not usually restricted from eating food supplied or cooked by women, but women were subject to certain restrictions regarding food provided by men. In addition, a woman's *mahi* could not be touched by a man unless he was acting as her servant, and if it was handled incorrectly by any other man then it became unfit for her to eat. This happened on one occasion when Banks inadvertently touched some leaves on a fermenting pile and the woman overseeing it immediately tore down the pile in front of him (see *Banks's Journal*, vol. 1, p. 345).

41 Breadfruit specimen

The British Museum, London. Oc,TAH.71. (Edge-Partington registration slip and C.H. Read registration slip, both 'Cook collection'.) Length 15 cm.

Dried breadfruit specimens like this one are rare. It probably dates to the late eighteenth century, and has a separate, damaged label reading 'A dryed bread[fruit] Otahaite', which was probably written by Daniel Solander while he was employed at the British Museum. James Edge-Partington registered the breadfruit as a 'Cook collection' item (C.H. Read agreed). It is hollow with a single hole at the top.

The breadfruit, *Artocarpus altilis* (Parkinson) Fosberg, was already known in Banks's day. William Dampier had described it as far back as 1697 in his *New Voyage Round the World* (London: James Knapton). Banks saw it for himself during the *Endeavour* voyage, and noted its utility as a food source requiring little effort to cultivate. To many Europeans it almost seemed that such a naturally abundant food embodied one of the many blessings available on an island paradise like Tahiti (see *Banks's Journal*, vol. 1, p. 341). Tahiti was, of course, no Arcadian idyll, as its frequent wars and sometimes brutal religious practices showed, but there is nevertheless a deep irony in the fact that after the voyage there was considerable interest in transferring the breadfruit from the Pacific to the British West Indies to be used as a staple for plantation slaves. Calls for its transfer became more urgent because the French were moving useful plants to their own West Indian possessions and, more importantly, because vital trade between the British West Indies and America had been severed following the American Revolution. Banks was consulted by government about mounting a Pacific mission to obtain breadfruit, for which he drew up plans and appointed Kew gardeners.

Thus commenced the *Bounty* mission under William Bligh in 1787 to gather breadfruit and other useful plants from the Society Islands for delivery to the West Indies, a voyage that famously ended in mutiny in April 1789. Bligh was therefore sent back to the Pacific in the *Providence* and the *Assistant* in 1791–3 to complete the task. This he did, delivering nearly 700 breadfruit saplings to the West Indies along with a variety of other plants gathered during the voyage, after which he returned to England in August 1793 with some 1,283 plants that he collected before reaching the West Indies or there received as gifts for the Royal Gardens at Kew. This was one of the largest single accessions to the Royal Gardens that Banks ever engineered. It may be added that although breadfruit was not initially successful as a food among the slaves in the West Indies it has since become a staple in the region.

42 Breadfruit splitter

The Museum of Archaeology and Anthropology, University of Cambridge. D.1914.19. Handle 32.5 and head 13 cm. Wood.

The breadfruit (*uru*) is an important food staple in many parts of Polynesia, where it provides a valuable source of carbohydrates (see number 41 opposite). It is usually split open and the starchy contents then removed to be baked or boiled into soup for human consumption. Several types of splitting implement were devised for this purpose. This is a one-piece wooden adze tool that was used for opening breadfruit. Undecorated, it has a long, slightly bent handle and a head with a chopping edge for breaking into the fruit.

43 Three food pounders *penu*

1. The Museum of Archaeology and Anthropology, University of Cambridge. D.1914.16. Height 14.5 and base diameter 12.5 cm. Basalt, faceted.

2. The Museum of Archaeology and Anthropology, University of Cambridge. D.1914.18 Height 15.5 and base diameter 11.8 cm. Basalt, cross-bar.

3. The British Musem, London. Oc,TAH.15. (Edge-Partington registration slip, 'Cook collection'.) Height 19 cm and base diameter 11.5 cm. Basalt, forked top.

Important household items, food pounders were used for preparing the starchy soup (*poi*) that was made from breadfruit and a range of fruit and vegetables such as taro, sweet potato and banana. These items were pounded to the required consistency on a wooden pounding table. Formally, food pounders have been classified into three basic types according to their tops, the 'forked top', the 'faceted top' and the 'cross-bar top'. A number of Society Island food pounders now held in modern collections were gathered during Cook's voyages, although which voyage is not always clear. At Cambridge there are three food pounders from the *Endeavour* voyage, D.1914.16–18. Pounder D.1914.17, not shown here, is of the forked top type, but one of the forks is broken off. These pounders were included in a collection of *Endeavour*-voyage objects given by James Cook to Lord Sandwich, who then passed part of the collection to Trinity College, Cambridge, which later deposited everything at the Museum of Archaeology and Anthropology.

Another first-voyage example of a forked top pounder is to be found at the Pitt Rivers Museum, Oxford, although its handle is also damaged, 1887.1.391. This pounder was part of a collection of artefacts donated by Banks to Christ Church, Oxford, and later deposited at the museum. A further early voyage example, again of a forked top pounder, is held at the British Museum, Oc,TAH.15. This has an attached label, probably by Daniel Solander, who was in charge of Natural and Artificial Curiosities at the museum from 1773, and the label reads 'Stone Pestles. Otahaite'. James Edge-Partington registered the pounder as a 'Cook collection' object. Which Cook voyage it might have come from is not certain, although see King, *Artificial Curiosities from the Northwest Coast of America*, pp. 16–17, where labelled items such as this one are linked to Cook's final voyage. The pounder has been likened to one featured in an illustration by J.F. Miller for Joseph Banks dating to about 1771, BL Add. MS 15508, f. 31, but an even closer if still not exact resemblance exists with the pounder depicted by Miller in number 39 (see Kaeppler, *Artificial Curiosities*, p. 149).

43.1

43.2

43.3

44 Barkcloth *tapa*

The Pitt Rivers Museum,
University of Oxford.
1886.21.16.
192 × 190 cm. Bark.

This piece of Tahitian barkcloth is thin, cross-ribbed and dyed brown, and was donated to Christ Church, Oxford, by Joseph Banks. Barkcloth, widely known as *tapa* cloth, a name originating from Tahiti and the Cook Islands, was used for clothing, bedding, screens and wrapping objects and it could be decorated with paint or print patterns. The quality of barkcloth and its designs varied, and more valuable pieces would be worn by higher-status individuals. Gifts of pieces of barkcloth were traditionally made in Polynesia for ceremonial reasons and as a peace offering or show of goodwill. Members of the *Endeavour* expedition would have been presented with such gifts in the Society Islands. They later found that barkcloth was highly valued by the Māori of New Zealand, and so pieces gathered earlier in the voyage featured as a trade item during first contacts there. A great deal of barkcloth was gathered during the voyages of this period, and specimens of it were placed in barkcloth books that contained many sample pieces, although this tends to make the origin of each piece difficult if not impossible to determine. Barkcloth books demonstrate the range of purposes, decoration and fineness of this material.

The most famous barkcloth books are those by Alexander Shaw, who in 1787 compiled a number of bound volumes with a printed introduction. In them he included Cook-voyage specimens from Hawai'i (where such cloth is known as *kapa*), the Society Islands, Tonga and possibly Rurutu, not all of which, of course, were visited on the *Endeavour* voyage. The date of publication indicates that the barkcloth contained in these books must have come from Cook's voyages. In full, the title is *A catalogue of the different specimens of cloth collected in the three voyages of Captain Cook, to the Southern Hemisphere; with a particular account of the manner of the manufacturing the same in the various islands of the South Seas; partly extracted from Mr Anderson and Reinhold Forster's observations, and the verbal account of some of the most knowing of the navigators: with some anecdotes that happened to them among the natives*. No two barkcloth books by Shaw are quite the same. Specimens have been removed in some while others have more than the 39 pieces stipulated in his preface. A written note in the barkcloth book held at the Textile Museum, Washington, indicates that Shaw intended to dedicate this work to Banks, but that possibility is unconfirmed.

Other collectors compiled barkcloth books, in some of which are to be found a mix of barkcloth, feathers and shells. There are a considerable number of barkcloth books in repositories around the world today, providing evidence both of the popularity of such collections and their varied content. A recent survey of those made by Shaw estimates that there are no less than 63 copies held worldwide, a figure likely to grow with time. As explained, individual pieces of barkcloth were also collected in large numbers and exist in modern collections (see numbers 45–6 immediately below). The Sandwich Collection at the Museum of Archaeology and Anthropology, Cambridge, contains a selection of eleven separate pieces of barkcloth gathered by James Cook during the *Endeavour* voyage. The museum also holds a unique prototype version of the Shaw barkcloth book, once in the possession of what was formerly the Royal Canterbury Museum, Kent, now the Beaney House of Art and Knowledge, 1976.159. Similarly, the Pitt Rivers Museum holds a representative series of four pieces of barkcloth from the voyage, originally donated by Joseph Banks to his old college, Christ Church. These are made from different sorts of tree bark and display a variety of colour, quality and thickness for this material.

44

45

45   Barkcloth beater *i'e*

The Pitt Rivers Museum, University of Oxford. 1887.1.383. Length 33 and width 3.8 cm (maximum dimensions). Wood.

Barkcloth was made by the women of Polynesia mostly from the cleaned inner bark of the paper mulberry tree (*Broussonetia papyrifera*), which was soaked and then beaten on a wooden anvil with mallets or *i'e*. Barkcloth may also be fabricated from the bark of the breadfruit tree and that of *Ficus tinctoria* G. Forst., the dye-fig, both of these and *Broussonetia papyrifera* belonging to the mulberry family, Moraceae. The cloth was probably originally fabricated from the bark of the dye-fig, which is native to Oceania, but this was supplanted by the much better paper mulberry that was introduced from South East Asia during early voyages of migration.

The women traditionally made barkcloth by beating and broadening strips of bark. Textured patterns resulted from the beaters, which had grooved lines carved into their faces. These increased in number and hence fineness on each face of the beater, the most fine being used when the cloth itself was beaten to a thin state. After the barkcloth was made it might be decorated with paint or print designs or be left plain.

A number of barkcloth beaters from the Society Islands that were gathered during Cook's voyages are to be found in modern, mainly European collections. The present barkcloth beater from the Banks collection at the Pitt Rivers Museum was acquired in the Society Islands, probably at Tahiti while the *Endeavour* expedition was there. It is almost precisely 13 inches long, and on its four beating faces has grooves of increasing fineness numbering 8, 16, 24 and 36. The square end of the beating head has been carved in relief with a z-shape. Another known first-voyage beater is held in the Sandwich collection at the Museum of Archaeology and Anthropology, Cambridge, D.1914.24. This is nearly fifteen inches in length (14$^{13}/_{16}$ inches) and has grooves on its faces numbering 10, 21, 38 and 46. The beating head end is also carved, with roughly incised zig-zags.

A further early voyage beater held at the British Museum, Oc,TAH.20, was registered by James Edge-Partington as 'Cook's Voy:'. He and later Adrienne Kaeppler noted an apparent resemblance between this one and a drawing of an upright beater for Banks by J.F. Miller, BL Add. MS. 15508, f. 30, but documentary evidence to prove a direct link does not exist (see Kaeppler, *Artificial Curiosities*, p. 151). Miller gave a length measurement on his drawing of 15 inches, to which the museum beater approximates at 14¾ inches, but it has 12 rather than the 11 grooves seen on the Miller beater. In all there are 12, 27, 45 and 62 grooves on the museum beater, and v-shaped relief carvings on its beating head end. The Cambridge beater approximates to a length measurement of 15 inches, but again the number of its grooves does not correspond exactly with those in Miller's drawing.

Number 39 above, also drawn by Miller, includes an upright beater with a plan outline of the beating head end showing 11, 23, 43 and 56 grooves, but no length measurement and, as in Miller's folio 30 beater illustration, no indication of additional carving on the head end. The beater in number 39 differs in a number of respects from that seen at folio 30 and indeed from each of the surviving beaters described here. None of these particular beaters appears to have been illustrated by Miller. Note that a barkcloth beater appears in the bottom right-hand corner of Benjamin West's portrait of Banks as viewed (see number 143).

46 'Woman scraping bark to make cloth SS' and 'Women beating cloth'

By Sydney Parkinson.

The British Library, London. BL Add. MS 23921, f. 50(a,b). 16.2 × 20.4 cm; 20.3 × 23.2 cm.

These are pencil drawings by Sydney Parkinson of women making barkcloth, perhaps sketched in April 1769, when he described this process in his journal (see *Journal of a Voyage*, p. 18). Alexander Buchan died of a fit on 17 April and so thereafter Parkinson undertook figure and landscape illustrations in addition to his usual work depicting natural history specimens. The present drawings are separately annotated in pencil on the recto as above, with 'SS' presumably standing for South Seas. Recto titles have also been added by an unknown hand in ink on the folio beneath the drawings, respectively 'A Girl scraping the Bark to make Cloth' and 'Girls beating out the Bark with their Cloth beaters'. The girl scraping bark in the first drawing is wearing a sunshade or *taumata* on her head. In Polynesian society it was only the women who made barkcloth.

The scraping and beating of bark into barkcloth was recorded by Banks: 'When the trees are arrivd at a sufficient size they are drawn up and the roots and topps cut of[f] and strippd of their leaves; the best of the *Aouta* are in this state about 3 or 4 feet long and as thick as a mans finger but the *ooroo* are considerably larger. The bark of these rods is then slit up longitudinaly and in this manner drawn off the stick; when all are stripd the bark is carried to some brook or running water into which it is laid to soak with stones upon it and in this situation it remains some days. When sufficiently soakd the women servants go down to the river, and stripping themselves set down in the water and scrape the pieces of bark, holding them against a flat smooth board, with the shell calld by the English shell merchants Tygers tongue *Tellina Gargadia*, dipping it Continualy in Water untill all the outer green bark is rubbd and washd away and nothing remains but the very fine fibres of the inner bark. This work is generaly finishd in the afternoon; in the evening these peices are spread out upon Plantain leaves. In doing this I suppose there is some some dificulty as the mistress of the family generaly presides, all that I could observe was that they laid them 2 or 3 layers thick, and seemd very carefull to make them every where of equal thickness; so that if any part of a peice of Bark was scrapd thinner than it ought, another peice of the same thin quality was laid over it, in order to render it of the same thickness as the next. When laid out in this manner the size of a peice of cloth [is] 11 or 12 yards long and not more than a foot broad, for as the longitudinal fibres are all laid leng[t]hwise they do not expect it to stretch in that direction tho they well know how considerably it will in the other.'

Much of the water in the bark had drained away by morning and the pieces were starting to adhere. Banks continued: 'It is then taken away by the women servants who beat it in the following manner: they lay it upon a long peice of wood one side of which is very Even and flat, which side is put under the Cloth; as many women then as they can muster or as can work at the board begin; each is furnishd with a battoon made of a very hard wood calld by the natives *Etoa* (*Casuarina equisetifolia*); these are about a foot long and square with a handle; on each side of the 4 faces of the square are many small furrows of as many different fineness, in the first or coarsest not more than [15] in the finest one [56] which cover the whole face of the side. With the coarsest then they begin, keeping time with their strokes in the same manner as smiths or Anchor smiths, and continue until the Cloth which extends itself very fast under these strokes shews by the too great thinness of the Grooves which are made in it that a finer side of the beater is requisite; in the same manner they proceed to the finest side with which they finish, unless the Cloth is to be of that very fine sort which they call *Hoboo* which is almost as thin as muslin' (*Banks's Journal*, vol. 1, pp. 354–5).

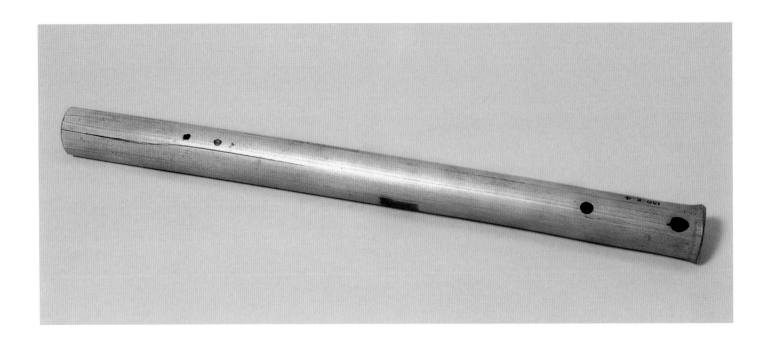

47 Noseflute *vivo*

The Pitt Rivers Museum, University of Oxford. 1903.130.20. Length 41.3 and diameter 3.4 cm (maximum dimensions). Bamboo.

Probably from Tahiti, this is a plain bamboo noseflute (some have bands of plaited coconut fibre wrapped around them or are otherwise decorated), in which the nostril-hole used for blowing is next to the closed end, with a large fingerhole close by. One of the two small fingerholes towards the other, open end has been plugged. These fingerholes would have been used to vary pitch when playing. Noseflutes were played for pleasure as well as at special events, festivities, dances or other performances. They were often accompanied by drums, the other main musical instrument used by the islanders (see numbers 48–50 below and 39 above). The present example was a gift by Joseph Banks after the *Endeavour* voyage to his old college, Christ Church, Oxford.

A further first-voyage noseflute, D.1914.27, also plain, was given by Lord Sandwich to Trinity College and is currently deposited at the Museum of Archaeology and Anthropology, Cambridge. Only a few noseflutes thought to be of Cook-voyage origin exist today. The Museum of Archaeology and Anthropology holds another one with bands of plaited coconut fibre that may have come from a Cook voyage. Part of the Pennant collection, 1925.385, this noseflute may have been a gift from Banks to his fellow naturalist Thomas Pennant, although that is not certain and nor is the particular voyage from which it came.

48 'The Lad Taiyota, Native of Otaheite, in the Dress of his Country.'

Stanfield Parkinson, ed., A Journal of a Voyage to the South Seas in his Majesty's Ship, The Endeavour (London, 1773; Fothergill reissue 1784). Plate 9. Drawn by Sydney Parkinson. Engraved by R.B. Godfrey.

The British Library, London. L.R.294.c.7. Rebound in half brown leather and brown cloth. Title, place and date of publication, gilt lettering on spine. Author given. 29 × 24.2 cm (closed). Plate 27.5 × 22.7 cm.

This is a print engraving from Sydney Parkinson's posthumously published account of the *Endeavour* voyage for which the original drawing is lost. Taiato or Tayeto was the young Tahitian boy servant of Tupaia. He is shown in this illustration wearing a poncho. Both Polynesians sailed with *Endeavour* when it departed the Society Islands, but neither reached England as they died of diseases contracted at the notoriously unhealthy Dutch port of Batavia (now Jakarta). While exploring New Zealand with James Cook in October 1769, Taiato survived a kidnap attempt by Māori off what Cook therefore chose to call Cape Kidnappers.

103

49 Drum *pahu*

The Museum of Archaeology and Anthropology, University of Cambridge. D.1914.26. 65 × 16 cm. Wood, fish skin, coconut fibre.

In the Society Islands drums were used for musical entertainment as well as in dances and religious ceremonies. The present example is typical of its kind. It was formed by hollowing out a log, setting it on end, and covering the upper opening with a stretched fish-skin tympanum lashed by coconut fibre cords to a base of six short legs standing on a circular pedestal. Here some of the cords have detached, which is not unusual in objects of this construction and age.

Drums were played with the palms of the hands rather than with sticks. Their size and hence tone varied. Taller and more slender drums produced higher pitches, whereas shorter, wider drums had a lower pitch. Tone might also be varied by altering the place on which the drum skin was struck. Tall drums were about twice the height and half the width of short ones.

A small number of drums thought to have been collected during the Cook voyages survive today. The present drum is a taller type, probably from Tahiti. Another tall example resides at the British Museum, measuring 63.5 cm in height, Oc,TAH.22. James Edge-Partington registered this drum as a 'Cook collection' item, and both he and later Adrienne Kaeppler noted an illustration by J.F. Miller of a comparable drum (see Kaeppler, *Artificial Curiosities*, p. 140). Completed for Banks after the *Endeavour* mission, Miller's illustration appears to be of a tall drum like those at Cambridge and the British Museum (see number 61).

50 Musicians of Tahiti

By Tupaia.

The British Library, London. BL Add. MS 15508, f. 10(b). 26.7 × 36.8 cm.

This is a pencil and watercolour drawing by Tupaia, to whom a small quantity of expedition artwork previously thought to be by an unknown 'Artist of the Chief Mourner' is now attributed. Joppien and Smith previously suggested that the apparently naive manner of this artwork indicated that it was by Banks or perhaps even Cook, but this is now known not to be the case (see Joppien and Smith, *The Art of Captain Cook's Voyages*, vol. I, p. 73). In number 48 above, Tupaia's travelling companion Taiato is shown playing a noseflute. Compare the hand positions in that print engraving and in this drawing. Tupaia here shows how the noseflute was played, with a finger closing one of the nostrils at the blowing end. The other hand is placed further down the flute to play fingerholes situated towards the far end. The drums appear to be played sitting down or squatting, and two sorts are shown by Tupaia, one short and one tall. When played with the palms of the hands, these would make lower and higher pitched sounds respectively. Tahitian music mainly consisted of singing to the accompaniment of a noseflute, with drums and clapping providing a rhythmic beat (see numbers 47–9, 51–2 and 58).

In the Society Islands a special order called the *arioi*, whose members were drawn from the upper and middling ranks, travelled in groups from place to place performing musical dances and plays for gifts of provisions and *tapa* cloth. The purpose of the *arioi* was obscure to early British voyagers, who sometimes likened them to the strolling players of England, but appears to have been linked to fertility rituals and worship of the god Oro. Perhaps, too, the sect gave a role to those offspring of the nobility who could not inherit property after the birth of the first male heir, and who might therefore become a costly burden or even a disruptive influence in society, while also providing a reserve of well-conditioned warriors. Their performances were lively events, and often contained erotic dance routines. Dramatic elements may have served to satirize the shortcomings of individuals, including those of tribal and religious leaders. *Arioi* members did not marry and this and their promiscuity, coupled with the fact that they were not permitted children and so practised infanticide, probably as a way of regulating numbers and avoiding confusion over paternity, meant that the cult was rapidly stamped out once missionaries arrived.

Banks encountered a company of *arioi* at Tahiti on 12 June 1769. He recorded the event in his journal: 'In my mornings walk today I met a company of travelling musicians; they told me where they should be at night so after supper we all repair to the place. There was a large concourse of people round this band, which consisted of 2 flutes and three drums, the drummers accompanying their musick with their voices; they sung many songs generaly in praise of us, for these gentlemen like Homer of old must be poets as well as musicians. The Indians seeing us entertaind with their musick, asked us to sing them an English song, which we most readily agreed to and receivd much applause, so much so that one of the musicians became desirous of going to England to learn to sing' (*Banks's Journal*, vol. 1, p. 290). Earl Morton gave Cook and Banks 'Hints' before they departed, suggesting that music might be used to win the trust and goodwill of indigenous peoples (see number 6). Music and other types of performance were used by Polynesian and voyager alike to entertain one another and to try to bridge gaps in mutual understanding.

51 'Sketches of Inhabitants'

By Sydney Parkinson.

The British Library, London. BL Add. MS. 23921, f. 36(a-e).

a. 23.2 × 23.2 cm. Dancing girl.

b. 18.9 × 15.4 cm. Man tying sash.

c. 19.7 × 16.5 cm. Bare-breasted dancer.

d. 17.2 × 14.1 cm. Man holding club.

e. 16.5 × 15.4 cm. Young woman.

Annotated in ink on the folio as above, these five pencil sketches by Sydney Parkinson are of three girls and two men (one of the men is tying a sash, and the other is holding a club). One of the drawings, that of the bare-breasted dancing girl, is labelled in pencil on the recto by Parkinson 'Heiva Dancing girl SS.' These drawings were probably completed on or after 7 August 1769, when, according to his journal, Banks took Parkinson to make illustrations at a *heiva* on the west coast of Ra'iatea: 'in the afternoon took Mr Parkinson to the *Heiva* that he might scetch the dresses' (*Banks's Journal*, vol. 1, p. 328). See also Parkinson's sketches of female dancers at folios 37 and 38, in which that at folio 38(b) resembles the folio 36(a) dancing girl seen here with her almost-clasped arms raised to shoulder height.

The figures depicted in folio 36 were later used by Giovanni Battista Cipriani to produce an idealized scene of girls dancing to drum and flute music in a Ra'iatean house (see number 52 below). This scene was then engraved by Francesco Bartolozzi and published in Hawkesworth's 1773 edition of the *Endeavour* voyage account, vol. 2, plate 7. Another sheet of preparatory sketches, in this case of male figures featured in the Cipriani scene, exists at the National Library of Australia, Canberra, Acc. no. R7219. Denis Carr, in *Sydney Parkinson*, p. xiv, attributes these sketches to Parkinson, but in terms of content and style they are far more likely to be by Cipriani. It would appear that Cipriani based some of the Canberra sketches on material in folio 36, for example the man tying his sash at 36(b). The others he may have designed himself or taken from further drawings by Parkinson that have since been lost. On the reverse of the Cipriani sketch is a plan and a three-dimensional view of a house similar to that in which his final dance scene is set. The Cipriani sketches and those by Parkinson at folio 36 are, then, combined in the final Cipriani scene.

A *heiva* was not the performers, as Banks thought, but the performance of a dance or theatrical event, usually by groups of travelling *arioi*. *Heiva* might be simple entertainments provided for a small group, or dramatic or ceremonial displays for a formal occasion. Elements of music, dance and acting were combined in a single performance as required. The dress, facial contortions and erotic gyrations of *heiva* dancers are what most struck early voyagers.

Banks had already witnessed a *heiva* performed by a group of Ra'iatean *arioi* on 3 August, and at the time he noted differences in social organization between the *arioi* of this island as compared with those of Tahiti: 'In the course of our walk we met a set of stroling dancers Calld by the Indians *Heiva* who detain us 2 hours and during all that time entertaind us highly indeed. They consisted of 3 drums 2 women dancers and 6 men; these Tupia tells us go round the Island as we have seen the little *Heivas* do at Otahite, but differ from those in that most of the people here are principal people, of which assertion we had in the case of one of the women an undoubted proof.

I shall first describe their dresses and then their dances. The women had on their heads a quantity of *tamou* or plaited hair which was rolled and between the interstices of it flowers of Gardenia were stuck making a head dress truly Elegant. Their shoulders arms and breasts as low as their arms were bare, below this they were coverd with black cloth and under each shoulder was placd a bunch of black feathers much as our ladies nosegays or Bouquets. On their hips rested a quantity of cloth pleated very full which reachd almost up to their arms and fell down below into long peticoats reaching below their feet, which they managd with as much dexterity as our opera dancers could have done; these pleats were brown and white alternately but the peticoats were all white.

In this dress they advancd sideways keeping excellent time to the drums which beat brisk and loud; they soon began to shake their hips giving the folds of cloth that lay upon them a very quick motion which was continued during the whole dance, they sometimes standing, sometimes sitting and sometimes resting on their knees and elbows and generaly moving their fingers with a quickness scarce to be imagind. The chief entertainment of the spectators seemd however to arise from the Lascivious motions they often made use of which were highly so, more indeed than I shall attempt to describe …. Between the dances of the women (for they sometimes rested) the men acted a kind of interlude in which they spoke as well as dancd. We were not however sufficiently vers'd in their language to be able to give an account of the Drama' (*Banks's Journal*, vol. 1, p. 325–6).

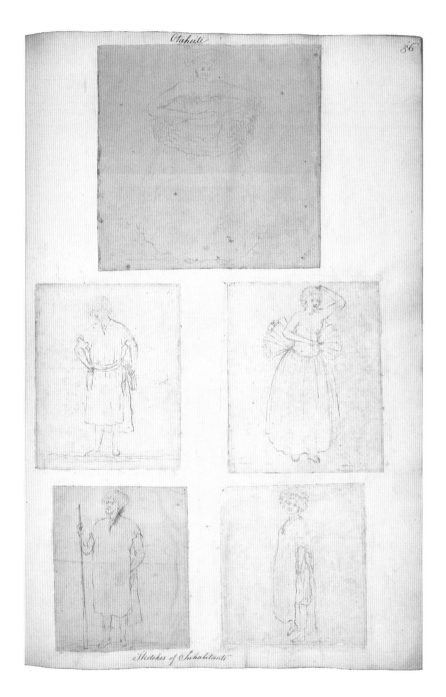

The *heiva* that both Parkinson and Banks subsequently witnessed on 7 August concluded with another dramatic display, which Parkinson thought showed the conquest of Ra'iatea by the forces of Borabora 'in which they exhibited the various stratagems used in the conquest, and were very vociferous, performing all in time with the drum. In the last scene, the actions of the men were very lascivious' (*Journal of a Voyage*, p. 74). Enacted by the victors, this particular performance took place not far along the coast from lands at Rautoanui that had been seized from the unfortunate Tupaia by the Borabora invaders. On 8 August, Banks took Solander to see another *heiva*, and in his journal Banks described a dramatic interlude by the men featuring the theft of a basket of meat from negligent sleeping servants.

52   A Dance in Ra'iatea

By Giovanni Battista Cipriani after Sydney Parkinson.

Trustees of the Goodwood Collection. 266.

33 × 73.7 cm.

The London-based Italian artist Giovanni Battista Cipriani produced this oil painting of a dance on the island of Ra'iatea, probably some time in 1772. It was developed from a pen and wash drawing by him (Dixson Library, Sydney. DL PXX2, 14), which was in turn based on original sketches by Sydney Parkinson (see number 51 above). Cipriani's drawing was engraved by his great friend and countryman Francesco Bartolozzi for publication in the 1773 official account of the *Endeavour* voyage, edited by Dr John Hawkesworth. It appears in volume 2 as plate 7, and represents, in effect, a reinterpretation of the Parkinson materials in keeping with the norms of neo-classical taste then prevailing in London's artistic circles. The Ra'iatean house itself may have been derived from a lost drawing by one of the voyage artists, although a sketch for it by Cipriani exists in Canberra. Joppien and Smith note that, 'Cipriani taught at the third Duke of Richmond's sculpture gallery at Whitehall, and it seems likely that the painting came into his family's possession during the fourth Duke's lifetime' (Joppien and Smith, *The Art of Captain Cook's Voyages*, vol. 1, p. 152). They further observe that the present oil painting is slightly extended on the left-hand side as compared with the Dixson Library version. This has the effect of introducing an extra column, making nine on this side rather than the original eight. Goodwood House, where Cipriani's oil painting is held, is the family seat of the Dukes of Richmond.

After *Endeavour* had departed the Society Islands and then the Austral Islands, Banks wrote an essay account of the 'Manners & customs of S. Sea Islands'. Fond of music and dance himself, and able to play the flute and guitar, Banks must have found the *heiva* performances particularly engaging. Of them he observed: 'Their Drumms they manage rather better: they are made of a hollow block of wood coverd with sharks skin, with these they make out 5 or 6 tunes and accompany the flute not disagreably; they know also how to tune two drums of Different notes into concord which they do nicely enough. They also tune their flutes if two play upon flutes which are not in unison, the short one is leng[t]hned by adding a small roll of leaf which is tied round the end of it and movd up and down till their ears (which are certainly very nice) are satisfied. The drums are usd cheifly in their *heivas* which are at Otahiti no more than a set of musicians, 2 drums for instance two flutes and two singers, who go about from house to house and play; they are alway[s] receivd and rewarded by the master of the family who gives them a peice of cloth or whatever else he can best spare and while they stay, 3 or 4 hours maybe, receives all his neighbours who croud his house full. This diversion the people are extravagantly fond of most likely because like concerts assemblys &c. in Europe they serve to bring the Sexes easily together at a time when the very thoughts of meeting has opend the heart and made way for pleasing Ideas. The grand Dramatick *heiva* which we saw at Ulhietea [Ra'iatea] is I beleive occasionaly performd in all the Islands but that I have so fully Describd in the Journal of that Island Augst ye 3^d 7th and 8th that I need say no more about it.

Besides this they dance especialy the young girls whenever they can collect 8 or 10 together, singing most indecent words using most indecent actions and setting their mouths askew in a most extrordinary manner, in the practise of which they are brought up from their earlyest childhood; in doing this they keep time to a surprizing nicety, I might almost say as true as any dancers I have seen in Europe tho their time is certainly much more simple. This excercise is however left off as soon as they arrive at Years of maturity for as soon as ever they have formd a connection with a man they are expected to leave of[f] Dancing *Timorodee* as it is calld' (*Banks's Journal*, vol. 1, pp. 350–1).

By way of contrast with Cipriani's romanticized illustration, see the later and more credible depiction of a Tahitian *heiva* by John Webber, artist of the third Cook voyage, BL Add. MS 15513, f. 19 (plate 28 in the Atlas to the third-voyage official account).

53 'An Heiva, or kind of Priest of Yoolee-Etea, & the Neighbouring Islands.'

Stanfield Parkinson, ed., A Journal of a Voyage to the South Seas in his Majesty's Ship, The Endeavour (London, 1773; Fothergill reissue 1784). Plate 11. Drawn by Sydney Parkinson. Engraved by T. Chambers.

The British Library, London. L.R.294.c.7. Rebound in half brown leather and brown cloth. Title, place and date of publication, gilt lettering on spine. Author given. 29 × 24.2 cm (closed). Plate 28.4 × 22 cm.

This published print engraving of a man wearing a wicker helmet or *fau* and a breast gorget or *taumi* appeared in the posthumously published edition of Sydney Parkinson's voyage account, where it is linked in a footnote to a visit to a *marae* at Opoa Valley, Ra'iatea, on 21 July 1769 (see *Journal of a Voyage*, p. 70). The footnote mistakenly suggests that such dress was worn by a priest acting as the Chief Mourner, who in fact usually appeared in different ritual attire, for which see number 58. Situated on the eastern coast of Ra'iatea, Opoa was a site of enormous spiritual and historical significance (see number 72). A pencil sketch by Parkinson of the figure and helmet is at BL Add. MS 23921, f. 39(a), and his annotation on the sketch reads 'Whow SS.'. The sketch, or a lost Parkinson drawing developed from it, was used by Thomas Chambers to produce the impressive head-and-shoulders print engraving featured here.

Banks later described a *fau* in his journal when writing about the welcome given him and Daniel Solander on 2 August at the houses of the principal people in the Rautoanui area on the island's western side. In one of the dwellings a *heiva* or dance was performed for the naturalists: 'Gratefull possibly for the presents we had made to these girls the people in our return tryd every method to Oblige us; particularly in one house the master orderd one of his people to dance for our amusement which he did thus:

He put upon his head a large cylindrical basket about 4 feet long and 8 inches in diameter, on the front of which was fastned a facing of feathers bending forwards at the top and edged round with sharks teeth and the tail feathers of tropick birds: with this on he dancd moving slowly and often turning his head round, sometimes swiftly throwing the end of his headdress or *whow* so near the faces of the spectators as to make them start back, which was a joke that seldom faild of making everybody laugh especialy if it happned to one of us' (*Banks's Journal*, vol. 1, p. 324).

Banks took Parkinson to illustrate another *heiva* on 7 August (*Journal of a Voyage*, p. 74). No *fau* is mentioned in Parkinson's journal for this date, but one is described in an editorial footnote to his earlier entry for 21 July, as explained above. See number 54 below for discussion of *fau*, their illustration and collection during the Cook voyages.

An Heiva, or kind of Priest of Yoolee-Etea, & the Neighbouring Islands.

54 Feathered helmets *fau*

1. The British Museum, London. Oc,TAH.9. (Edge-Partington registration slip, 'Cook collection'.) Height 164 cm. Wicker, barkcloth and feathers.

2. The Pitt Rivers Museum, University of Oxford. 1886.1.1683 (Forster 5). Height 137.5 cm. Wicker, barkcloth and feathers.

Only two examples of this rare type of headdress exist today and both are now extremely fragile. One is held at the British Museum, Oc,TAH.9, where it was registered in the late nineteenth century by James Edge-Partington as part of the 'Cook collection', but this provenance was assigned belatedly and gives no indication as to a particular Cook voyage. The other is held at the Pitt Rivers Museum, 1886.1.1683 (Forster 5), and is known to have been collected on the second of Cook's Pacific voyages. Both comprise a central cylinder with an attached frontal shield made of screwpalm (*Freycinetia* sp.) basketry over which a covering of various feathers was placed.

Such ceremonial headdresses were worn by high-ranking individuals, war leaders and priests for example, often with a breast gorget, making for an impressive visual effect. They embodied the authority and spiritual potency or *mana* available to the wearer. Cook-voyage illustrations of Tahitian warriors in their war canoes tend to set a figure so attired in an imposing central position. At the time when the Pitt Rivers *fau* was obtained, probably 28 April 1774, Cook and the Forsters witnessed a formidable Tahitian warfleet mustering prior to battle with the island of Mo'orea. This fleet numbered well over 300 craft assembled from just two districts of Tahiti Nui, according to Cook. Some *fau* headpieces were worn by fleet members, but the voyagers noted that these were soon removed and set aside on the vessel fighting platforms. Large and elaborate, *fau* were not well suited to actual combat, but would have been immediately visible at such a gathering. According to Banks they also served in entertainment and dance activities seen during the *Endeavour* voyage. Additionally, in the posthumously published account of that voyage by Sydney Parkinson, a priest is depicted wearing a *fau*. This variation of use in the written and illustrated records pertaining to *fau* may simply indicate that they traditionally had more than one function and probably form. Alternatively it might be that the place of such headdresses in island life and ritual was in a state of flux during the late eighteenth century. This was a time of upheaval in the islands due to war and religious change. The cult of Oro was becoming increasingly prominent and, as it did, *fau* may have diminished in importance along with older forms of worship to which they related. Perhaps their presence in dances like the one that Banks witnessed was a derogation from former high-status use in sacred or other contexts. During the second Cook mission the voyagers traded using highly prized Tongan red feathers that helped them obtain such valuable artefacts as mourning dresses, *taumi* and presumably *fau*. Such exchanges may also have helped to fuel the pace of change since red feathers were central to the rites performed for Oro and were therefore readily accepted by chiefs and other eminent persons in return for items of this kind. No *fau* was, it seems, gathered on the third voyage.

For many years it was assumed that the British Museum *fau* was originally collected by Banks and included by Benjamin West in the famous portrait of him as the pioneering Pacific collector (see number 143). However, it is today believed that the British Museum *fau* is an altogether different *fau* to that featured in the West portrait, one supposed to have belonged to Banks since it was painted alongside him, but which is now thought to be lost (see the 2007 article by Stevenson and Hooper, 'Tahitian *Fau* – Unveiling an Enigma'). Various differences of structure and decoration have been noticed between the museum *fau* and the West portrait *fau*. The frame of the central cylinder in the former has five hoops visible above the frontal shield, which has the remains of two rows of white feathers inside its rim. The latter, by contrast, appears shorter with only three hoops in the central cylinder being visible above the shield, which has a single row of white feathers and shark teeth inside its rim. Stevenson and Hooper draw attention to *fau* illustrations by Sydney Parkinson at BL Add. MS 23921, f. 39(b), and by Banks's London artist J.F. Miller at BL Add. MS 15508, f. 18, which appear to be of the same or a similar example to that painted by West. Each has a shark-tooth border around the shield rim and this and the visual similarity between the three *fau* depictions, Stevenson and Hooper suggest, may mean that they all show a *fau* seen by Banks during *heiva* performances on Ra'iatea on or about 2 August 1769 (see Banks's journal quoted in number 53 above). The Parkinson and the Miller illustrations both identify the *fau* shown as coming from that island and being used in dances.

To judge from illustrations alone, it appears possible that Banks possessed at least one other *fau*.

54.1

For after the voyage Miller recorded what may be a second *fau* at BL Add. MS 15508, f. 17, one annotated by him as being from Ra'iatea. The *fau* in question is shown with a single feather not a shark-tooth border inside the shield rim, no fringe of white-tailed tropic bird feathers adorning the shield edge and three hoops visible in the central cylinder above the shield top. Stevenson and Hooper observe that Sydney Parkinson had previously illustrated a similar-looking *fau* at BL Add. MS 23921, f. 39(a), which was later used for the published head-and-shoulders plate of a Ra'iatean priest in number 53 above. They add that this second *fau* may be the same as that featured by Parkinson in his illustration of a Ra'iatean war canoe (see number 71). None has a shark-tooth border on its frontal shield. Three cylinder hoops are also visible above each shield with, perhaps, the exception of the *fau* in the illustration of a war canoe, where just two such hoops appear to be visible. This, then, is a plausible interpretation, except that the second Miller illustration includes a pale patch of feathers towards the top left side of the shield, while Parkinson's war-canoe *fau* has a central patch of pale feathers immediately beneath its shield curve and Parkinson's priest has no such patch on his shield at all. A further possibility might be that the two Miller diagrams show the same *fau*, both with and without a shark-tooth border and fringe of tropic bird feathers around the shield. Such a pair would better convey the construction of the *fau* by revealing more of its make-up, in much the way that drawings by natural history artists often showed different or concealed parts of the same plant to reveal more of its taxonomic characters. It may be worth noting in this regard, that Banks entitled folio 17 'Dancing Cap or Whow from Ulhietea without its feathers', and further that his title on folio 18 states the headdress to be a 'Dancing Cap calld Whow from Ulietea'. Each includes six hoops in the central cylinder, with three hoops visible above the shield. The presence of a shark-tooth border within the shield rim for the folio 18 *fau* suggests that it was collected at a *heiva* on the west coast of Ra'iatea. But a more likely explanation is that Miller did indeed draw two separate *fau*, displaying various differences in shape, shield decoration (the patch of light-coloured feathers seen on folio 17 but not on folio 18) and different cord attachments.

It would therefore seem that at least two now lost *fau* were collected by Banks or members of his party during the *Endeavour* voyage as distinguished in surviving illustrations. All the *fau* featured in contemporary depictions appear to be in sound physical condition. Those by Miller and Parkinson lack the feather covering seen on the central cylinder of the British Museum *fau*, as do the *fau* included in Parkinson's drawing of a Ra'iatean war canoe and that in West's portrait. This may show that unfeathered cylinders were an accepted indigenous form for some *fau*.

Then there is the impressive *fau* held at the Pitt Rivers Museum. This is known to be from the second Cook voyage to the Pacific, in which Banks did not participate due to a clash with the Admiralty over the fitting out of the ships for that mission. Two German naturalists went in his place, a father and son team, Johann Reinhold Forster and Johann Georg Adam Forster. They obtained this second precious headdress at Tahiti, probably in late April 1774, and the Forsters donated it to the University of Oxford in 1776 along with a valuable collection of Pacific objects from the voyage. The Forster collection was housed at the Ashmolean Museum until 1886, when it was transferred to the Pitt Rivers Museum. The Forster *fau* has two rows of feathers inside the shield rim, one white and one black, and the shield itself is bordered by projecting feathers of the white-tailed tropic bird, some of which were added during a restoration in 1970. Shorter than the British Museum *fau*, it has just four of the hoops in its central cylinder visible above the shield. Moreover, its cylinder is open-topped whereas the British Museum's *fau* was, although now sprung, once joined at its peak. On the shield front there is a patch of centrally placed yellow feathers, but this is situated at a lower point than the patches seen in the *fau* illustrations mentioned above. Evidently, some *fau* incorporated this element but its position might vary. No other examples of *fau* are known to survive, although one, possibly of second or less likely third Cook-voyage origin, was once held by Trinity College, Dublin. On this reasoning some five *fau* may be identified as having been brought to Britain from Cook's voyages: the British Museum example; two with Banks; the Pitt Rivers example, and the *fau* formerly in Dublin. However, there is insufficient evidence to be sure how many headdresses were brought back with Cook, and comparisons between the surviving examples and contemporary illustrations and journal accounts are by no means always convincing.

The provenance of the British Museum *fau* remains unclear. Stevenson and Hooper speculate as to whether it was collected during the second Cook voyage (pp. 189 and 192), which could have been the case. Among the Admiralty papers there is a note from Cook, dated 13 August 1771, summarizing the bulk of his collections as passed to the Admiralty after the *Endeavour* voyage, and this includes 'a head Ornament worn at the Hievas at Ulietea', Adm. 1/1609. This item is unlikely to be a *fau*. Perhaps it refers to a *tamau*, a headdress made of many strands of finely plaited human hair that was adorned with flowers and worn wrapped around the head by dancers. Several parts were necessary to achieve the required length for such attire (see number 51). A *tamau* ascribed to Tahiti appears to have been among the artefacts given by Cook to Lord Sandwich and then passed by him to Trinity College, Cambridge (see MS Add. a. 106, ff. 108r–9v and MS Add. a. 106, ff. 211r–12v). A small bundle of many strands of plaited black hair resides in the Sandwich collection now held at the Museum of Archaeology and Anthropology in Cambridge, D.1914.5. At the British Museum there is another small bundle of strands of plaited human hair, designated by James Edge-Partington as a 'Cook collection' item of Tahitian origin although the voyage from which it came is unconfirmed, Oc,TAH.56. At the Etnografiska Museet, Stockholm, there is yet another bundle, probably from the first or second of Cook's Pacific voyages, since it was originally included in a gift of natural history and artefact specimens given by Banks to his Swedish friend Johan Alströmer during a visit that Alströmer made to London in 1777–8, 1848.01.0047. These are all likely to be parts of historic *tamau* headdresses. A variety of headdresses existed in the Society Islands in the late eighteenth century and other types were observed and doubtless also collected during Cook's voyages.

54.2

55 Breast gorget *taumi*

The Pitt Rivers Museum, University of Oxford. 1887.1.392. Height 40 and width 51 cm. Cane, plant fibre, coconut fibre, shark teeth, feathers, dog hair and pearl shell.

This Society Island breast gorget, or *taumi,* is part of the Banks collection at the Pitt Rivers Museum, Oxford. It is a known *Endeavour*-voyage object. Breast gorgets are thought to have been worn by warriors and others of high rank. They are formed upon a horseshoe-shaped cane framework, to which are fixed bands of plaited coconut fibre covered in feathers. Curving rows of shark teeth are attached to the gorget and, together with a fringe of fair dog hair around the sides and top edges, such objects give the wearer a striking appearance. It has been suggested that the shark teeth, arranged as they are, represent a shark's gaping jaws and further that a warrior sought to take on the fearsome nature of a shark when wearing a gorget. It also seems to be the case that such gorgets were sometimes worn in pairs, back and front (see number 71). In the present example there are two circular mounts on each shoulder. Designed to be edged with feathers, two of these mounts still retain their original pearl shell discs in the centre. Typically *taumi* have two or three mounts on each shoulder.

Breast gorgets appear to have been fairly numerous at the time explorers started to arrive in the Society Islands. They have always been highly valued by European collectors, and a number of possible Cook-voyage examples survive in modern collections, mainly as individual pieces, but sometimes also in pairs. Known breast gorgets from the *Endeavour* voyage are, however, rare. This is perhaps explicable by the fact that only during the second and third Pacific voyages could Cook and his men trade prized Tongan red feathers for these and other precious Society Island objects. Red feathers were among the most sacred objects within island tribal society and were essential in prayer and worship of its many gods. Cook did not visit Tonga on his first Pacific mission and so had no red feathers to offer in 1769.

In the British Museum collections there are two splendid breast gorgets at Oc,TAH.57 and 57*, the first of which was registered by James Edge-Partington as a 'Cook collection' item. A further museum gorget exists at Oc,1986,Q.1. Possibly this is a lost gorget also registered by Edge-Partington as 'Cook collection', Oc,TAH.537, and likened by him to a post-voyage gorget illustration by J.F. Miller for Banks, BL Add. MS 15508, f. 7 (see Kaeppler, *Artificial Curiosities*, p. 129). As with many comparisons of this kind the match is not a good one. Indeed, none of these three gorgets resembles this or other illustrations of such objects arising from the *Endeavour* voyage: by Miller at BL Add. MS 15508, f. 6; by James Roberts at BL Add. MS 23921, f. 52a, for plate 8 in volume II of the official mission account, and by Charles Praval at BL Add. MS 7085, f. 9. Nor is there documentary evidence to confirm that they resulted from a Cook voyage although this remains a possibility.

The Pitt Rivers Museum gorget was definitely not illustrated by Miller. Another known first-voyage gorget resides in the Sandwich collection at the Museum of Archaeology and Anthropology, Cambridge, D.1914.10. That gorget resembles the illustrations in folios 7 and 52a above but not exactly.

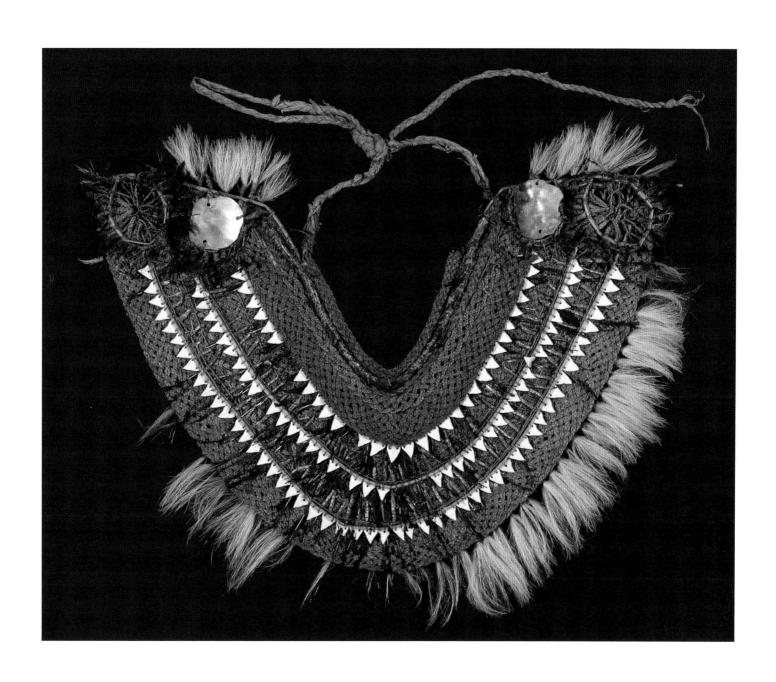

56 Deity figure *to'o*

The British Museum, London. Oc,TAH.64. (Edge-Partington registration slip, 'Cook collection'.) Length 61.9, width 6.6 and depth 6.1 cm. Wood and coconut fibre.

God images took various forms in the Society Islands. This is an example of a *to'o* deity figure. *To'o*, literally meaning a baton or staff, typically comprise a central wooden or coir shaft wrapped with a binding of coconut fibre to which feathers were attached. It has been suggested that early Cook-voyage examples are distinctive in that the ends of the core were left unbound. Here the coconut binding references a figure, with the knots as nipples, a navel and penis, while the crossed cords appear to suggest arms. *To'o* did not represent a deity as such, but enabled a dialogue with the gods or *atua*. In *pa'iatua* ceremonies associated with the god Oro they would be unwrapped and the bindings and feathers, including sacred red feathers, circulated among lesser chiefs and priests, to whom they imparted not only spiritual but also political power. Such ceremonies were conducted by ritual experts and chanting, like the manner in which the *to'o* were bound and unbound, helped to harness the potency or *mana* of the gods for the benefit of humans. *Mana* was a potentially dangerous spiritual force and therefore had to be managed with great care. This particular example was depicted in a drawing by J.F. Miller along with another *to'o* now held in the David King Collection, San Francisco (see the third item in number 57 opposite). Banks probably gave the present *to'o* to the British Museum while that now in America originally went to his old friend, the mineralogist and son of the Earl of Warwick, Charles Greville.

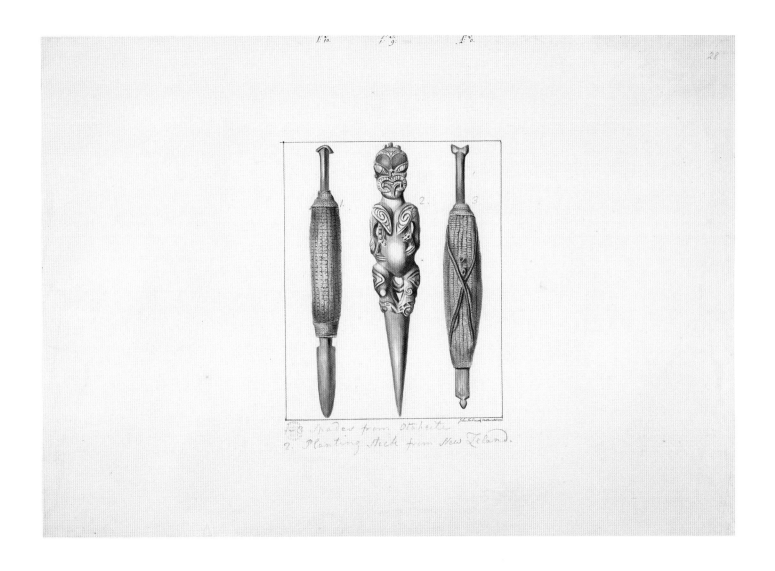

57 Artefacts from Tahiti and New Zealand

By John Frederick Miller.

The British Library, London. BL Add. MS 15508, f. 26.
16.5 × 20.3 cm.

The pen and wash illustration is signed on the recto 'John Frederick Miller del 1771'. A key in pencil in Banks's hand appears on the recto: '1–3 Spades from Otaheite. 2. Planting Stick from New Zeland'. The dimensions of the objects, in sequence from 1 to 3, are given in ink at the top of the folio, '1ᶠ 10ᵇ', '1ᶠ 9ᵇ' and '2ᶠ 0ᵇ'. The central figure is of a Māori *whakapakoko rākau* or 'god stick', which was used to mark an area that was *tapu*, or sacrosanct, for religious reasons. The original of this object as drawn by Miller now resides at the Museum of New Zealand (Te Papa Tongarewa), Wellington, WE001838, and is believed to have been a gift from Banks to the collector and museum proprietor William Bullock. For the third object, see number 56 opposite.

58 Tahiti – A dancing girl and a chief mourner

By Tupaia.
The British Library, London. BL Add. MS 15508, f. 9(a,b).
27.9 × 39.4 cm.

These pencil and watercolour illustrations by Tupaia of Tahitian figures in their ceremonial dress are annotated anonymously on the recto in pencil 'Otaheite', 'Dancing Girl' and 'Chief Mourner'. For dancing girls, see number 51 by Sydney Parkinson, which depicts dancers seen at a *heiva* on Ra'iatea early in August 1769. Similarities between Parkinson's illustrations and Tupaia's dancing girl here suggest that the latter's may also have been seen at Ra'iatea, not 'Otaheite'. If this is so, Tupaia drew his dancer in early August 1769, perhaps around the 7th. Note the angle of the dancer's mouth, an attempt to show the facial contortions adopted during performances.

The second illustration is of the chief mourner wearing his striking ritual attire, the *heva tupapau*. This was probably drawn in June 1769, for it was on 10 June that Banks took part in a mourning procession at Tahiti. The chief mourner was the leading protagonist in mourning rites performed for high-ranking dead. After the deceased had lain in state for a suitable period the chief mourner, a prominent person or a relative of the deceased, appeared wearing impressive formal garments and carrying mother-of-pearl clappers and a shark-tooth staff or *paeho*. Moving as a spirit being, the chief mourner led a group of followers through the country. These followers had beforehand undressed and blackened their skins with charcoal, and they acted in an outlandish and terrifying fashion as if mad with grief. In this state they might become susceptible to possession by demonic spirits and hence dangerously violent, but as such could not be held responsible for their actions. The chief mourner used his clappers to warn of the approaching procession and anyone caught in its way could expect to be beaten with his staff. People hid or fled in horror from the chief mourner and his followers.

Banks participated in such a ritual while at Tahiti, in which he acted as one of the assistants to the chief mourner, removing his clothing, wearing only a strip of barkcloth about his waist and being 'smutted' with charcoal and water. On the orders of the chief mourner, he and his fellow assistants easily dispersed a crowd of over one hundred Tahitians that was encountered at Fort Venus. Taking part in this important ceremony provided Banks with a unique opportunity to gain insights into local religious beliefs, but the complex displays of grieving and ritual that marked the death of individuals in Tahitian society remained difficult for him and other European visitors to comprehend. Early accounts of mourning ceremonies convey different and sometimes conflicting impressions of what was seen. In particular, early observers struggled to understand rapid switches of emotion on the part of the bereaved from extremes of distress to apparent levity, as well as the bloody self-mutilation of the women by striking their heads and shoulders with shark teeth or shells, which was also part of the mourning rites. All too often the insights gained by explorers and, later on, by missionaries were incomplete, not least where the mourning rituals of the poor were concerned. For social reasons these might be less conspicuous than those of senior figures and probably as a result were less well documented.

Mourning dresses were highly valued by Tahitians, who used various precious materials in their manufacture and devoted enormous time and care to their production. None appear to have been collected on the *Endeavour* voyage, although parts of at least ten were obtained on the second Cook mission, when the voyagers were able to offer red parrot feathers gathered at Tonga, these being greatly prized by Tahitians. Each dress was an extremely elaborate creation, comprising long inner robes of barkcloth tied about the waist by a cord, with a netted shawl covered in pigeon feathers over the back. The head mask was made of two pieces of mother-of-pearl fitted together into a disk that covered the face, with a brow ornament of mother-of-pearl, tortoiseshell and, radiating around it, the tail feathers of the sacred tropic bird. A crescent-shaped breast ornament of wood, pearl oyster slices and feathers, and hanging below it an apron arrangement of finely worked mother-of-pearl plates, all added to the spectacular appearance of the chief mourner.

Herman Diedrich Spöring also drew a mourning dress, BL Add. MS 23921, f. 32, but his illustration does not agree as closely with Tupaia's depiction as does that shown in the admirable scene by Sydney Parkinson (see number 59 below). The present illustration is important as an example of indigenous art since Tupaia, who produced it, was the first Polynesian to learn to draw on paper. As a priest he

well understood the forms and significance attached to mourning rites. He was probably tutored in drawing by Parkinson, and his illustrations reflect what he considered significant in scenes and objects from his own culture. Previously it was thought that a mysterious 'Artist of the Chief Mourner', named after this particular drawing and sometimes even thought to be Banks himself, was responsible for a small quantity of distinctive expedition artwork by the same amateur hand, but Tupaia is now understood to have produced this valuable series. It mostly concerns ceremonial and religious subjects or those relating to navigation, in which Tupaia was a native authority (see numbers 50, 58, 67, 69, 77 and 109).

For Banks's journal account of the mourning ceremony in which he participated, one held for the mother of a young woman called Hoona, see *Banks's Journal*, vol 1, pp. 288–9; see also Banks's comments in his journal essay account of the Society Islands (August 1769), pp. 376–9. The chief Tupura'a was chief mourner on this occasion, and he required Banks to act as one of his followers in order to be able to witness the event. It is interesting that Tupura'a thereby engaged one of the most influential voyagers in his train, something that would, presumably, have served to enhance his own standing in Tahitian circles. Tupura'a was deeply embroiled in island politics. He was paramount chief of the Faaa district, the brother of the prominent noblewoman Purea, and a close ally of another district chief, Tutaha. Both he and his friend were slain in battle with the forces of Tahiti Iti in 1773.

59 'A Tupapow in the Island of Otaheite'

By Sydney Parkinson.

The British Library, London. BL Add. MS 23921, f. 31(a). 23.8 × 37.4 cm.

This is a wash illustration by Sydney Parkinson, annotated on the folio in an unknown hand in ink 'Otaheite' and as given above, and on the verso in pencil by Parkinson 'Ewhatta no te tuobapaow'. It was probably composed from sketches made during or after May 1769, as in the first week of that month Parkinson described a Tahitian lying in state: 'In walking through the woods we saw the corpse of a man laid upon a sort of bier, which had an awning over it made of mats, supported by four sticks; a square piece of ground around it was railed in with bamboos, and the body was covered with cloth' (*Journal of a Voyage*, p. 26).

In this illustration we see the funerary enclosure erected to hold a high-ranking corpse with the chief mourner in attendance. During mourning rituals the body was covered in *tapa* cloth and placed on a specially constructed raised bier sheltered by a roofed structure, the *fare tupapau* or house of the dead. A protective fence of bamboo demarcated the area held to be *tapu* when approaching the body. It was thought that the spirit of the dead lingered in this world while flesh still hung about the body, witnessing the mourning of relatives and friends, and perhaps also causing mischief. Various rituals were peformed to guard against the latter eventuality, and the body's flesh was allowed to decay in order to release its spirit before being committed for burial or sometimes cremation. Complex mourning ceremonies were conducted that saw the women express their grief through weeping and self-mutilation, to be followed by the men and sometimes also women in a ritual enactment led by the chief mourner (see number 58 above and *Banks's Journal*, vol 1, pp. 265-6). Such rituals were intended to placate the spirit of the deceased and enable it to rest in peace. Gifts were also made to the gods, including of food, lest during the mourning period a hungry deity should light upon and consume the dead.

There is some disagreement as to the length of time that a corpse would be displayed on its bier before burial, and why in Tahitian society this period varied from one individual to another. Just one or two days were allotted in some cases, while in others up to five days were thought necessary, and for senior chiefs a number of months might elapse. The smell of the decaying body was one obvious incentive for its fairly rapid interment, and it may well be that in most cases the deceased was not exposed for very long. Not all bodies were kept in the open and there allowed to rot until only the bones remained. Some, however, did receive elaborate treatments over an extended period of time while laid out in structures of the kind depicted by Parkinson. Their viscera were removed and scented oils applied to the remaining flesh until it was dessicated in the Sun and eventual decomposition had taken place. Then the remains were committed for final burial in an appropriate *marae*. Tending bodies in this way was a costly process, so high-status individuals were most likely to undergo protracted mourning rituals, but there is also evidence that less important persons were sometimes accorded lengthy rites, and this is perhaps explicable by the strong attachment felt for them by their kin. Note the Tahitian hiding from the chief mourner in a palm tree. He would have been beaten with the chief mourner's shark-tooth staff or assaulted by his dreaded followers if apprehended by them.

John James Barralet later used Parkinson's illustration to produce a pencil and wash depiction of the same scene (Dixson Library, Sydney, DL PXX2, 45), making various alterations in order to open up a central view of the chief mourner and corpse. In particular, Barralet moved to the viewer's left the palm tree that Parkinson had originally positioned in the middle of his scene (as a botanical artist might perhaps be expected to do). Barralet also brought forward a breadfruit tree situated to the right of Parkinson's background, thus ensuring that his own version was framed on both sides. Barralet's hiding Tahitian is placed in this breadfruit tree, not in the palm tree as Parkinson's had been. Barralet's version was engraved by William Woollett for Hawkesworth's 1773 *Endeavour* voyage account, vol. 2, plate 5. Woollett was a master engraver, widely acknowledged in his own time as one of the finest exponents of his craft. He famously engraved Benjamin West's painting of *The Death of General Wolfe*, published in 1776, and he also contributed the most expensive plate to the official account of Cook's last voyage. This was entitled 'A Human Sacrifice, in a Morai, in Otaheite' (after John Webber, BL Add. MS, 15513, f. 16), and it recorded a scene witnessed by Cook at Tahiti (see plate 15 in the Atlas to the third-voyage official account, published in 1784).

60 Bow and quiver with twelve arrows

The Museum of Archaeology and Anthropology, University of Cambridge. D.1914.97/83. Bow 164 cm; quiver 95.5 and 6 cm; arrows average 80 cm. Bamboo and wood.

In the Society Islands bows and arrows were not used for war, but were instead employed in ritual contests and occasionally for hunting birds. They were simple in design. Bows were made of softwood and bark string. Arrows were made of bamboo, were wooden-tipped and unfeathered. Quivers were made of bamboo and sometimes decorated. Archery contests were a pursuit of the ruling elite, the *ari'i*, and appear to have had religious significance. Prayers and ceremonies usually accompanied a match. Contests were undertaken from a special platform and were decided on distance rather than accuracy. The loser supplied his fellow competitors with food and dancing. That a primary weapon like the bow and arrow was not employed in warfare is noteworthy in islands where rivalry and conflict were commonplace. One reason given for this is that by making the use of the bow *tapu* for their subjects, while continuing to practise its use themselves, the *ari'i* were better able to safeguard their position at the head of society. Throwing lances and hurling stones with slings were the other main throwing sports, but with these accuracy as well as distance mattered, and both were used in war. Bows and arrows from Cook's voyages survive in several European collections. J.F. Miller depicted a set for Banks, BL Add. MS. 23921, f. 57 (see number 61).

In his journal for 11 June 1769, Banks recorded a not untypical cultural misunderstanding, when Third-Lieutenant John Gore and the chief Tupura'a agreed to a bow and arrow contest: 'This evening Tubourai came to the tents bringing a bow and arrows, in consequence of a challenge Mr Gore had given him sometime ago to shoot. This challenge was however misunderstood, Tubourai meant to try who could shoot the farthest, Mr Gore to shoot at a mark and neither was at all practisd in what the other valued himself upon. Tubourai to please us shot in his way; he knelt down and drew the bow and as soon as he let slip the string droppd the bow from his hand, the arrow however went 274 yards' (*Banks's Journal*, vol. 1, pp. 289–90). The simultaneous dropping of the bow as the arrow was released was to avoid the backlash of the bowstring since chiefly archers wore no wrist guard. It would have been, of course, a wholly inappropriate technique in any battle situation.

Gore rated himself a good marksman with gun and bow, and not without reason (see number 116).

61 Artefacts from Tahiti

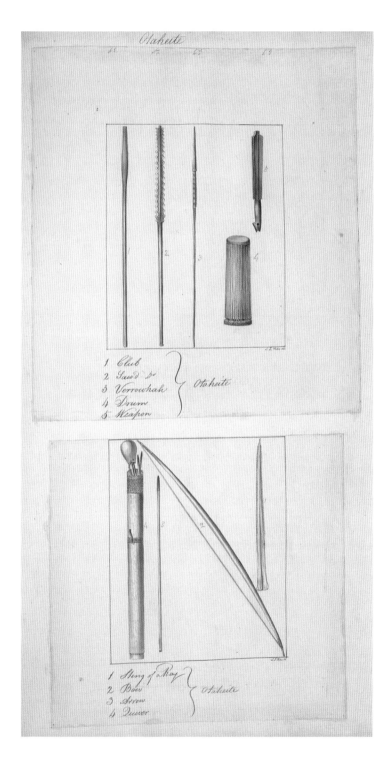

By John Frederick Miller.

The British Library, London. BL Add. MS 23921, f. 57(a,b).
20.4 × 16.5 cm;
20.6 × 16.5 cm.

This is a pair of pen and wash illustrations, probably produced in 1771, of weapons and a drum, (a), and of a quiver, bow and arrows with a stingray point, (b). Both by J.F. Miller, (a) is inscribed in ink with a key, '1 Club 2 Saw'd D° 3 Verrowhah 4 Drum 5 Weapon', all given as from 'Otaheite'. Then (b) is also inscribed in ink with a key, '1. Sting of a Ray 2. Bow 3. Arrow 4 Quiver', all given as from 'Otaheite'. Miller signed each illustration on the recto 'J.F. Miller del'. The object for number 5 in (a) is at the British Museum, Oc,TAH.65, 'Cook collection' (Edge-Partington and O.M. Dalton). Its original purpose is not clear, but it may well have been a deity figure and not a weapon at all.

James Cook observed in his journal: 'I shall next describe the Arm[s] with which they attack their enemies both by sea and land, these are Clubs, spears or Lances, Slings, and stones which they throw by the hand, the Clubs are made of a hard wood and are about 8 or 9 feet long, the one half is made flatish with two edges and the other half is round and not thicker than to be easily grasped by the hand; the lances are of various length some from 12, 20 or 30 feet and are generaly arm'd at the small end with the stings of stingrays which makes them very dangerous weapons. Altho these people have bows and arrow[s] and those none of the worst, we are told that they never use them in their wars which doubtless is very extraordinary and not easily accounted for' (*Cook's Journals*, vol. 1, p. 132).

In his journal essay account of the Society Islands, Banks commented: 'Their weapons are Slings which they use with great dexterity, pikes headed with the stings of sting Rays, and Clubbs of 6 or 7 feet long made of a very heavy and hard wood. With these they fight by their own account very obstinately, which appears the more probable as the Conquerors give no Quarter to Man Woman or Child who is unfortunate enough to fall into their hands during or for some hours after the Battle, that is till their Passion is subsided' (*Banks's Journal*, vol. 1, p. 386).

62 A tattooing comb and a tattooing mallet

1. The Museum of Archaeology and Anthropology, University of Cambridge. D.1914.35. Length 15.5, head length 3.4 and head width 3.1 cm. Wood, bone and plant fibre.

2. The British Museum, London. Oc,TAH.118. (O.M. Dalton registration slip, 'Cook collection'.) Length 39 cm. Wood.

Tattooing was a widespread cultural practice throughout Polynesia at the time of the *Endeavour* voyage, but it was not, of course, a new technique to Europeans. In the Society Islands tattoos were mainly applied to the arm, leg, buttock and lower back areas, while the head and shoulders were avoided due to the belief that these parts were sacred. Elsewhere in Polynesia, such as New Zealand, the face and most of the body might be adorned. Tattoos were not merely decorative. In the Society Islands they could indicate the descent, social status and coming to maturity of an individual. They were applied by a priest or *tahu'a* and had religious significance by venerating the god Ta'aroa. The process was a painful one in which a needle-comb or *ta* was dipped in pigment, and then a mallet used to strike its handle, pushing the pigment into the skin. The modern English word 'tattoo' derives from the Tahitian *tatau* meaning 'to tap'. Typically needle-combs comprised a serrated blade made of shell or bone secured to a wooden handle by coconut fibres. In the nineteenth century missionaries discouraged Polynesian tattooing, but recently it has undergone a revival. Banks appears to have had a tattoo placed on his arm as did Parkinson (see DTC I, 54–5, and *Journal of a Voyage*, p. 25). A number of tattooing instruments gathered on Cook's Pacific voyages survive, and Banks's artist J.F. Miller illustrated a first-voyage selection, BL Add. MS 23921, f. 55(a). The present comb is a known first-voyage object originally given by James Cook to Lord Sandwich, and the mallet is thought to be from a Cook voyage, although which one is uncertain.

In his journal essay on the Society Islands, Banks remarked that tattoos were rarely if ever applied to the face. Other parts of the body, however, received distinctive patterns: 'yet all the Islanders I have seen (except those of Ohiteroa) agree in having all their buttocks coverd with a deep black; over this most have arches drawn one over another as high as their short ribbs, which are often ¼ of an inch broad and neatly workd on their edges with indentations &c.' As to the method of application, a painful process that Banks witnessed, he recorded that: 'The colour they use is lamp black w[h]ich they prepare from the smoak of a kind of oily nutts usd by them instead of candles; this is kept in cocoa nut shells and mixt with water occasionaly for use. Their instruments for pricking this under the skin are made of Bone or shell, flat, the lower part of this is cut into sharp teeth from 3 to 20 according to the purposes it is to be usd for and the upper fastned to a handle. These teeth are dippd into the black liquor and then drove by quick sharp blows struck upon the handle with a stick for that purpose into the skin so deep that every stroke is followd by a small quantity of Blood, or serum at least, and the part so markd remains sore for many days before it heals' (*Banks's Journal*, vol. I, p. 335–7; see also p. 309, a description of a girl being tattooed on her buttocks).

The nut referred to above by Banks is the seed of the candlenut tree, *Aleurites moluccana* (L.) Willd. The seed has a very high oil content, which makes it useful as a candle. Producing ink for tattoos is only one of its multifarious uses in cooking, cosmetics and traditional medicine.

62.1

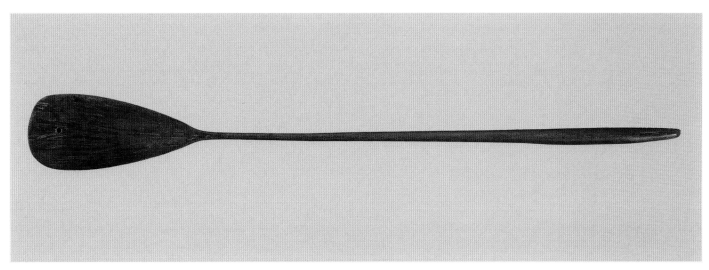

62.2

63 Society Island tattoos

By Sydney Parkinson.

The British Library, London. BL Add. MS 23921, f. 51(d) verso.
29.7 × 42 cm.

Folio 51 is a set of sketch drawings on three sheets by Sydney Parkinson of four male faces showing the distortions of the mouth employed in Society Island dances, anonymously annotated 'Otaheite' and 'Distortions of the Mouth used in Dancing'. On the verso of the last two faces, which appear together on a single sheet, are these three pen-and-ink studies of tattoo designs (see Joppien and Smith, *The Art of Captain Cook's Voyages*, vol. 1, p. 148, who suggest that the faces were completed on Ra'iatea on 7 August 1769; see also number 51 regarding the *heiva* that Parkinson attended with Banks on this island on that day).

In the verso tattoo sketches Parkinson notices the placing of tattoos on the buttocks and the use of the important Polynesian crescent motif. Number 71 below features a central figure in a Ra'iatean war canoe with similarly tattooed buttocks. Parkinson regularly sketched details of figures and objects to be developed into or featured in finished composite scenes. Likewise he produced sketch outlines of plants and animals with notes on them as to colouring, intending to complete these illustrations later (see numbers 122–6). Following Parkinson's death during the voyage, his annotated plant sketches were worked up by others in London for inclusion in Banks's planned *Florilegium*. Similarly, some of Parkinson's drawings of scenes and figures were engraved to illustrate the official account of the voyage and the edition of his voyage papers that was published by his brother, Stanfield. In reproducing the latter scenic drawings, however, many were drastically altered by artists in London who had never visited the Pacific and who portrayed life there in the classical styles with which they were most familiar. Another Parkinson sketch of tattooed buttocks, probably from Tahiti, is at BL Add. MS 9345, f. 1 verso.

64 Canoe steering paddle

The British Museum, London. Oc,TAH.87. (Edge-Partington and C.H. Read registration slips, designated 'Cook collection' by Read.) Length 182.5, width 32.5 and thickness 3.5 cm. Wood.

Two British Museum registration slips associated with this paddle describe it as, first, coming from the 'Tahitian Group' (Edge-Partington) and, second, as a 'Steering paddle from Ulietea [Ra'iatea]' assigned to the 'Cook collection' (Read). It closely resembles a paddle depicted in an illustration for Joseph Banks by J.F. Miller, BL Add. MS. 15508, f. 29 (see Kaeppler, *Artificial Curiosities*, p. 155). The paddle is well designed for its function. Its shaft terminates in a simple cross-grip handle, which is semi-oval in section, and this and the broad, flat blade are unadorned. In his journal essay on life in the Society Islands, Banks remarked: 'These boats are paddled along with large paddles which have long handle and a flat blade resembling more than any thing I recollect a Bakers peel; of these generaly every one in the boat has one except those who set under the houses and with these they push themselves on pretty fast through the water' (*Banks's Journal*, vol. 1, pp. 366–7; see also number 71).

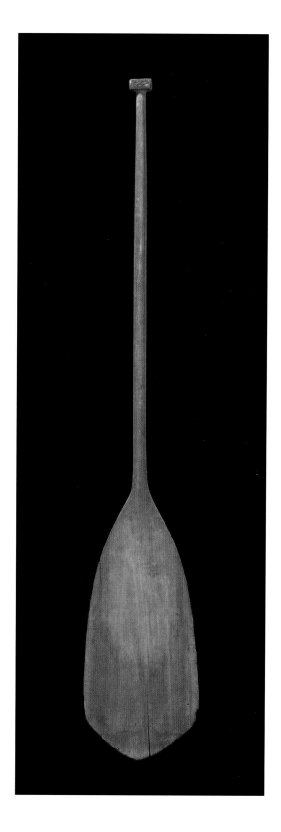

65 Canoe bailer *tata*

The Museum of Archaeology and Anthropology, University of Cambridge. D.1914.28. 41 × 24 × 11 cm. Wood.

This bailer is from the Society Islands, probably Tahiti. It is a known *Endeavour*-voyage object because in 1771 it was part of a collection of artefacts given by James Cook to Lord Sandwich, who passed a selection of this material to Trinity College, Cambridge, which later deposited everything at the Museum of Archaeology and Anthropology. The bailer is carved from a single piece of wood, with an integral handle displaying no embellishments. Its handle is formed by a central bar running parallel with the sides, which connects with a transverse bar at the mid-point of the bailer. Plain and functional, it is similar in design to other Society Island bailers brought back by Cook's missions. Among these is one held at the Pitt Rivers Museum, Oxford, again a known first-voyage object. This bailer was part of the *Endeavour* collection donated by Joseph Banks to his old university college, Christ Church, and later passed to the Pitt Rivers Museum. In that collection it formed a pair with another bailer from New Zealand, showing Banks's organized approach to collecting since his gift included more than one object of the same type from different Pacific locations.

A further early-voyage bailer is held at the British Museum, Oc,TAH.6, 'Cook collection', with a label affixed, 'Boat-Scoop from Otaheitee.' This label is thought to be in the hand of Daniel Solander, who was employed at the museum from 1763 and in 1773 became under-librarian (keeper) of the department of Natural and Artificial Curiosities. As such he took charge of the increasing quantity of Pacific material that was directed to the museum during and after Cook's voyages, not least by Banks. This bailer may well have been collected on one of Cook's voyages, but which one is not certain, although see King, *Artificial Curiosities from the Northwest Coast of America*, pp. 16–17, who suggests that labelled items like this one probably derive from Cook's last voyage.

J.F. Miller illustrated a Society Island bailer for Banks following the *Endeavour* voyage (see number 93). It is not possible to determine on appearance alone whether any of the above bailers was the one drawn by Miller, but of them that at Cambridge bears the closest general resemblance to Miller's illustration, particularly in the curved edge of its handle bridge and those of its underside and mouth, the bailers at Oxford and the British Museum having straighter lines by comparison. See at number 71 an impressive composite set-piece illustration drawn by Sydney Parkinson, showing such a bailer in use alongside a range of other objects gathered or observed by the explorers (see also number 92).

66 Shark hook and fish hook

1. The Pitt Rivers Museum, University of Oxford. 1887.1.378. (Society Islands shark hook.) Length 35.5 and width 12.5 cm. Wood, coconut fibre.

2. The Pitt Rivers Museum, University of Oxford. 1887.1.379. (Māori fish hook.) Length 18 and width 8 cm. Wood, bone, New Zealand flax and Freycinetia banksii.

The maritime culture of Polynesia was of particular interest to the voyagers on *Endeavour*, not least because they hailed from a nation whose reliance on the sea was for so many reasons also pivotal. Various kinds of hooks, lures and other fishing gear were collected at the Society Islands and elsewhere during Cook's voyages, and drawings and accounts were compiled of the methods and equipment used by Polynesians when fishing. Fishing provided an essential source of food and natural products. Freshwater as well as saltwater fishing was skilfully practised throughout the region. Polynesian peoples principally relied on spear, net and line to catch their prey. This ranged from jellyfish, octopus and squid to shelled creatures including oysters, mussels, periwinkles, sea urchins, lobsters and crabs. Many of these could be gathered inshore, in pools, lagoons, reefs, sheltered coves and elsewhere about the coast. There, too, and further out at sea was a large assortment of fish, ranging from small fry and eels to bonito and albacore, with dolphin, porpoise, swordfish, shark and even whale (when stranded) numbering among the largest prey taken.

66.2

The hooks used in line fishing were of different sizes and shapes, and could be assembled from more than one part and material. They might be used inshore or in deeper waters, and their design reflected the size and type of catch being sought. Most fish hooks were made of shell, bone or wood, or some combination of them, with wood being preferred for hooks employed to catch larger prey. The first of the present hooks is probably from Tahiti and was used in bait fishing for bigger fish such as shark, hence its greater size and strength. A two-piece wooden hook, it is fitted with a separate wooden point bound with coconut fibre to the main shaft. The second fish hook is from New Zealand. It is a two-piece fish hook, made of wood with a bone point and attached cord. Both hooks were part of an *Endeavour*-voyage collection donated by Joseph Banks to his old college, Christ Church, Oxford, some time before mid-January 1773, which was later passed to the Pitt Rivers Museum. As a pair, the hooks show how Banks selected examples of the same type of object from different Pacific locations to represent a range of Pacific production. See illustrations of very similar if not actually the same hooks as these ones by J.F. Miller for Joseph Banks, dating to 1771 or 1772, BL Add. MS 15508, f. 27. A selection of first-voyage fish hooks and a line and sinker donated in 1771 by Lord Sandwich to Trinity College, Cambridge, are deposited at the university Museum of Archaeology and Anthropology.

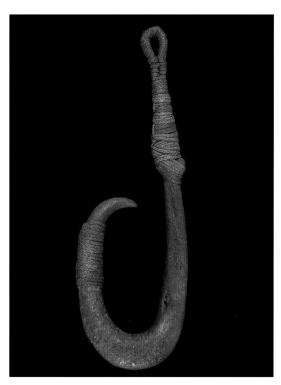

66.1

67   A scene in Tahiti

By Tupaia.

The British Library, London. BL Add. MS 15508, f. 12.

26.7 × 36.8 cm.

This is a pencil and watercolour illustration by Tupaia of a scene in Tahiti, most probably completed between April and July 1769. Tupaia was an influential *tahu'a* or priest when Samuel Wallis first arrived at Tahiti in the *Dolphin* in mid-1767. He was originally from Ra'iatea, the religious centre of the Society Islands, but was driven from his home when that island was overrun by the warlike men of Borabora in about 1760, narrowly escaping with his life after sustaining a serious spear wound during the fighting. Tupaia took up residence in Tahiti, but he yearned to return to Ra'iatea. *Tahu'a* communed with gods and spirits, often in the sacred precincts of the *marae*, and some were distinguished by possessing special skills and knowledge. In Tupaia's case, these related to his knowledge of Polynesian navigation, and so this illustration depicts elements of island life about which he knew a great deal. It shows a typical scene featuring a longhouse with, either side of it, pandanus, breadfruit, banana, coconut trees and the taro plant. Three vessels appear, one under sail (*va'a*), the other two (*pahi*) with fighting platforms on which are armed figures, one holding a quarterstaff and the other a spear. The illustration is unfinished, with some of the figures being in pencil outline only.

Tupaia was a close advisor to Purea and also one of her lovers (see number 68 immediately below), and when Wallis was at Tahiti their power was in the ascendant. Purea's standing had greatly diminished since then owing to a war that she and her partner, Amo (formerly Tevahitua), the chief of Papara district, fought and lost in December 1768 against chief Vehiatua from Tahiti Iti. Consequently, Tupaia, who was heavily embroiled in the power politics of the island, had good reason to be concerned about his future by the time *Endeavour* was at Tahiti. He may also have hoped to enlist British help against his old Borabora enemies. He expressed a desire to leave on *Endeavour* and, supported by Banks, was allowed aboard by Cook along with a boy servant, Taiato. Tupaia proved a useful addition to the voyage due to his knowledge of Pacific geography, which was displayed on an impressive map showing the relative positions of islands ranging from the Marquesas in the north-east to Rotuma in the west (see number 77). He was also able to communicate with the Māori of New Zealand where, given the difficult nature of contacts, his assistance proved invaluable. Some have even suggested that the Māori thought Tupaia was in charge of *Endeavour* because he was a high-ranking priest and the one with whom they had the most dealings. Tupaia and Taiato died in December 1770 of fevers contracted at Batavia (Jakarta) as *Endeavour* sailed for home.

68 'Oboreahs Canoe Otaheite'

By Herman Diedrich Spöring.

The British Library, London. BL Add. MS 23921, f. 23(a). 18.7 × 32.5 cm.

This pencil drawing by Herman Diedrich Spöring is annotated anonymously in pencil on the recto as given above, with a verso ink note by Spöring, 'Canoe about 30 feet in length belonging to Öbalhoea'. Spöring was born and educated at Åbo (Turku), at the time in Sweden but now in Finland, where his father was professor of medicine at the university. He may have practised medicine for a while, but in 1755 moved to Britain, where he worked in London first as a watchmaker and then as a clerk to Daniel Solander at the British Museum. He travelled on the *Endeavour* mission in the latter capacity, to fair copy the plant and animal descriptions of the naturalists and to label artwork, but, after the death at Tahiti of Alexander Buchan, Spöring started to produce landscape and ethnographic illustrations as well as some natural history drawings and coastal profiles. He worked in pencil without applying colour, but achieved a high degree of technical accuracy. It has been said of Spöring, not unfairly, that the precise eye of a watchmaker is apparent in his drawings (see, for example, numbers 88, 94 and 111–12).

Samuel Wallis preceded James Cook to Tahiti, being the first European navigator to visit and describe the island, during six weeks from June to July 1767. At that time it was incorrectly assumed that Purea (Oboreah to the British) was ruler of the island, it not being understood that Tahiti was divided into a series of separate districts, each ruled by a chiefly family or group called the *ari'i*. These rulers were venerated and alliances could be formed among them, including by intermarriage, but war between tribal districts and indeed islands was also common. Under the *ari'i* were the *tahu'a*, an important priest class who held spiritual power. Their role was central because spiritual entities were seen as active in every aspect of the human and natural world. *Tahu'a* made offerings to and sought help from the gods, who could enter and speak through them, and their rituals helped underpin the authority of the *ari'i*. They might possess special knowledge and skills, for example in navigation as with Tupaia. Then came the *ra'atira*, nobility of lesser status but who were able to own land, and beneath them were the ordinary people or *manahune*. Tahitian society was highly structured, with developed customs and beliefs and was not, in fact, ruled by one individual at all.

At Matavai Bay it was Tutaha who governed, not Purea, whose district was Papara at the south-west corner of the island. Tutaha had been a prominent figure in the fall of Purea and her partner Amo, ruler of Papara, in December 1768. Tutaha's base was Paea on the west coast, but by the time of Cook's arrival his control extended to include the north-west corner of the island and hence Matavai Bay. Purea was nevertheless recognized when Cook anchored at the bay in April 1769, and treated as an island queen, much to the resentment of Tutaha. Only slowly did the voyagers realize that their European-based conceptions of Tahitian society were an inadequate means of comprehending its actual make-up. Indeed, the political, religious and economic life of the Society Islands was complex and volatile. Although the voyagers could not have known it, they had arrived in the islands during a period of change characterized by tribal wars between the ruling families of each district and religious upheaval. This was partly the result of the recent invasion of Ra'iatea and Tahaa by neighbouring Borabora, causing the cult of Oro to shift its base from the great Marae Taputapuatea (see number 72) to Tahiti, where increased worship of Oro only served to fuel religious and dynastic conflict. Nor could the voyagers have known that their advent in turbulent times may have been foretold in native prophecies based on earlier encounters with European vessels, giving the voyagers divine associations that would certainly help explain the importance attached by local rulers to securing their support, as Purea had tried to do in her dealings with Wallis in 1767 (see numbers 69 and 77, and the 1982 article by H.A.H. Driessen, 'Outriggerless Canoes and Glorious Beings'). On arriving back at Tahiti in August 1777 during his third voyage, Cook discovered that Purea had died, probably in 1775 or 1776. Before that Tutaha was himself slain in battle after attacking the forces of Tahiti Iti in 1773.

It was during a visit to see Tutaha on 28 May 1769 that Purea offered her canoe as a sleeping place for Banks. Tupaia stayed close to Banks, while Cook and Daniel Solander, who had accompanied him on his trip, found sleeping places elsewhere. It being hot Banks undressed for the night, during which his jacket, waistcoat, two pistols and a powder-horn were stolen. When he discovered his loss in the night

there was a commotion in which Purea, Tutaha and Tupaia were all involved, but Banks never retrieved his possessions and suspected that Purea and Tutaha were implicated in their theft. Following his return from the voyage satirists in England made much of Banks's supposed liaison with the 'Queen of Tahiti', although Banks had not been physically attracted to his royal hostess (see number 137).

On the same night Cook was also robbed of his stockings, while Solander escaped without loss. The trip was not wholly unprofitable, however, for the next morning the party witnessed surfing, as practised by a group of Tahitians using an old canoe stern. What to Banks appeared fatally high breakers, were evidently an enjoyable playground for experienced Tahitian swimmers: 'We stood admiring this very wonderfull scene for full half an hour, in which time no one of the actors attempted to come ashore but all seemd most highly entertaind with their strange diversion' (*Banks's Journal*, vol. 1, pp. 281–3).

69 *Marae*, Tahiti

By Tupaia.
The British Library, London. BL Add. MS 15508, f. 14(16).
31.7 × 38.7 cm.

This pencil drawing by Tupaia of a *marae* is perhaps based on the large one constructed at Mahaiatea by Purea and Amo in honour of their son Teri'irere. As a priest and a follower of Purea, Tupaia would have known the *marae* at Mahaiatea especially well since he had assisted in its design and consecration under the god Oro. The actual *marae* featured in this illustration remains, however, unclear but Tupaia represents those features typically found at such sites. *Marae* were the open-air religious structures where Tahitians commemorated the dead, made sacrifices (including human) and worshipped the many gods of the Society Islands. Their confines were sacred and entry was forbidden without permission. *Marae* varied in size and importance, but commonly consisted of a stepped-stone platform or *ahu* that formed one side of a low-walled court. Tupaia here depicts an *ahu* comprising eleven steps. He includes a paved court area and a centrally placed *fare atua*, or deity house, where the *to'o* deity figures were kept. A sacrificial pig appears on an altar table, or *fata*, forward of the deity house.

The transit of Venus having been observed, Cook and Banks travelled around Tahiti in a clockwise direction from 26 June to 1 July. They saw various *marae*, but by far the most impressive example to be found anywhere on the island was the pyramidal structure raised at Mahaiatea in the Papara district on Tahiti Nui. Built in the years 1766–8, this was one of the greatest works of architecture in the Society Islands and a conspicuous monument to Purea's pride. Through it she hoped to establish the power of her son Teri'irere above all rivals by proclaiming him paramount chief, or *ari'i rahi*, of the island at a ceremony held within the *marae*. Here the explorers learned of the war made on Purea and Amo by their enemies, who resented the couple's audacious schemes, and they witnessed numerous human bones lying on a nearby shore. These were the discarded remains of those defeated in or sacrificed after the decisive battle, which was followed by the looting and burning of villages in the area.

Banks was astonished by this grandiose structure, recording that: 'Its size and workmanship almost exceeds beleif, I shall set it down exactly. Its form was like that of *Marais* in general, resembling the roof of a house, not smooth at the sides but formd into 11 steps, each of these 4 feet in hight making in all 44 feet, its leng[t]h 267 its breadth 71. Every one of these steps were formd of one course of white coral stones most neatly squard and polishd, the rest were round pebbles, but these seemd to have been workd from their uniformity of size and roundness. Some of the coral stones were very large, one I measurd was 3½ by 2½ feet. The foundation was of Rock stones likewise squard, one of these corner stone[s] measurd 4^{ft}: 7^{in} by 2^{ft}: 7^{in}. The whole made a part of one side of a spatious area which was walld in with stone, the size of this which seemd to be intended for a square was 118 by 110 paces, which was intirely pavd with flat paving stones. It is almost beyond beleif that Indians could raise so large a structure without the assistance of Iron tools to shape their stones or mortar to join them, which last appears almost essential as the most of them are round; it is done tho, and almost as firmly as a European workman would have done it, tho in some things it seems to have faild. The steps for instance which range along its greatest leng[t]h are not streight, they bend downward in the middle forming a small Segment of a circle: possibly the ground may have sunk a little under the greatest weight of such an immense pile, which if it happend regularly would have this effect. The labour of the work is prodigious: the quarry stones are but few but they must have been brought by hand from some distance at least, as we saw no signs of quarry near it though I lookd carefully about me; the coral must have been fishd from under water, where indeed it is most plentifull but generaly coverd with 3 or 4 feet water at least and oftenest with much more. The labour of forming them when got must also have been at least as great as the getting them; they have not shewn us any way by which they could square a stone but by means of another, which must be most tedious and liable to many accidents by the breaking of tools. The stones are also polishd and as well and truly as stones of the kind could be by the best workman in Europe, in that particular they excell owing to the great plenty of sharp coral sand which is admirably adapted to that purpose and is found everywhere upon the seashore in this neighbourhood. About 100 yards to the west of this building was another court or pavd area in which were several *ewhattas*, a kind of altars raisd on wooden pillars about 7 feet high, on these they offer meat of all kinds to the gods; we have seen large Hogs offerd and here were the Sculls of above 50 of them besides those of dogs, which the preist who accompanied us assurd us were only a small part of what had been here sacrafisd. This *marai* and aparatus for sacrafice belongd we were told to Oborea and Oamo. The greatest pride of an inhabitant of Otahite is to have a grand *Marai*, in this particular our freinds far exceed any one in the Island' (*Banks's Journal*, vol. 1, pp. 303–4).

70 'Society Islands. Canoes of Ulietea'

By Sydney Parkinson.

The British Library, London. BL Add. MS 23921, f. 20. 29.8 × 48.2 cm.

This unsigned wash drawing by Sydney Parkinson is annotated anonymously in ink on the folio as given above and with a verso pencil note by Banks 'Ulietea', meaning Ra'iatea. *Endeavour* visited Ra'iatea and its immediate neighbour Tahaa from 20 July to 9 August 1769, when this illustration must have been made. It shows in the foreground a two-masted double canoe of the *pahi* type with carved figures on its tapering bow and stern. Behind it to the right as viewed another double canoe may be seen, with further craft visible in the far distance. The foreground canoe has adults, children, a baby, some chickens and a dog on it. It is provided with one cabin situated well forward and another placed amidships. Pacific reef herons appear to the right.

The people of Ra'iatea were highly accomplished canoe builders. They lodged their canoes in large boathouses, considerable numbers of which were observed by Banks and Solander in the Opoa area on Ra'iatea's east coast. Voyage drawings of Society Island canoes and their houses focus on those seen at Ra'iatea and Tahaa. Canoes were essential to island life for travel, trade, fishing and warfare. They came in two main hull types, the *pahi* as seen here, and the *va'a*.

Pahi canoes were constructed on Huahine, Ra'iatea and other Leeward Islands, but not it would seem on Tahiti Nui and, as a result, few were seen there during the first voyage. A number appear to have been observed by Banks and Cook on Tahiti Iti when they journeyed around Tahiti before finally departing the island. Banks noted that those he saw were visitors from Ra'iatea. On later voyages more *pahi* canoes were seen on Tahiti Nui, implying that their construction had by then spread there, perhaps due to greater contacts with the Leeward Islands and almost certainly via adjacent Tahiti Iti, where they were probably already being made. *Va'a* canoes were, by contrast, built throughout the Society Islands.

Timber for canoes was taken from felled trees using stone adzes and not bent but carved to the required shape. The parts of a canoe were bound firmly together using coconut fibre thredded through pierced holes, and the joins between them were sealed by caulking made of beaten coconut husk and resin. Double canoes were formed by lashing spars athwart two hulls of the same type. Decking could then be placed between the connected hulls, on which one or two cabins or open-sided shelters might be situated. When used in warfare such canoes were necessarily larger in size and could be fitted with a raised fighting platform, for which see number 71. All Society Island canoes could be paddled, with a sail or sails made of strong mats sewn together being employed on those of a larger size or that were undertaking a longer journey. Canoes could also be used individually, in which case an outrigger would be added for stability. Single outrigger canoes of less than 25 feet in length were usually paddled without a sail. Basic rafts might be employed for short trips and ferrying cargoes.

Va'a canoes were of a simpler design than the *pahi* type, which might help explain their wider use, being hollowed-out from a log or logs to make a hull that was flattened or gently curved at its bottom. In their most basic form *va'a* might be nothing more than a single hollowed-out log. Their sides were upright and could be heightened in larger examples by the addition of one or two planks or strakes. Typically the *va'a* stern was raised in a curve, although for smaller examples this might only amount to a slight upward trend in the hull. Hollow carved pillars could be fitted to the stern to increase its height to about 14 feet in the air in some cases, limiting the vessel's capabilities far from shore, but displaying the importance of its occupants. *Va'a* canoes were for use in shallow and, if large enough, ocean waters and were the preferred vessel for a range of purposes, among them inshore fishing and short trips. For Oboreah's *va'a*-type double canoe, with a cabin in which she could sit when being carried from place to place or occasionally sleep during an overnight stay, see number 68.

Pahi canoes were a more sophisticated 'keel'-style assembly, with sides made from carved planks that in cross-section resembled the shape of an inverted spade in modern European playing cards. Interior timber ribs strengthened such hulls. *Pahi* incorporated a vertically raised prow and stern, the stern always being higher than the prow. They were better adapted for longer, deep-water voyages than *va'a* canoes and were able to carry the extra stores needed for such journeys in their rounded hulls. They were of greater general length too, ranging from about 30 to over 75 feet, against which *va'a*

canoes varied in length from 10 to a maximum of about 70 feet. According to Banks, local people could spend up to twenty days voyaging in a *pahi*, visiting many islands on extended journeys that lasted several months. Both canoe types required frequent bailing as their open hulls easily shipped water. For a bailer and paddle, see numbers 64–5.

At Opoa on 21 July 1769, Banks and Daniel Solander witnessed a *pahi* being made. Banks noted the dimensions of the craft and detailed its method of construction, remarking on the skilful use of carpentry tools by the locals, including their chisels incorporating a blade of human bone: 'these they grind very sharp and fix to a handle of wood, making the instrument serve the purpose of a gouge by striking it with a mallet made of a hard black wood, and with them would do as much work as with Iron tools was it not that the brittle Edge of the tool is very liable to be broke' (*Banks's Journal*, vol. 1, pp. 319–20; see also pp. 363–8). For an example of a chisel, see number 38. Adzes were also important in the construction of canoes, for which see number 37.

71 'Society Islands. A War Canoe'

By Sydney Parkinson.

The British Library, London. BL Add. MS 23921, f. 21.
29.8 × 48.3 cm.

Annotated anonymously in ink on the folio as given above, this wash drawing by Sydney Parkinson shows a Ra'iatean war canoe of the *pahi* type, and was probably completed in August 1769. The warriors are armed with a range of weapons, and one man displays tattoos on his buttocks that Parkinson also drew in separate sketches (see number 63). This man is using a bailer to remove water from the canoe. Garments of barkcloth are featured, such as the turbans on the heads of three of the canoe occupants. Note, too, the central figure on a fighting platform wearing a breast gorget or *taumi* and a tall feathered helmet or *fau*. The two men either side of the central figure clearly wear their gorgets back and front.

These and other types of objects collected or observed during the voyage are portrayed by Parkinson to enrich further what is in fact an elaborate composition. In scenes such as this one Parkinson combined a range of subject matter, some of which he had drawn or sketched beforehand, to be shown in a single set-piece view of local life and action. Joppien and Smith suggest that 'This appears to be the earliest drawing (in terms of subject matter, if not also in time) by Parkinson in which he assembles figures in an ordered composition' (Joppien and Smith, *The Art of Captain Cook's Voyages*, vol. 1, p. 154).

72 'a Morai with an offering to the Dead'

By Sydney Parkinson.

The British Library, London. BL Add. MS 23921, f. 28.
23.8 × 37 cm.

This wash drawing by Sydney Parkinson is annotated anonymously on the folio in ink as given above and with a verso pencil note by Banks 'Huaheine'. In fact, it is thought to show a scene on Ra'iatea, not Huahine, dating to July 1769. The scene is one of enormous significance for it features the ancient Marae Taputapuatea at Opoa Valley on the southeast coast of Ra'iatea, the most sacred of all the royal *marae*. Marae Taputapuatea stands among a complex of ceremonial structures only recently beginning to be excavated and more fully understood (see the 2010 article by A. Smith, 'Archaeology, local history and community in French Polynesia'). The great creator Ta'aroa was believed to have first entered the Earth at Opoa, creating there the ancestral homeland of East Polynesians now known as Ra'iatea. At the time of European contact Opoa was the religious, ceremonial and navigational heart of Central Polynesia, from which great voyaging canoes are said to have departed in the past for islands as distant as the Cook Islands and New Zealand. From at least the sixteenth century it was the centre for worship of Oro, the god of war and fertility, led by members of the influential *arioi* cult. The *arioi* spread Oro's worship to other islands, carrying on their travels stones from Marae Taputapuatea with which to establish a powerful physical and spiritual network of new *marae* across East Polynesia.

Parkinson depicts a hog on a raised platform, or *fata rau*, as well as what appear to be two fish and perhaps some breadfruit. Offerings to the gods might also include dog, turtle and plants such as bananas or coconuts. Rendering the local language as best he could, Banks recorded a visit to Taputapuatea on 20 July: 'After this we walk together to a great *Marai* calld *Tapodeboatea* whatever that may signifie; it is different from those of Otahite being no more than walls about 8 feet high of Coral Stones (some of an immense size) filld up with smaller ones, the whole ornamented with many planks set upon their ends and carvd their whole leng[t]h. In the neighbourhood of this we found the altar or *ewhatta* upon which lay the last sacrafice, a hog of about 80 pounds weight which had been put up there whole and very nicely roasted' (*Banks's Journal*, vol. 1, p. 318). The planks to which Banks makes reference are called *unu*, carved upright wooden boards that, along with the presence of the Oro deity symbols or *to'o*, denoted the high status of this *marae*. The sombre mood of Parkinson's illustration captures some of the awe associated with this site.

71

72

73 Fort Venus, Tahiti

Copy drawing by Charles Praval after Herman Diedrich Spöring.

The British Library, London. BL Add. MS 7085, f. 8(a–d). 62.5 × 59.5 cm.

This is a pen drawing by Charles Praval after Herman Diedrich Spöring, completed between January and July 1771.

It is in four parts showing at the top, in folio a, 'A VIEW of a part of the West Side of GEORGES ISLAND taken from the Ship at Anchor in ROYAL BAY'.

Folio b is 'THE WEST ELEVATION of the FORT'. Here the key identifies various parts of the fort, in which Banks's tents are located at 'a', with the observatory being located at 'b'. The key runs 'a M^r· Banks's Tents b The Observatory c Officers Tent d. Mens Tent and Guard-room e. Cook-room and Smith Forge f. Coopers and Sailmakers Tent'.

Then bottom left is folio c, 'A PLAN of ROYAL or MATAVIE BAY in GEORGES ISLAND.' Scale 1 mile is 5 inches.

Bottom right is folio d, 'A PLAN of FORT VENUS in ROYAL BAY.' Here a key again identifies Banks's tents at 'a' and the observatory at 'b'. The key runs 'a. M^r· Banks's Tents b. The Observatory c. The Clock d. Officers Tent e. Mens Tent and Guard-room f. Magazine g. Oven and Cook-room h. Smiths Forge i. Necessary House k. Carriage Guns l. Swivels m. Coopers and Sailmakers tent'. Scale 100 ft is 3⅝ inches.

Praval was entered as a supernumerary in the muster book of *Endeavour* on 19 December 1770, while the ship was still at Batavia (Jakarta). He was one of nineteen mainly British hands taken on board to strengthen the crew due to losses sustained at that port through disease. Both Spöring and Sydney Parkinson died at sea of fevers after departing Batavia, on 24 and 26 January 1771 respectively, thus robbing the mission of its two remaining illustrators. Praval possessed modest skills as a draughtsman and copyist that were soon recognized by Banks and Cook. It would seem that he was therefore enlisted as an AB for Wages and Victuals on 6 February. On the way back to England his time was devoted to copying drawings of charts, coastal views, landscapes and some artefacts (BL Add. MS 7085) for an account of the voyage that Cook was apparently preparing for possible later publication. The present illustrations of Fort Venus form part of this presumed work. Fort Venus featured in other manuscript illustrations, see BL Add. MS 23921, f. 2 (Spöring) and, similar in various respects to it, BL Add. MS 15508, f. 5 (Praval after Spöring). The west elevation of the fort in folio b resembles the latter illustration in a number of ways. A published view of the fort appeared in Sydney Parkinson's *A Journal of a Voyage*, plate 4, as engraved by S. Middiman. This is thought to have been derived from a lost drawing by Parkinson.

Endeavour anchored in Matavai Bay on 13 April 1769 and, after contacts with the local people had been established, it was possible to negotiate the use of a site for a base ashore. This was located on the headland at what was called Point Venus, where, on 15 April, tents (starting with those brought by Banks) were erected and the construction of a defensive wooden stockade begun. Once finished, Fort Venus was some 55 yards long by 30 yards wide, and in it was the observatory containing the instruments for observing the forthcoming transit of Venus. A trade rapidly developed with the Tahitians for supplies, and Banks showed himself to be a confident and sociable intermediary in dealings with them. This was demonstrated when the astronomical quadrant, essential to the transit observation, was stolen by a local man, and Banks, the astronomer Charles Green and others ventured into the countryside to retrieve it. Spöring, who had earlier in his career worked as a watchmaker in London, was able to repair minor damage to the instrument (see number 14). Fort Venus was the land base of operations for the *Endeavour* expedition until 13 July, when the ship departed Tahiti to gather supplies among the Leeward Islands and refresh the crew before heading southwards. Cook gave these islands the collective name of the Society Islands because they 'lay contiguous to one a nother', although it is sometimes incorrectly stated that he did so in tribute to the Royal Society, which initiated this voyage (see *Cook's Journals*, vol. 1, p. 151).

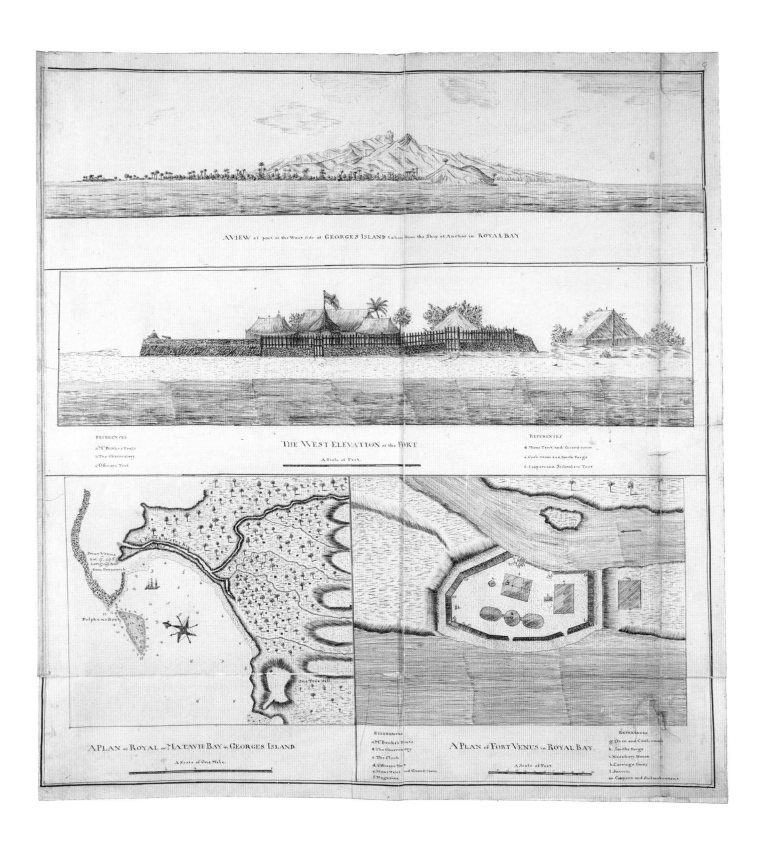

74 'Observations made, by appointment of the Royal Society, at King George's Island in the South Sea; by Mr. Charles Green, formerly Assistant at the Royal Observatory at Greenwich, and Lieut. James Cook, of his Majesty's Ship the Endeavour', by Charles Green and James Cook, *Philosophical Transactions*, vol. 61 (1771), pp. 397–421

The Royal Society of London. Executive Secretary Sequence. Half bound in red leather and tan cloth. 23.3 × 18.3 cm (closed).

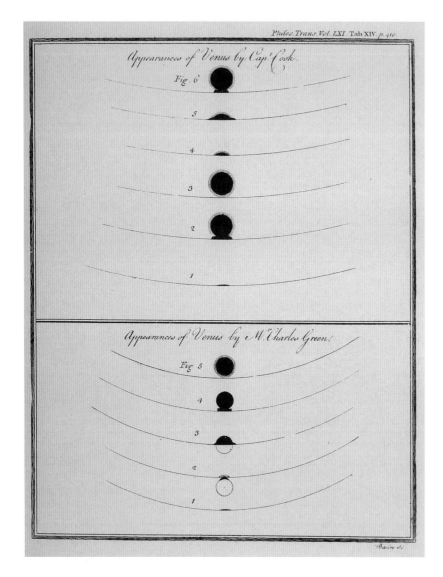

Observation of the transit of Venus took place on Saturday 3 June, with Charles Green and James Cook conducting it at Point Venus. Using the information gained from this and other observations taking place around the world, the mean distance of the Earth from the Sun could be calculated. During the voyage Green trained a number of the *Endeavour*'s officers in astronomical observation. Taking no chances, Cook arranged for two other parties to make observations elsewhere in case Matavai Bay was cloudy, but in the event the weather was clear. Second-Lieutenant Zachary Hicks went east to Motu Taaupiri and Third-Lieutenant John Gore took a party to Motu Irioa on Mo'orea. Banks and Herman Diedrich Spöring accompanied Gore. Later, as befitted a mission launched at the instigation of the Royal Society, the astronomical results were published in the society's journal, the *Philosophical Transactions*.

The present illustration from the *Philosophical Transactions*, table XIV at page 410, shows Venus passing across the disc of the Sun as seen by Cook and Green. It reveals an unexpected optical problem with the observations that made it extremely difficult to time the event with the required degree of accuracy. As Venus approached the Sun a shadow appeared to join the two bodies, making it impossible to distinguish the exact moment of contact. This phenomenon became known as the 'black drop' effect. Nevertheless, by combining data from various transit observations made in 1761 and 1769, the French astronomer J.J. Lalande in 1771 obtained a value for the distance from the Earth to the Sun equivalent to 153 million kilometres. The value determined by modern radar is 149.6 million kilometres. Note that the *Philosophical Transactions* observations are dated 2 June because the astronomical day did not begin until noon in order that an entire night's observations could be recorded under a single date.

75 'Observations made by appointment of the Royal Society at King Georges Island in the South Sea by Mr Chas Green formerly Assistant at the Royal Observatory at Greenwich and Capt. James Cook of his Majesty's Ship the Endeavor.'

The Royal Society of London.
RS L&P V 257, ff. 1–10.
30 × 19 cm.

This is the first folio side of the manuscript report of the observation of the transit of Venus on 3 June 1769 by Charles Green and James Cook. It was read as a paper before the Royal Society on 21 November 1771 and then published in the society's journal (see number 74 opposite). The main hand appears to be that of Charles Green with annotation by the astronomer royal Nevil Maskelyne.

76 'CHART *OF THE ISLAND* OTAHEITE, BY *LIEUT. J. COOK* 1769.'

John Hawkesworth, ed., An account of the voyages undertaken by the order of His Present Majesty for making discoveries in the southern hemisphere, and successively performed by Commodore Byron, Captain Wallis, Captain Carteret, and Captain Cook, in the Dolphin, the Swallow and the Endeavour: drawn up from the journals which were kept by the several commanders, and from the papers of Joseph Banks, Esq, 3 vols (London: Printed for W. Strahan and T. Cadell, 1773). Vol. II, facing page 79. By J. Cook and I. Smith. Engraved by J. Cheevers.

The British Library, London. 455.a.22. Rebound in half blue leather and blue cloth. Title, place and date of publication, gilt lettering on spine. Editor and volume given. 29 × 24.2 cm (closed). Plate 31 × 48 cm. Scale ca. 1:250,000 (lat.).

Tahiti is the administrative centre of present-day French Polynesia and its capital, located on the north-west side of the island, is Papeete. The island comprises two parts, the larger Tahiti Nui lying to the north-west of the smaller Tahiti Iti, joined by a narrow isthmus. This print engraving of a chart of Tahiti was published in 1773 in the official three-volume account of the *Endeavour* voyage, as edited by Dr John Hawkesworth. It shows Tahiti, with its two main island sections named Opoureonu and Tiarrabou. A number of charts of Tahiti were compiled by James Cook, with help from his draughtsman and relation Isaac Smith, which are variously based on one another. That at BL Add. MS 7085, f. 7, resembles the present printed chart. Those at BL Add. MS 7085, f. 6, BL Add. MS 21593B and Dixson Library, Sydney, Safe 166, appear to be related to one another. Andrew David, in *The Charts & Coastal Views of Captain Cook's Voyages*, vol. 1, pp. 108–12, dates the development of the manuscript charts mentioned here to the period April to July 1769, when *Endeavour* was at Tahiti.

On his arrival at Tahiti in April 1769, Cook selected Matavai Bay as a suitable anchorage and Tutaha, the chief of the local district, allowed him to make a headland base from which to observe the imminent transit of Venus. *Endeavour* did not depart immediately after the observation on 3 June (see numbers 74–5), and this allowed the island to be explored and its people to be observed in more detail. On 26 June, Cook and Banks set out to travel round the island in a clockwise direction, a journey that took six days walking or afloat in the pinnace. On the way they saw the bay where the French navigator Louis Antoine de Bougainville had anchored in the previous year. They discovered, too, that Tahiti is formed of two parts connected by an isthmus, which they crossed in order to complete the round trip. Back on Tahiti Nui they visited the great *marae* at Mahaiatea in the Papara district, which was built by Purea and Amo for their son. Cook and Banks then continued along the west coast to the districts of Faaa and Pare, and finally back to Matavai Bay, where preparations to depart were commenced. This trip enabled Cook to complete his outline of the island showing its bays and harbours.

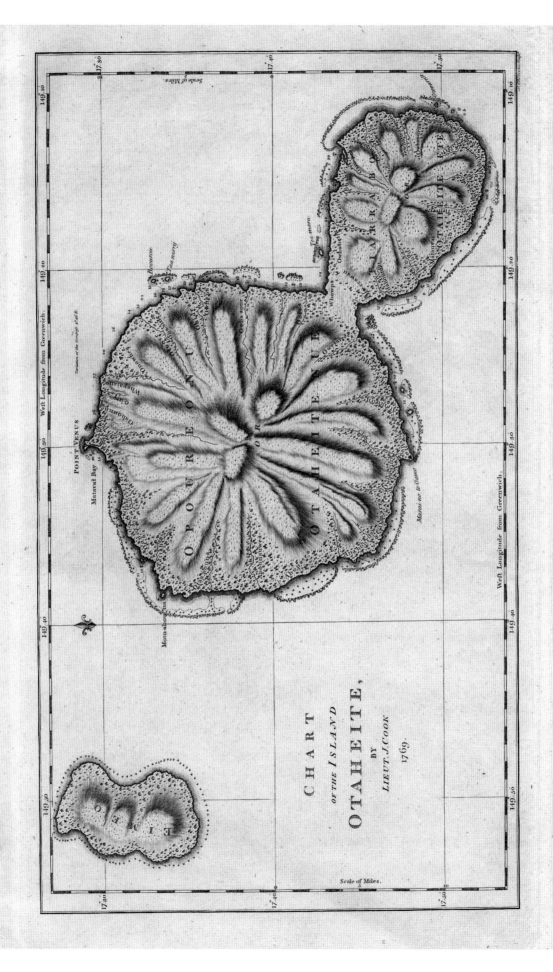

77 The Society Islands copied by James Cook from an original chart by Tupaia

Copy chart by James Cook after Tupaia.

The British Library, London. BL Add. MS. 21593C. 20.2 × 33.6 cm. Scale approx. 1:10,000,000.

This ink and wash chart of the Society Islands and wider Pacific is believed to be a copy by James Cook of a now lost original chart by Tupaia, the Ra'iatean priest and navigator who joined the *Endeavour* expedition when it departed Tahiti in July 1769 to explore for a southern continent. A note in pencil reads 'Drawn by Lieut Jas Cook 1769'. Andrew David, in *The Charts & Coastal Views of Captain Cook's Voyages*, vol. 1, p. 131, ascribes this note to Banks, but it does not appear to be in his hand. The chart shows the many Pacific islands known to Tupaia. Five captions relate to individual islands and past events concerning them, and suggest that Tupaia was aware of earlier contacts with European vessels, one of which may have been *De Africaansche Galey*, a ship from Jacob Roggeveen's squadron that was wrecked at Takapoto in the Tuamotus in 1722 and some deserters from it possibly murdered by locals, if so probably at Anaa. The chart resides among collections formerly belonging to Sir Joseph Banks that are now held at the British Library. It was first published in 1955 by R.A. Skelton in a portfolio of charts and maps from the *Endeavour* voyage accompanying J.C. Beaglehole's landmark edition of Cook's journals.

Cook was doubtful about the idea of bringing Pacific islanders back to Britain, but Tupaia wished to depart on *Endeavour* and, at Banks's request, he and his boy servant Taiato were taken on board when the vessel left Tahiti. Banks notoriously recorded that he wanted Tupaia more or less as a curiosity to impress his friends in England, 'I do not know why I may not keep him as a curiosity, as well as some of my neighbours do lions and tygers at a larger expence than he will probably ever put me to', in effect making Tupaia another object for collection and display (*Banks's Journal*, vol. 1, pp. 312–13). Notwithstanding this comment, Banks had led the way in studying Tahitian society and befriending its people. He well understood that Tupaia's knowledge of island habits and produce was a valuable asset to the mission, in addition to which Tupaia had a fair grasp of English by this time, and so could help interpret during any future encounters, a critical factor as it turned out in dealings with the Māori of New Zealand. A sometimes sceptical Cook also came to appreciate Tupaia for the geographical knowledge that he possessed and, as the voyage progressed, both men turned to the Ra'iatean for insights and information. Banks, for example, was able to draw on Tupaia's knowledge when, having left them, he sat down to write his journal essay account of life in the Society Islands. Similarly, Cook benefited from Tupaia's capabilities as a pilot among the reefs and bays of the Leeward Islands. Having at Huahine ascertained the ship's draught, Tupaia warned against entering depths of less than five fathoms, understanding that *Endeavour* might be at risk in such waters. He indicated suitable harbours at Ra'iatea, his home island. He predicted, too, the arrival of *Endeavour* at Rurutu after the ship left the Society Islands, this being an island that he had apparently visited before. Using his skills as a navigator Tupaia was thereafter always able to indicate the direction in which Tahiti lay.

More significant still, later in the voyage Tupaia produced a chart showing his extensive knowledge of Pacific island geography. Although the original is now lost, the surviving copy by Cook is an impressive testament both to Tupaia's knowledge and to the maritime heritage of his people. It stands, too, as a monument to the complex and sometimes confused nature of cross-cultural exchanges on missions of this kind. Having evidently been drawn at Cook's request to help him better comprehend Tupaia's geographical grasp, interpretations of what the chart actually shows and why remain inconclusive. Its broad visual arrangement of the various islands into concentric circles centered on Tahiti has been thought to correspond to their distances from that location in respect of sailing times (see David, *The Charts & Coastal Views of Captain Cook's Voyages*, vol. 1, pp. 130–1). Interestingly, a number of islands are located incorrectly, but David Turnbull has argued that this is not due to inadequacies in Tupaia's knowledge, but perhaps arose from misapprehensions between him and Cook as to how to lay down such a chart (see Turnbull, in Lincoln, ed., *Science and Exploration in the Pacific*, pp. 126–8). Basing his interpretation on comments in the mid-nineteenth century by the American ethnographer Horatio Hale, Turnbull suggests that in places the chart appears to be upside-down, possibly because Cook misconstrued Tahitian terms indicating the direction in which the north and south winds blow for actual north and south. Those islands with which the

Tupaia's Map

(A chart of the islands surrounding Tahiti, drawn by Tupaia, c. 1769)

Island names visible on the chart include:

Oahoatroa, Oryvavai, Olematerra, Oateeu, Oruruitu, Oahoo-ahoo, Opatoerow, Ohevapoto, Oheva roa, Ocito, Tebooi, Teposuhah, Whatterrero, Ouraaitua, Toutepa, Oureu, Moluhea, Whennua ouda, Oura, Oo-ahe, Teohaow, Oryroa, Whateretuah, Whanganea, Temauno, Tetinoheva, Oweha, Opoporea, Opopooa, Maatiah, Oahohe, Oinah, Whareva, Orumaroa, Orivavie, Oroluma, Bola-bola, Ohahah, Tupi, Maurua, Ohevatoutouai, Uietea, Huaheine, Whaow, Oirotah, Tetupatappa eahow, Mocratayo, Ohevavane, Ohete maruiru, Ourappe, Tetrypoepooma'riteha, Oheavie, Teavrooroatina-roa, Opoorvo, Ooaticw, Otaheite, Myrea, Ilonae, Oheteroa, Mannua, Noutou, Tencu'hanmeatane, Teaineovohete, Ohetepolo, Ohotohoutou-atu, Ohetetouton-nii, Ohetetoutoureva, Ohetehizaure, Ohooteerra, Onowhea, Teatewhete, Tereai Tsooftera, Etatahieta Oheitoottera, Opatoa, Tinuna voyagers were familiar are added correctly despite the inversion. Those islands that Tupaia knew, but which the voyagers did not, are situated according to the Ra'iatean's sense of wind and hence sailing direction. Some errors may be Tupaia's, of course, while certain islands are given twice and various name spellings remain obscure even today, nevertheless over half of the 74 islands included in the chart were unknown to Europeans at the time Cook entered the Pacific, and the area it covers is vast. New Zealand, Easter Island and Hawai'i apparently do not feature. More recently, it has plausibly been suggested that the chart represents a mosaic of plotting diagrams by Tupaia showing the sailing directions from more than one island centre, each diagram being orientated differently and on a different scale, and not a single statement of Pacific geography set out on a grid of fixed co-ordinates with Tahiti at its centre as Cook and his contemporaries would have understood such a document to be (see the 2007 article by Di Piazza and Pearthree, 'A New Reading of Tupaia's Chart').

Over the centuries humans spread in stages from South East Asia across the ocean, but by the late eighteenth century the long-distance migrations of earlier periods had ceased. The real extent of voyaging by Society Islanders at the time of Cook's visits remains unclear, but the present map and the journal accounts of the voyagers indicate that it could on occasion have been far wider than might be supposed. Tupaia's knowledge of Pacific geography must have derived in part from the discoveries of earlier generations of Polynesian voyagers since he was a priest well versed in his people's history and an expert in their various navigation methods. Doubtless trading networks and other contacts between islands, including those occurring by chance, must also have been a source of information for him and his kin. All this made him a living repository of information that Banks and Cook were only partially able to tap. In March 1770, some nine months after Tupaia joined the expedition and as it was about to depart New Zealand, Cook's thoughts turned to the possibility of a future Pacific mission. If attempted, such a mission would probe a final time for any hidden southern continent, which an increasingly sceptical Cook felt must lie in high latitudes if anywhere at all (Tupaia had never heard of one), but it might also explore the many islands of the southern tropics. With this latter aim in mind Cook compiled a journal list of 73 islands with their directions from Tahiti as known to Tupaia. Of these Tupaia claimed to have visited twelve according to Cook, all but two of which lie within the cluster that forms the Society Islands group, indicating a probable limit to Tupaia's normal voyaging of this vicinity (the most accurately rendered part of his chart), although it would seem that much longer journeys were still possible, with return destinations of up to 400 leagues to the west lying within comfortable reach (ten to twelve days' sailing) according to Tupaia. Cook noted that: 'The above list was taken from a Chart of the Islands Drawn by Tupia's own hands, he at one time gave us an Account of near 130 Islands but in his Chart he laid down only 74 and this is about the Number that some others of the Natives of Otaheite gave us an account of, but the Accounts taken by and from different people differ sencibly one from another both in names and Number' (*Cook's Journals*, vol. 1, pp. 288–94; see also pp. 138–9, pp. 153–4 and pp. 156–7, the latter pages referring to longer voyages westwards in canoes that Cook believed sailed much faster than *Endeavour*.)

In 1778 an edited and updated version of Tupaia's chart was published in Johann Reinhold Forster's account of Cook's second Pacific voyage. Forster participated in this mission as a naturalist, along with his son Johann Georg Adam Forster. Engraved by William Faden, the chart appeared facing page 513 of *Observations made during a voyage round the world, on physical geography, natural history, and ethic philosophy*. This work emerged, however, against a background of disagreement and acrimony concerning the official voyage account by Cook. A fine naturalist and a keen observer of the peoples encountered during the voyage, the elder Forster claimed that he had been promised authorship of the official account, apparently by the antiquary Daines Barrington, but having returned to England in 1775 he was excluded following a dispute with the Admiralty and Cook over who would head that work. After the disappointing first-voyage account edited by John Hawkesworth, Cook evidently had no intention of relinquishing control a second time, and when Admiralty chief Lord Sandwich's patience with the obstinate Forsters eventually ran out there was little the hapless naturalists could do to alter the situation (see number 32). Yet a prohibition laid on the elder Forster preventing him from independently publishing an account did not apply to his son, who produced one using his father's journals that appeared ahead of Cook's by only six weeks, *A voyage round the world, in His Britannic Majesty's sloop Resolution* (1777). A German translation of this work, also by Georg, was very well received on the Continent.

To print Tupaia's chart in 1778, the elder Forster obtained a manuscript copy version of the original from Richard Pickersgill, master on *Endeavour*, and a copy was also provided by Banks. He therefore had access to two copy versions, only one of which now survives, and perhaps even to the lost original, although he does not state this. Using these as well as lists provided by Cook, Forster updated the published version with more islands discovered by other European explorers, or that he himself saw or ascertained from local reports during the *Resolution* voyage, and he also collated the various spellings of island names then available. The result was a chart with 79 islands on it as compared to the 74 shown in the surviving British Library copy. In the accompanying text Forster commented briefly on each island, including five more islands not shown on his published chart. He also made appreciative comments regarding Tupaia's knowledge of Pacific geography, calling him 'the most intelligent man that ever was met with by any European navigator in these isles', pp. 509–13. Following publication of their accounts father and son departed Britain for academic posts on the Continent, in 1780 and 1778 respectively. For comparison see a chart of Cook's Pacific discoveries, number 139.

The Material History of the Endeavour: Joseph Banks at the British Library

PHILIP J. HATFIELD

The British Library

THIS ESSAY BEGINS not with the *Endeavour* expedition but with a detail, Banks's bookplate-cum-library stamp (see number 15). 'Jos: Banks', his typical signature and the form his stamp took, is frequently to be found in books originally from Banks's personal library, now housed in the British Library at St Pancras. It is not just a marker of history but, often, a benchmark of quality.[1] While curating the recent British Library exhibition *Lines in the Ice: Seeking the Northwest Passage*, I was struck by the regular occurrence of Banks's stamp. It is found in works by explorers such as Samuel Hearne[2] and it is also in works containing thanks for the use of Banks's library. One example is Thomas Pennant's *Arctic Zoology*. Pennant made use of Banks's library at Soho Square and paid tribute to him in the early pages of this book, stating 'Sir Joseph Banks, Baronet, will, I hope, accept my thanks for the free admittance to those parts of his cabinet which more immediately related to the subjects of the following sheets'.[3] These are some of the many references still to be found at the British Library that suggest the enduring impact of Banks and the effect of his activities on how we see our scientific history and, indeed, the world as it now is.[4] It is this impact, as recorded in and in part produced by the Banksian collections held at the British Library, which this chapter sets out to consider.

Banks's role on the *Endeavour* expedition is perhaps the most widely known way in which a modern audience understands his contribution to science and culture, and this expedition too is represented in the collections that he bequeathed to the nation at his death in 1820. Banks not only gave his main library and natural history collections to the British Museum at the end, but throughout his life he also gave books and papers to its library, effectively pruning his own collection as it expanded and his interests increasingly focused on botany. That this range of material should now be stored in such large concentrations in London's museums and libraries is testament to how deeply Sir Joseph Banks was embedded in the very fabric of Enlightenment London. Banks and his family were ardent supporters of the British Museum and its library,[5] donating their various collections to it. Sir Joseph was also a direct participator in the early administration of the museum, most notably through his role as a trustee.[6] The British Museum library and the extensive private collection that Banks developed both shared a similar core goal – to assemble information about the world at a time when Britain's political horizons were expanding and to make this information available to a broad community of scholars, authors and colonial and other officials.

Banks's collections, composed of books, maps, illustrations, natural history specimens, manuscripts and much else, have been dispersed as relevant across the collections of the British Museum, the Natural History Museum and the British Library, to name the three main national institutions. As such his collections have had a profound influence on some of Britain's most significant historical and scholarly collections, not least in reflecting the history of the *Endeavour* expedition. For the British Library this has meant the presence of a significant quantity of manuscripts, maps, topographical views and printed books related to the expedition and these are found in the manuscript, map and printed book collections of the library.[7] There are also items that derive from Banks's other travels. For example, he developed an interest in Norse sagas and started to collect them in manuscript form as a result of his trip in 1772 to Iceland. However, almost certainly due to his focus on botany, he later pruned these particular materials from his collection and donated them to the British Museum.[8]

When the *Endeavour* expedition was under preparation, Banks's library was of course small in comparison to the size and coverage that it would subsequently attain, but it still provided significant reference works to aid his research and that of the team he assembled for the voyage. Books and charts already possessed by Banks would have been consulted in order to develop his prior understanding of the areas to be encountered. During the voyage a number of them endured hard conditions, as Banks noted in his journal when the ship crossed the equator: 'all the books in my Library became mouldy so that they were obliged to be wiped to preserve them' (see numbers 15, 16 and 106 and *Banks's Journal*, vol. 1, p. 78). Such an approach was by no means new to Banks for it is also recorded that a number of his books accompanied him on his earlier tour of Newfoundland and Labrador in 1766. The work of description and classification on these voyages depended on such volumes. It was from these small but discerning beginnings that his library, in its day an unrivalled resource for natural historians and travellers alike, would later grow.

Another point worth making here is that the volumes from the voyage that survive at places like the British Library, as well as the manuscripts that were brought back, are a direct material link to the *Endeavour* expedition. As the *Endeavour* itself is lost and few artefacts from its structure remain, these book and manuscript materials provide some of the few direct physical links that still exist between our present and a voyage with significant scientific, cultural and colonial outcomes. Those *Endeavour* relics that do exist, such as the ship's guns abandoned to re-float the vessel from the Great Barrier Reef, are rightly accorded considerable historic status. See number 114. Such objects can acquire even greater significance than this when they also become involved in contemporary historical narratives, myth-making and even geopolitical debates. A recent example is the Canadian government's successful attempt to locate HMS *Erebus*, where the material heritage of a British expedition underpins contemporary nationalist and political claims to sovereignty in the Arctic.[9]

After the return of the *Endeavour* Banks's collections in Soho Square were the *de facto* authoritative source of information about the South Pacific, its peoples, flora and fauna.[10] As a result, these collections had a prominent effect on the scientific and cultural development of Britain at a crucial point in its expansion as an empire and global power into that region. Moreover, after the *Endeavour* expedition Banks became a significant force in setting the scientific boundaries of British voyages of exploration, especially to the Pacific region. His library underpinned his ability to act as an influencer, and its contents also had an intellectual impact on many of those going overseas, such as members of James Cook's later expeditions and those involved in the mounting of missions under William Bligh and George Vancouver. In short, the Banks books, manuscripts and other items held at the British Library are not just records of the history of science and exploration in Britain, but materials which directly connect us to a distant past. In so doing, they open up the opportunity to reflect critically on what such materials show us and their enduring impact, not just on the voyage and time during which they were produced, but on the way we and the world at large now see the globe around us.

While the library of Sir Joseph Banks mostly grew after his voyage to the Pacific with James Cook, it is the manuscripts, maps and objects arising from this voyage (and, to a lesser extent, future collecting influenced by them) which continues to elicit the greatest scholarly and popular interest. Banks travelled on Cook's *Endeavour* expedition as an innovative, youthful British naturalist with recent experience collecting abroad in Newfoundland, considerable resources and an extensive retinue. As elsewhere explained, this retinue consisted of the artists Sydney Parkinson and Alexander Buchan;

the botanist Daniel Solander; a clerk and draughtsman Herman Diedrich Spöring, and it was later supplemented with an indigenous expert in the shape of Tupaia, a Ra'iatean priest with extensive knowledge of local Pacific geography and an ability to communicate across Pacific Island dialects. This group was invaluable to the scientific and maritime record produced by the expedition. They recorded the geography, botany, zoology, ethnography and material culture not just of those areas commonly associated with the expedition – Tahiti, New Zealand and eastern Australia – but also places such as Tenerife, Brazil, Tierra del Fuego and Batavia (Jakarta).

The written and illustrative records created by these individuals were preserved in the collections of Sir Joseph Banks after the mission and valuable portions of them today reside in the British Library's manuscript collections. These manuscripts were used to produce various printed maps and published books which also later became part of Banks's library, and thus the British Library. Held in the department of manuscripts, these papers are not the only manuscripts relating to the *Endeavour* expedition at the British Library,[11] but they do provide a centrally important and detailed view of the places, peoples and environments that were observed by the *Endeavour* and its crew.

In particular, the manuscripts from Banks's library contain significant numbers of coastal views and charts. Buchan was originally intended to be the main producer of views during the voyage, until his untimely death from a fit at Tahiti. Those that he did make are usefully devoid of an overtly romanticized perspective. Given how little work the *Endeavour* crew and Banks's entourage were able to conduct in Brazil, Buchan's depictions of the coast, including Pao de Acucar (Sugarloaf Mountain), are a notable material record of this stage of the voyage (see number 27).[12] Some of Buchan's sketches and his chart of Tierra del Fuego also informed a plate found in John Hawkesworth's edition of the official *Endeavour* voyage account, 'A View of Part of the N.E. side of Tierra del Fuego, with three other views ...', vol. II, facing page 39 (see also numbers 28, 30, 33 and 34).[13] Hawkesworth's 1773 book made extensive use of Banks's journal and the illustrations produced by members of his party. After Buchan's death Parkinson took on many of the landscapes, in addition to undertaking his own botanical illustrations, and a number of the former are to be found in the manuscript department collections once belonging to Banks.[14] At the British Library there are also cartographic items that belonged to Banks, which are held in the map collection, such as a manuscript chart of New Zealand by the ship's master Robert Molyneaux.[15]

While the various manuscripts from Banks's library are a highlight of the *Endeavour* material, there are a number of interesting items in other collections that show the range of Banks's exploration, not least among them a set of four large, bound atlases that formerly belonged to him and are now held in the map department.[16] All differently bound, they contain 180 maps in total. These atlases are assembled geographically, with the voyaging and collecting interests of Banks apparent in some of the sections that they contain. For example, in one volume Iceland is a particular focus and there are a number of striking maps of the island. In the volume devoted to the Americas, Newfoundland and Labrador receive detailed interest. It is within this Americas volume that the *Endeavour* and subsequent voyages by Cook are also a key concern.

Fronting the atlas volume on Europe are a number of world maps in various projections. These world maps were bound chronologically, from 1750 to 1765, reflecting Europe's developing perception of the world over that period, especially of North America and the Pacific. Later, however, a 'definitive' chart has been added in front of them all, a four-part, double hemispheric projection (each part bound individually), which pays particular attention to Cook's voyages of exploration. This is Aaron Arrowsmith's 1794 'Map of the world on a Globular Projection, exhibiting particularly the nautical researches of Capn. James Cook, F.R.S.', which charts the routes taken during the three Cook voyages and notes the location and date of his death. It would appear that Banks situated these maps at the beginning as a comprehensive visual statement of the developed state of European geographical knowledge and also to mark his own increasing involvement with this expanded geography. Together, the atlases demonstrate his interest in cartography and map collecting, itself an important adjunct to his study of natural history and human society across the regions covered.

There are further noteworthy maps in the atlases, in particular one Antarctic projection chart of the southern oceans.[17] Showing the various lands charted by the *Endeavour* expedition and personally commissioned by Banks, this is possibly the first printed map arising from the expedition, noting the new Pacific islands, the shores of New Zealand and those of the east coast of Australia that were visited by Cook. Once thought to exist only as a copper plate (from which a twentieth-century press was made), this map predates by about a year those printed for Hawkesworth's *An account of the voyages undertaken by the order of His Present Majesty for making discoveries in the southern hemisphere*, published in 1773, and nestles among the various maps acquired by Banks detailing the Americas and the Pacific Ocean (see number 134.)[18] When looking at these maps it is important to keep in mind the extensive use to which they were put by visitors to Banks's library, and later the British Museum library. These atlas volumes strongly reflect the world as seen by Banks through his expeditions and interests, reminding us of his significant role in shaping how late eighteenth- and early nineteenth-century Britain perceived its empire and the world.

One historical gap in the British Library's holdings relating to Banks is the absence of his *Endeavour* journal. This was offered for sale in 1884 and deemed too expensive to justify acquisition, and as a consequence it is now held by the State Library of New South Wales, Sydney. Nonetheless, Banks's written perspective of the journey is apparent in other elements of the British Library's collections. Among the rare books is his own copy of Hawkesworth's edition of *An account of the voyages undertaken by the order of His Present Majesty for making discoveries in the southern hemisphere*, a work created using both Banks's and Cook's voyage journals. Despite the errors and other deficiencies caused by Hawkesworth's handling of these materials, the published account still provides intriguing insights into Banks's experiences on the *Endeavour* expedition. Moreover, both his manuscript journal in Sydney (as edited and published in 1962 by J.C. Beaglehole) and the Hawkesworth account contain illuminating details that supplement the manuscripts and charts populating the rest of his former library. For example, his notes on participating in a mourning ceremony at Tahiti may be read in conjunction with manuscript illustrations of the splendidly attired chief mourner who led this mysterious ritual, as drawn by Tupaia, himself a priest.[19] The same comparison could also be done with Parkinson's depiction of the event, which featured in Hawkesworth's edition (see numbers 58 and 59). Banks's direct involvement in such a ritual and his description of it was a method of anthropological study far in advance of his time, and it highlights the way in which he and his team greatly enriched the voyage record. Other details described by Banks provide first notes on the impact of Polynesian culture on subsequent European and colonial society, for example when he provides the first English record of a practice we would today call surfing.[20]

As well as this, other published accounts of the voyage are present in Banks's library, including that of Sydney Parkinson. Parkinson's work during the voyage can justifiably be described as prodigious, including various illustrations at all major locations visited until he died of a fever after departing Batavia. While Parkinson's primary work was on the plants that were illustrated during the journey, he also depicted the landscapes, people and the material culture encountered by the *Endeavour* and its crew. His journal and some plates were eventually published, albeit after a difficult and protracted argument over the rights to print this work that took place between Banks and Sydney's brother,

Stanfield. Parkinson's account adds a further perspective on the new lands, plants and animals encountered by Banks's entourage. These published accounts, along with the various manuscripts produced by Banks and those who accompanied him, provide a detailed view of some of the earliest historic European contacts with the Polynesian world, its peoples and its flora and fauna. As such they are an important record not only of the *Endeavour* voyage itself but also of expanding British scientific and cultural horizons.

These documents recorded for the first time landscapes, plants and animals new to Europeans, frequently documenting species that have now been lost and depicting locations which have now undergone considerable change. In particular, the scientific record is important as a reference not only for contemporary scholars of the history of science but also for those interested in ecological change. The botanical and zoological specimens collected by Banks are, in this regard, a historic benchmark against which any diminished biodiversity in these areas can be measured.[21] The contents of Banks's library offer still more than this, by acting as an opportunity to reflect on our colonial history and its effect on the societies and cultures encountered by Cook, Banks and those who travelled with and after them.

Many accounts of the voyage and its visual record dwell, appropriately, on the scientific illustrations, views and maps produced by Buchan, Spöring, Parkinson and Cook himself. However, there is also great value in the cultural record arising from the presence of Tupaia among the voyagers and his contribution to their works. A Ra'iatean priest and navigator, Tupaia accompanied Banks after Cook departed Tahiti, ostensibly because this remarkable man wanted to see more of the world. He was invaluable to Banks and Cook as a source of knowledge and a bridge between cultures, for example at New Zealand where he was able to communicate with the Māori. Like Parkinson, he too sadly died of dysentery contracted at Batavia. Tupaia became, in effect, a part of the *Endeavour*'s crew, and his knowledge of Pacific island geography and language was extremely useful to the mission, so it is particularly important that accounts of and various drawings by him live on in the collection of Joseph Banks held at the British Library.

The main source of nautical and geographical information provided by Tupaia is a chart, copied from a now lost original by Lieutenant James Cook (see number 77).[22] Tupaia's chart is an important repository of the Polynesian names of the islands encountered by Cook before the major charts of Cook's voyages replaced these with a European nomenclature. It is also a record of the Tahitian way of perceiving the relationship between widely separated Pacific islands, which is to say that the chart does not reflect how these places relate to each other in linear space as European maps do, but rather in terms of their sailing time from Tahiti. This results in a fascinating chart that shows the islands arranged into broadly concentric rings. A printed version of Tupaia's chart was later produced by Johann Reinhold Forster for his 1778 account of Cook's second Pacific voyage, but this later version changes the geography of the islands to one more recognizably like a European chart and thus loses some of the cultural and historical value of the original manuscript as copied by Cook. Banks's collections are rightly famed for their botanical, zoological and ornithological importance, but Tupaia's charts and the cultural depictions found elsewhere in Banks's *Endeavour* materials add a further element of significance.

Since the late twentieth century an increasing amount of interest has been paid to the value of European collections arising from expeditions such as that of the *Endeavour* in preserving and, sometimes, reviving elements of indigenous cultures eroded by the impact of colonialism.[23] In this regard the illustrations produced by Parkinson, particularly his well-known sketches of Māori people, such as 'Attitudes of Defiance: New Zealand',[24] as well as those by Buchan of Fuegian people and their material culture,[25] or those completed after the voyage by J.F. Miller of many artefacts, are all of great significance as documentary evidence of Polynesian and Aboriginal history. The scope and detail of the visual record is impressive and, notwithstanding the romantic touches occasionally apparent in some of the illustrations, in the main it constitutes a reliable source of information about places, peoples, material culture, dress and decoration of the body that, all too often, have been lost in the intervening centuries. This sort of record can be a resource of significance and meaning, as seen with Canada's first peoples and their developing relationship with North American heritage institutions. Indeed, descendants of the Nuu-chah-nulth people encountered on Vancouver Island by Cook on his third expedition are now involved in engagements with British Columbia institutions such as the University of British Columbia's Museum of Anthropology. One example is the 2012 repatriation of a Nuu-chah-nulth fish club gifted to Cook in 1778 (and until 2012 held in a private, non-First Peoples collection), which is now housed at the museum after a process of collaboration and research involving the contemporary Nuu-chah-nulth nation, for whom the object provides unique insights into pre-contact culture and history. Banks's collections arising from the *Endeavour* expedition can also offer insights into a wide range of contemporary issues stretching across the humanities and the sciences, a testament to his own insatiable curiosity and desire to comprehend the world at large.

The British Museum is the progenitor institution for a number of the capital's most eminent museums and libraries. Its collections have been used to develop its own world-leading reputation and to underpin those of the Natural History Museum and the British Library too, and so Banks has been a key founder to each of these institutions. His bequest was a significant gain to the British Museum at a time when funding was far from generous, but he had a long history of donating to the library beforehand. This was the result of what we might now call Banks's de-duplication and deaccessioning policy, where he weeded out material that was no longer wanted as his focus settled increasingly on the botanical world. Some of these donations were valuable ones that the British Library still counts among its Banksian collections. Banks was also a British Museum trustee and advised on the growth of the museum, thus having significant influence upon the development of the institution before it separated into various bodies in later centuries. The botanical and zoological specimens found their way to the Natural History Museum, where they continue to represent a significant element of its collections, particularly in the herbarium, while at the British Library there is Banksian material in the printed, manuscript and map collections discussed here. The continuing availability of this material at the British Library – to readers, exhibition visitors and many other individuals – ensures that the book stamp and signature 'Jos:Banks' will continue to be seen by those seeking insights into his activities and period, as well as by those interested in the natural sciences and, indeed, the world that surrounds us all.

NEW ZEALAND

In 1742 the Dutch navigator Abel Tasman visited what he called Staten Landt, now New Zealand, which he and afterwards others thought might be the fringe of a much larger landmass. It is possible that Spanish, Portuguese, Chinese and Malay ships had also visited New Zealand before the arrival there of James Cook in 1769. Once the transit of Venus had been observed, a subsidiary objective of the *Endeavour* mission was to look for a supposed southern continent. Cook's instructions therefore enjoined him to sail south from the Society Islands to the latitude of 40° and, if no land was found, to turn westwards towards New Zealand. This Cook did and, nothing but water intervening, by 8 October he was off the east coast of North Island.

Tupaia was a definite asset at this stage of the mission since he could communicate with the Māori inhabitants, a significant discovery in itself since it showed that their cultures were related, but Cook found it extremely difficult to gauge local reactions or establish mutual trust. Following bloody clashes involving a number of Māori casualties, he departed after just three days from Tūranganui-a-Kiwa, or what he chose to call Poverty Bay. The coastal conditions and unpredictable nature of relations with local tribes largely explain the limited number and relatively short duration of Cook's landings. In all, he made just ten stops around New Zealand on this voyage, but it nevertheless became a favourite haunt for the navigator, who returned on each of his ensuing Pacific expeditions and spent more than 300 days ashore or in its waters. Cook gradually learned from his experiences and developed policies for managing contacts. He was prepared to tolerate transgressions against his own people, most obviously when he did not exact *utu* (revenge) for the murder of eleven crew from *Adventure* during his second Pacific expedition, and he punished his men for stealing from or otherwise offending the Māori. On this his first visit Cook proved that New Zealand comprises two main islands, and is not part of a larger landmass, producing a masterly chart of the coasts that he surveyed.

Polynesians first arrived in present-day New Zealand (Aotearoa) between 1,000 and 1,200 AD, probably from the Society Islands, and over the next few hundred years they settled the country by a series of migrations in ocean-going canoes, or *waka*. They faced a cooler climate than the tropical one from which they came, and this is reflected in the more substantial clothing that they developed. Most of their vegetable staples such as the taro also proved difficult to grow in the temperate conditions, but the South American sweet potato or *kumara* thrived, suggesting return voyages by Polynesians as far eastward as that continent. Fishing and hunting provided food and materials. The birds that inhabited New Zealand were a valuable source of protein, most notably the flightless *moa*, but loss of habitat and over-hunting drove these giants to extinction. This, limited agricultural production and the pressures created by population growth led to a more warlike existence for which fortified settlements known as *pā* were established. But a complex society also emerged based on tribes or *iwi* and the smaller *hāpu* or clans. In the absence of writing, meaning was conveyed through oral traditions and art, including *moko* skin markings, and religious beliefs were strongly maintained. Skilfully carved and decorated ornaments, boats and dwellings attest to the remarkable ingenuity and visual beauty of Māori craftsmanship. At the time of Cook distinctive anthropomorphic and curvilinear patterns characterized Māori designs. New Zealand flax was used to make clothing and cord, and greenstone (nephrite jade), or *pounamu*, was especially prized for producing tools and pendants of different kinds.

78 'Portrait of a New Zeland Man'

By Sydney Parkinson.

The British Library, London. BL Add. MS 23920, f. 55. 38.7 × 29.5 cm.

This pen and wash drawing by Sydney Parkinson, annotated in ink on the folio as given above, may well be a study of one of the Māori men from the Poverty Bay area who visited *Endeavour* on 11 October 1769, three of that group remaining aboard overnight (see below). If so it represents one of Parkinson's earliest drawings in New Zealand, for which he appears to have made some preparatory sketches, BL Add. MS 9345, f. 12, and BL Add. MS 23920, f. 56. The final portrait was admirably engraved by Thomas Chambers, and included as plate 16 in the 1773 posthumous edition of Parkinson's voyage account by his brother, Stanfield. Parkinson carefully observes the curvilinear form of the man's facial markings (*moko*). He depicts an ornamental comb (*heru*), three feathers in a top-knot, a long ear pendant (*kuru*) and the iconic Māori *hei tiki* as well as a *pakepake* about the man's shoulders (see number 79 below).

On 8 October 1769 the voyagers entered what Cook later called Poverty Bay, site of present-day Gisborne, but known to the Māori as Tūranganui-a-Kiwa. There they had their first close view of a clearly inhabited landscape, and noted on the north-east headland the stockade of what was probably a fortified *pā*, thought by the voyagers to be an enclosure for keeping livestock. This was a definite misinterpretation, and not the only one during this unhappy first encounter, for, like Abel Tasman's before him, Cook's initial experience of the country's inhabitants ended in bloodshed. The voyagers sought trade and provisions, but were unable to make suitable overtures despite more than one attempt to do so, and they reacted in a heavy-handed fashion on being threatened. When four youths from the ship were left in charge of the yawl after landing, four armed Māori emerged from woods and attempted to cut off the boat. They then gave chase as it retreated downstream despite warning shots from the accompanying pinnace, and one of the assailants was shot dead by crew on the pinnace before he could throw a spear at the yawl. Worse, when it seemed the next day as though peaceful if tense contact had been made with a local group, another Māori snatched Charles Green's cutlass and there was yet more shooting (a mix of small shot and ball), with the Māori concerned being killed by the surgeon William Monkhouse. This despite Tupaia's ability to communicate with the locals, with whom he conversed, and his warnings about a possible Māori deception. Tupaia himself is recorded as having fired at these men during the clash. Evidently, Lord Morton's earlier injunction advising restraint in all dealings with local peoples was by now pushed to the back of British minds. The next day, while in their boats at the head of the bay, the voyagers got into another fight with the occupants of a canoe, resulting in the death of four more Māori. In their journals both Cook and Banks reflected with bitter regret on that day's events, the latter observing: 'Thus ended the most disagreable day My life has yet seen, black be the mark for it and heaven send that such may never return to embitter future reflection' (*Banks's Journal*, vol. 1, p. 403).

Yet there was some cause for hope that a better understanding would be reached in the light of harsh experience. Three young Māori boys from the captured canoe were brought aboard *Endeavour*, well treated and then, although initially reluctant to go, released again. Although Cook named the bay after his unproductive attempts to resupply there, as he departed southward several Māori canoes approached the ship, but they could not be enticed alongside until another canoe arrived from the Poverty Bay area and its four occupants bravely ventured aboard. The remaining canoes, carrying some fifty locals, then approached. This memorable visit provided a good chance for friendly trade on both sides, doubtless resulting in objects being acquired that were later brought home by the voyagers, some of which may even survive in collections today (see number 91). It allowed, too, the opportunity to meet and draw a Māori for the first time as here.

Banks recorded the encounter in his journal on 11 October, making reference to elements of local dress, skin markings, body ornaments, weapons and canoe construction: 'Weather this day was most moderate: several Canoes put off from shore and came towards us within less than a quarter of a mile but could not be persuaded to come nearer, tho Tupia exerted himself very much shouting out and promising that they should not be hurt. At last one was seen coming from Poverty bay or near it [perhaps Whareongaonga], she had only 4 people in her, one who I well rememberd to have seen at our first interview on the rock: these never stopd to look at

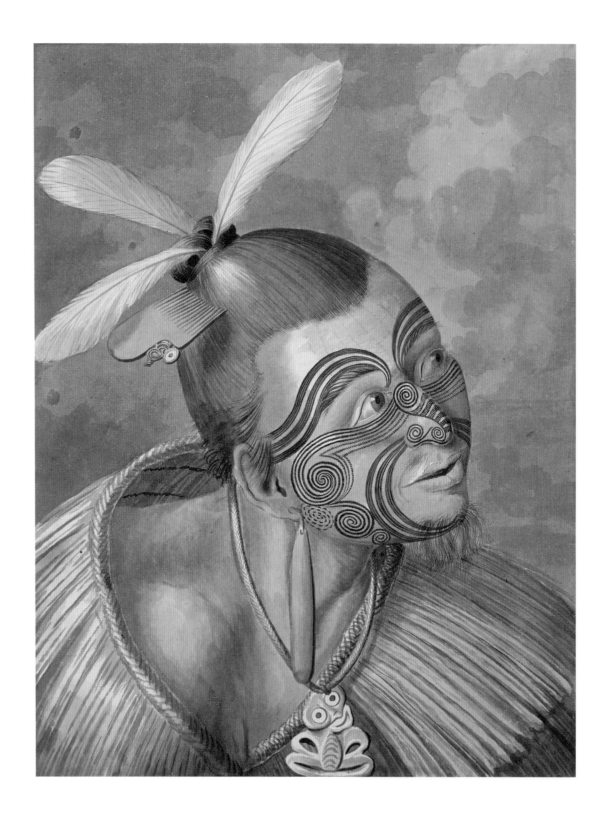

any thing but came at once alongside of the ship and with very little persuasion cam[e] on board; their example was quickly followd by the rest 7 Canoes in all and 50 men. They had many presents given to them notwithstanding which they very quickly sold almost every thing that they had with them, even their Cloaths from their backs and the paddles out of their boats; arms they had none except 2 men, one of whom sold his *patoo patoo* as he calld it, a short weapon of green talk [nephrite] of this shape intended doubtless for fighting hand to hand and certainly well contrivd for splitting sculls as it weigh[s] not less than 4 or 5 pounds and has sharp edges excellently polishd.

We were very anxious to know what was become of our poor boys, therefore as soon as the people began to lose their first impressions of fear that we saw at first disturbd them a good deal we askd after them. The man who first came on board immediately answerd that they were at home and unhurt and that the reason of his coming on board the ship with so little fear was the account they had given him of the usage they had met with among us.

The people were in general of a midling size tho there was one who measurd more than 6 feet, their colour dark brown. Their lips were staind with something put under the skin (as in the Otahite tattow) and their faces markd with deeply engravd furrows Colourd also black and forrnd in regular spirals; of these the oldest people had much the greatest quantity and deepest channeld, in some not less than $1/16$ part of an inch. Their hair always black was tied on the tops of their heads in a little knot, in which was stuck feathers of various birds in different tastes according to the humour of the wearer, generaly stuck into the knot, sometimes one on each side the temples pointing forwards which made a most disagreable apearance; in their Ears they generaly wore a large bunch of the down of some bird milk white. The faces of some were painted with a red colour in oil some all over, others in parts only, in their hair was much oil that had very little smell, more lice than ever I saw before! and in most of them a small comb neatly enough made, sometimes of wood sometimes of bone, which they seemd to prize much. Some few had on their faces or arms regular scars as if made with a sharp instrument: such I have seen on the faces of negroes. The inferior sort were clothd in something that very much resembled hemp; the loose strings of this were fastned together at the top and hung down about 2 feet long like a petticoat; of these garments they wore 2, one round their shoulders the other about their wastes. The richer had garments probably of a finer sort of the same stuff, most beatifully made in exactly the same manner as the S. American Indians at this day, as fine or finer than one of them which I have by me that I bough[t] at Rio de Janeiro for 36 shillings and was esteemd uncommonly cheap at that price. Their boats were not large but well made, something in the form of our whale boats but longer; their bottom was the trunk of a tree hollowd and very thin, this was raisd by a board on each side sewd on, with a strip of wood sewd over the seam to make it tight; on the head of every one was carvd the head of a man with an enormous tongue reaching out of his mouth. These grotesque figures were some at least very well executed, some had eyes inlaid of something that shone very much [abalone, *Haliotis* sp.]; the whole servd to give us an Idea of their taste as well as ingenuity in execution, much superior to any thing we have yet seen' (*Banks's Journal*, vol. 1, pp. 406–8; see also *Cook's Journals*, vol. 1, pp. 168–74).

79   Neck pendant *hei tiki*

By permission of Her Majesty Queen Elizabeth II. The Royal Collection RCIN 69263. 15 cm. Nephrite, abalone shell (Haliotis iris), flax and bone.

Hei tiki take the form of a human figure, which some scholars have likened to the female in birthing position, but precisely what such objects originally represented remains uncertain. They are nevertheless the most famous form of neck pendant made by the Māori and retain a wide popularity today. This one, made of prized greenstone, has shell inlaid eyes and a beautifully polished surface. It also has a bird-bone toggle and a woven flax cord. A particularly fine example, it was a gift from James Cook to George III and resides in the Royal Collection.

There is a possible but not identical illustration of this *hei tiki* by Herman Diedrich Spöring at BL Add. MS 23920, f. 76(a), which Spöring annotated in ink on the verso 'An Amulet.' and on the recto 'Hawke's-Bay New-Zeeland. Oct: 18 1769' (see Kaeppler, *Artificial Curiosities*, pp. 40 and 176–7). Banks described more than one friendly visit to *Endeavour* by Māori in the Hawke Bay area during which various items were traded, first on 14–15 October 1769 when the voyagers were sailing southwards along the coast, and then again on 18 October when they returned northwards (see *Banks's Journal*, vol. I, pp. 410–13 and 414 respectively, the former entry mentioning sight of a 'Green stone' object hanging around one man's neck, 'some of our people imagind it to be a Jewel'). A further depiction of a first-voyage *hei tiki* with a selection of other Māori objects was drawn for James Cook by Charles Praval, BL Add. MS 7085, f. 33.

Nephrite or *pounamu* is a type of jade found naturally in parts of the South Island, which derives its Māori name of Te Waipounamu from this substance. In English it is commonly called greenstone. Hard, tough and capable of being worked into a variety of forms with an exquisite finish, *pounamu* was highly valued by the Māori and was exchanged or traded throughout New Zealand. Since metal did not exist in New Zealand prior to the arrival of Europeans, *pounamu* was used not only for ornamental purposes but also to make tools and weapons. Later, as metal became increasingly available, many nephrite adze blades were converted to objects such as *hei tiki*, although as an *Endeavour*-voyage item the present example does not fall into this category.

80 *Phormium tenax* Forster & G. Forster, *Char. gen. pl.*: 48, t. 24 (1775)

By Frederick Polydore Nodder after Sydney Parkinson.

The Natural History Museum, London. Diment et al., Catalogue, Part 2 Botany (1987), NZ 4/176. Height 50.5 × width 33.5 cm; plant height 40.5 cm.

Based on an original outline drawing by Sydney Parkinson, this finished watercolour by Frederick Polydore Nodder is annotated in ink on the recto 'Fred$^{k.}$ Polydore Nodder pinx$^{t.}$ 1783'. There is a Banks herbarium specimen for this plant, which, like the drawings above, is held at the Natural History Museum, London. Banks first saw *Phormium tenax* when *Endeavour* arrived at New Zealand. The Māori extensively cultivated *Phormium tenax*, to them *harakeke*. The cooler temperate climate of New Zealand in comparison to the tropical regions of the central Pacific from whence the Māori came necessited the use of phormium for more substantial clothing. The Māori also used the plant's extremely strong fibres to produce not only cloth but ropes and lines, fishing nets, bags, mats and various utensils.

The British were quick to appreciate the commercial possibilities of phormium fibre. Its discovery as a new source of fibre for the production of sails and ropes was a potentially important one, because any interruption of the importation of flax for sails and hemp for ropes through the Baltic ports from the Russian dominions would have threatened Britain's sea power. Banks commented in his journal essay account of New Zealand (March 1770): 'But of all the plants we have seen among these people that which is the most excellent in its kind, and which realy excells most if not all that are put to the Same uses in other Countries, is the plant which serves them instead of Hemp and flax. Of this there are two sorts: the leaves of Both much resemble those of flags [*Iris pseudacorus*]: the flowers are smaller and grow many more together, in one sort they are Yellowish [likely *Phormium colensoi*] in the other of a deep red [*Phormium tenax*]. Of the leaves of these plants with very little preparation all their common wearing apparel are made and all Strings, lines, and Cordage for every purpose, and that of a streng[t]h so much superior to hemp as scarce to bear a comparison with it. From the same leaves also by another preparation a kind of snow white fibres are drawn, shining almost as silk and likewise surprizingly strong, of which all their finer cloaths are made; and of the leaves without any other preparation than splitting them into proper breadths and tying those strips together are made their fishing nets. So usefull a plant would doub[t]less be a great acquisition to England, especialy as one might hope that it would thrive there with little trouble, as it seems hardy and affects no particular soil, being found equaly on hills and in Valleys, in dry soil and the deepest bogs, which last land it seems however rather to prefer as I have always seen it in such places of a larger size than any where else' (*Banks's Journal*, vol. 2, pp. 10–11).

It is no surprise, then, to find that a drawing of a phormium is shown in an open folio at Banks's feet in the famous Benjamin West portrait of him (see number 143). Banks clearly had great hopes for the commercial uses of phormium fibre and suggested the name *Chlamydia tenacissima* for this plant. The publication of this name by Joseph Gaertner in 1788 postdates, however, the publication of the name *Phormium tenax* by J.R. and G. Forster in 1775. Between the 1820s and the 1860s a considerable trade in hand-dressed fibre was carried on between the Māori and Europeans. The dressed fibre was exported to Britain and also to British colonies in New South Wales mainly for rope making in ships. The original expectation that phormium would be widely employed in this capacity was never fully realized, as other fibres eventually proved more suitable, and its use became entirely superseded by the production of synthetic materials in the twentieth century. It should be noted, however, that in the nineteenth and twentieth centuries production continued for a range of purposes, including cordage, woolpacks and floor coverings.

81 Cloak *kaitaka*

*The Pitt Rivers Museum,
University of Oxford.
1886.21.20.
127 × 178 cm.
Flax, dog skin and hair.*

Originally given by Banks to Christ Church, Oxford, and now deposited at the Pitt Rivers Museum, this cloak is made of New Zealand flax with a *taniko* border formed of decorative panels of dyed flax, edged in places with strips of dog skin and some hair. It is a cloak of the type worn by Banks in his portrait by Benjamin West, so this one may be the portrait original, although other cloaks at the British Museum, Oc,NZ.137, and the Etnografiska Museet, Stockholm, 1848.01.0063, have in the past been noted as similar in some respects to that featured by West (see Kaeppler, *Artificial Curiosities*, p. 171; see also Rydén, *The Banks Collection*, pp. 81–2). In the portrait the cloak retains a white dog's hair fringe, and Banks is shown pointing at a piece of clothing made from what was in his view an extremely valuable plant for its useful properties (see number 80 immediately above). This point is emphasized in the West portrait by what appears to be a drawing in an open folio at Banks's feet of the New Zealand flax plant. Such a cloak would have been worn by a high-status individual, and there has been some speculation as to whether Banks was first given it or whether it was originally bestowed on Tupaia, an important figure in Māori eyes because he was a priest and because he could comprehend their language and therefore had more dealings with them. If the latter is the case, then the cloak probably came to Banks after Tupaia's death on the journey back to England (Tapsell in Hetherington and Morphy, *Discovering Cook's Collections*, pp. 102–7). Women had the specialist task of making such garments, using the technique of flax finger weaving in their manufacture, and decorating them in various ways (see number 83).

In his journal essay account of New Zealand, Banks commented: 'Besides this they have several kinds of Cloth which is smooth and ingeniously enough workd: they are cheifly of two sorts, one coarse as our coarsest canvass and ten times stronger but much like it in the lying of the threads, the other is formd by many threads running lengthwise and a few only crossing them which tie them together. This last sort is sometimes stripd and always very pretty, for the threads that compose it are prepard so as to shine almost as much as silk; to both these they work borders of different colours in fine stitches something like Carpeting or girls Samplers in various patterns with an ingenuity truly surprizing to any one who will reflect that they are without needles. They have also Mats with which they sometimes cover themselves, but the great pride of their dress seems to consist in dogs fur, which they use so sparingly that to avoid waste they cut into long strips and sew them at a distance from each other upon their Cloth, varying often the coulours prettily enough. When first we saw these dresses we took them for the skins of Bears or some animal of that kind, but we were soon undeceivd and found upon enquiry that they were acquainted with no animal that had fur or long hair but their own dogs. Some there were who had these dresses ornamented with feathers and one who had an intire dress of the red feathers

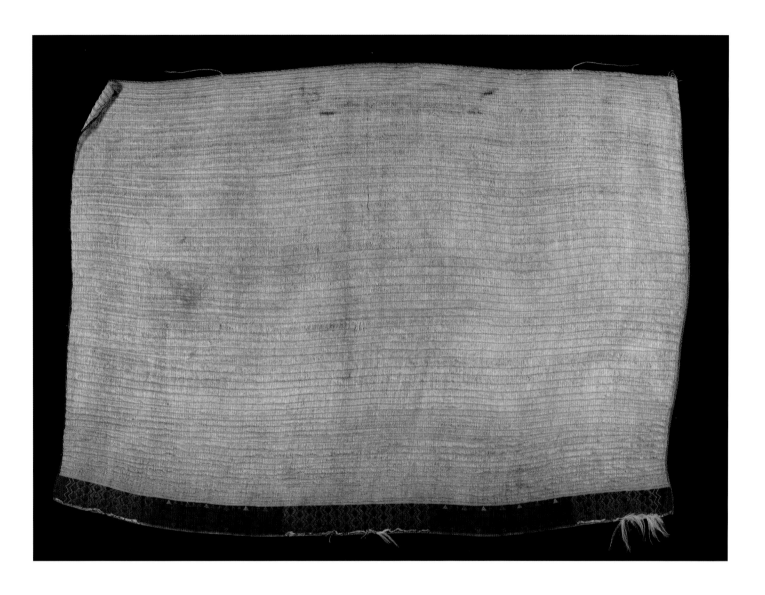

of Parrots, but these were not common' (*Banks's Journal*, vol. 2, p. 15).

Banks here recorded two types of Māori cloak, the *kaitaka* and the *kahu-waero*, the latter a chiefly garment of great value. The Pitt Rivers Museum holds what is thought to be the only surviving historic example of a *kahu-waero* cloak, 1886.21.19, as also included by Banks in the donation that he made to Christ Church after the *Endeavour* voyage. Banks thus described in his journal and arguably collected an example of each sort of cloak. There are six Māori cloaks from this voyage at the Museum of Archaeology and Anthropology, Cambridge, D.1924.80–5. These were collected by James Cook and passed to Lord Sandwich, who gave them to Trinity College, the college later depositing them at the museum. The Stockholm collection referenced above, originally a gift from Banks to his Swedish friend Johan Alströmer, probably contains cloaks from one or both of Cook's first two Pacific voyages if, as seems likely, it was obtained by Alströmer during a visit to London in 1777–8. The cloaks are five in number and, like the others mentioned here, are uniquely valuable. Indeed, that at Stockholm, 1848.01.0063, is currently believed to be the only one of its kind in existence. Some forty Māori cloaks believed to have been gathered during the Cook voyages survive in present-day collections.

82 Three belts *tatua*

1. The Pitt Rivers Museum, University of Oxford. 1886.21.2. 180 × 13 cm (excluding ties). Flax, plain; dog skin; plaited tying cords.

2. The Pitt Rivers Museum, University of Oxford. 1886.21.3. 136 × 5.5 cm (excluding ties). Flax, in two colours; plaited tying cords.

3. The Pitt Rivers Museum, University of Oxford. 1886.21.4. 125 × 7.5 cm (excluding ties). Flax, plain; plaited tying cords.

These belts are a set of three that were donated by Joseph Banks to Christ Church, Oxford. They are all made from flax using the same plaiting technique and each has ties. The first belt is plain in colour, of the finest weave of the three and has edges stitched with strips of dog skin. It is the widest of the set and shows evidence of patched repairs due to wear. It must have been valued by its original owner and used fairly frequently for this to be so. The second belt is the narrowest, and is woven in two colours. The third belt is slightly broader than the preceding one and of a finer weave. These belts were almost certainly made by women, in keeping with cultural norms throughout Polynesia, where women produced men's clothing of finely beaten barkcloth or plaited fibres.

An interesting aspect of such a set is the way in which Banks appears to have tried to include various Māori belts in his Oxford donation. These belts are not duplicates as such, but rather examples of different types that together represent a range of Māori production and design. Other Banks objects at Oxford appear to have been chosen with a similar aim in mind, and this can certainly be seen in the selection of Māori cleavers held there, or the four types of barkcloth from Banks that were also part of his donation (see numbers 65, 66, 92 and 99; see also Coote, *Curiosities from the Endeavour*, p. 8).

Another group of Māori belts known to have been gathered on Cook's first Pacific voyage exists in the Museum of Archaeology and Anthropology, Cambridge, D.1914.43–6. They were a gift by Lord Sandwich to Trinity College, which later deposited them at the museum. Other belts considered to be from Cook's voyages are held in various European collections, including, notably, an impressive set of five at the Etnografiska Museet, Stockholm, 1848.01.0036–40. These were part of an important collection of artefacts and natural history specimens that Banks presented to Johan Alströmer, a Swedish merchant and university friend of Daniel Solander. Alströmer probably received this material during a visit to London in 1777–8, and if so it would most likely have included items from the first and also the second Cook voyages. Three belts at the British Museum were designated 'Cook collection' by James Edge-Partington, and may have been gathered on one or more of Cook's Pacific voyages, Oc,NZ.128, 131 and 132.

83 Weaving pegs *turuturu whatu*

1. The Pitt Rivers Museum, University of Oxford. 1887.1.715. Length 45.5 cm. Wood.

2. The British Museum, London. Oc,NZ.68. (Edge-Partington registration slip, 'Cook collection'.) Length 40.5 cm. Wood.

Weaving pegs, *turuturu whatu*, were used in pairs. They were placed in the ground by their pointed end so that prepared flax fibres could be suspended between them for the finger weaving of cloaks and other clothing. The central section of the peg incorporates a carving similar in form to the *tiki wananga* deity figures in order that when it was made spiritual power or *mana* might, in effect, be woven into the garment. The first of the present examples was a donation by Banks to Christ Church, Oxford, and is now held at the Pitt Rivers Museum. The second example is held at the British Museum in its 'Cook collection', and was probably gathered on one of the Cook voyages. James Edge-Partington noted on his registration slip for this peg that it resembles one drawn after the voyage by J.F. Miller for Banks, see number 93, and there is indeed a resemblance.

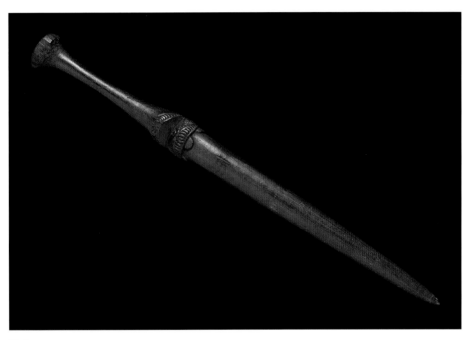

83.1

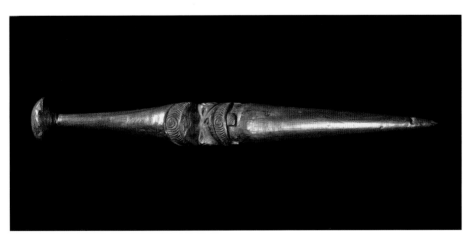

83.2

84 'Portrait of a New Zeland Man'

By Sydney Parkinson.

The British Library, London. BL Add. MS 23920, f. 54(a). 39.4 × 29.8 cm.

This pen and wash drawing by Sydney Parkinson of Otegoowgoow (Te Kuukuu) – the son of a chief of the Bay of Islands, who was wounded in the thigh by musket fire during a clash with a party from *Endeavour* on Motuarohia Island on 29 November 1769 – is annotated in ink on the folio as given above. In it Parkinson recorded the subject's elaborate facial markings (*moko*) and a series of chiefly ornaments. These include a bone comb, nephrite ear pendant and a whalebone *rei puta* neck pendant (see number 85). In his journal Parkinson described turbulent encounters in the days leading to the clash in which Otegoowgoow was wounded, referring to the 'somewhat unruly' behaviour of the Māori on the 29th as well as to firing by the British. Conflict escalated when locals attempted to seize two of *Endeavour*'s boats that were on the beach. At this point 'Mr. Banks got on shore, he had like to have been apprehended by one of the natives, but happily escaped' (*Journal of a Voyage*, p. 109; see also *Banks's Journal*, vol. 1, pp. 440–2). Fortunately, peaceful relations were eventually established.

John James Barralet appears to have produced a wash copy of this drawing at BL Add. MS, 15508, f. 21. That copy is annotated on the recto 'pl 13 in Hawkesworth', and may well have been the source for the engraving by an unknown hand that was published in John Hawkesworth's 1773 official account of the *Endeavour* voyage, vol. 3, plate 13. Thomas Chambers engraved a further version of this image for Stanfield Parkinson's 1773 edition of his deceased brother's voyage account, plate 21. It is curious that this latter engraving lacks significant details present in each of the other versions, such as the comb and neck pendant, and one wonders whether the Chambers engraving might have been prepared using the version in the official account. Such an explanation would help explain the deficiencies in it on the assumption that Banks or Hawkesworth held the original Parkinson drawing and did not permit Stanfield access to it as a result of the disputes surrounding these publications. When, however, the official version appeared, and Hawkesworth's injunction against Stanfield's book was finally lifted, the latter could hurriedly have included an engraved copy in his own edition. Chambers had already prepared a much more complete Māori head for plate 16 in Stansfield's edition, number 78 above, a head not found in the official account. Possibly Stanfield possessed his brother's illustration of that head, to which Chambers therefore had access while Banks and Hawkesworth naturally did not. Interestingly, the manuscript original and copy versions of both Māori heads today reside in collections once belonging to Banks. John Fothergill may have obtained a number of illustrations not in Banks's possession when he subsequently purchased the remaining copies of the deceased Stanfield's edition, and passed this material back to Banks as rightfully his property (see number 5).

Banks described the skin markings of the Māori, which, strictly speaking, are not a form of tattoo since the decoration is carved into the skin using chisel implements rather than being introduced through small puncture holes. In his journal essay account of New Zealand, Banks commented: 'Both sexes stain themselves with the colour of black in the same manner and som[e]thing in the same method as the South Sea Islanders, introducing it under the skin by a sharp instrument furnish'd with many teeth, but the men carry this custom to much greater leng[t]hs and the women not so far, they are generaly content with having their lips black'd but sometimes have patches of black on different parts of their bodies. The men on the contrary seem to add to their quantity every Year of their lives so that some of the Elder were almost coverd with it. There faces are the most remarkable, on them they by some art unknown to me dig furrows in their faces a line deep at least and as broad, the edges of which are often again indented and most perfectly black. This may be done to make them look frightfull in war; indeed it has the Effect of making them most enormously ugly, the old ones at least whose faces are intirely coverd with it. The young again often have a small patch on one cheek or over an eye and those under a certain age (may be 25 or 26) have no more than their lips black. Yet ugly as this certainly looks it is impossible to avoid admiring the immence Elegance and Justness of the figures in which it is form'd, which in the face is always different spirals, upon the body generaly different figures resembling something the foliages of old Chasing upon gold or silver; all these finishd with a masterly taste and execution, for of a hundred which at first sight you

would judge to be exactly the same, on a close examination no two will prove alike; nor do I remember to have seen any two alike, for their wild imaginations scorn to copy as appears in almost all their works. In different parts of the coast they varied very much in the quantity and parts of the body on which this *Amoco* as they call it was placd, but in the spirals upon their faces they generaly agreed, and I have generaly observd that the more populous a country was the greater quantity of this *Amoco* they had; possibly in populous countreys the emulation of Bearing pain with fortitude may be carried to greater lengh[t]s than where there are fewer people and consequently fewer examples to encourage. The Buttocks which in the Islands was the principal seat of this ornament in general here escapes untouched: in one place only we saw the contrary [off Cape Brett, 26 November 1769]: possibly they might on this account be esteemd as more noble, as having transferrd the seat of their ornament from the dishonourable cheeks of their tail to the more honourable ones of their heads' (*Banks's Journal*, vol. 2, pp. 13–14).

85 Neck pendant *rei puta*

The British Museum, London. Oc,NZ.159. (Edge-Partington registration slip, 'Cook collection', and C.H. Read registration slip.) Pendant length 13.5 cm. Sperm-whale tooth, gum, flax, bird bone.

This is a rare form of Māori neck pendant. It comprises a finely plaited flax neck cord with a loop at one end and an albatross-bone toggle at the other. The toggle is undecorated and one end of it is filled with a reddish clay. Three evenly spaced holes enable the neck cord to be attached to the pendant by loops of cord. The pendant is made from a single sperm-whale tooth, which has been fashioned into a flat top that curves to an enlarged end. The enlarged end has two slanting eyes carved on it that are inlaid with gum, but it lacks the nose sometimes seen in other *rei puta*. The surface is otherwise highly polished. Banks noted the *hei tiki* and *rei puta* pendants of the Māori in his journal essay account of New Zealand, commenting that 'these they wore about their necks and seemd to Value almost above everything else' (*Banks's Journal*, vol. 2, p. 17).

The British Museum holds four *rei puta*, but the other three have no Cook-voyage association. The one included here has been likened to that appearing in a contemporary copy of Sydney Parkinson's illustration at number 84, the copy concerned being by John James Barralet, BL Add. MS 15508, f. 21 (C.H. Read registration slip; see Kaeppler, *Artificial Curiosities*, p. 177). However, the similarity between these illustrations and the present *rei puta* is not exact, for the drawings show a pendant with a V-shaped nose, and eyes that are more angular in shape with clearly defined pupils in them, as well as lattice carving on the toggle. The drawings do, however, also show loop ties or rings attaching the pendant to its neck cord, and these better approximate to those seen on the museum *rei puta*. Other illustrations made during and after the voyage do not resemble the museum *rei puta*, particularly in the number and position of the holes by which the pendant is attached to its cord (see Herman Diedrich Spöring, BL Add. MS 15508, f. 32(b); J.F. Miller, BL Add. MS 23920, f. 76(b); and Charles Praval, BL Add. MS 7085, f. 33).

A further *rei puta* is held at the Great North Museum: Hancock, NEWHM C765. An admirable example of its kind, it too has three evenly spaced holes for its flax neck cord, a faint nose and eyes that better approximate to Parkinson's illustration than do those of the British Museum's *rei puta*, but its bone toggle is also undecorated and multiple strands of fibre fan out from each hole to connect with the neck cord. Additionally, some natural marks are visible about the pendant's face. This *rei puta* was formerly in the private collections of the antiquary George Allan (1736–1800), which passed to the Literary and Philosophical Society of Newcastle in 1822, and later formed the basis of the Hancock Museum.

86 Carved wooden object

The Museum of Archaeology and Anthropology, University of Cambridge. D.1914.65. 31.5 cm. Wood and abalone shell (Haliotis iris).

Experts do not know what this object was intended to be and various suggestions have been made as to its purpose. These include that it was an ornament for the prow of a boat (as seen, perhaps, in Sydney Parkinson's drawing of Māori fishing, number 96), a form of decoration for a latrine or other building component, or even a bird snare. Number 103, a plate from Sydney Parkinson's posthumously published journal, depicts a number of Pacific artefacts, including the large *wheku* face of such an object. The plate key states that this sort of object varied in size, with the smaller ones sometimes having carved handles, was coloured red (with ochre), and was frequently held up to *Endeavour* when it was approached by Māori. The suggestion is made that 'perhaps it may be the figure of some idol which they worship'.

Whatever their original purpose, such objects are rare. This one displays a stylized *wheku* face at one end of an elaborately carved wooden shaft with a smaller carved head at the other end. The eyes in the large face are open wide, and a tongue protrudes from the gaping mouth. The shell inlay of one of the eyes is missing. A rectangular recess on both sides of the shaft is pierced by two holes, one roughly square in shape and the other with five irregular sides, these holes presumably being used for fixing the object into place. The present example is a known *Endeavour*-voyage artefact, originally given by James Cook to Lord Sandwich and passed by him to Trinity College, Cambridge. Three more examples of this type of object exist, two at the British Museum, Oc,NZ.66 (Edge-Partington, 'Cook collection') and 7361 (presented by A.W. Franks, 1871), and one at the University of Göttingen, OZ.323. J.F. Miller illustrated such a carving in a side view for Joseph Banks after the *Endeavour* voyage, BL Add. MS 23920, f. 73.

87 'Carved Plank from New Zeland'

By John Frederick Miller.

The British Library, London. BL Add. MS, 23920, f. 75.

22.6 × 16.5 cm.

This pen and wash illustration of a wall-panel carving was made for Joseph Banks following the return of the *Endeavour* expedition, and is signed on the recto 'John Frederick Miller. del. 1771' and annotated in ink on the folio as given above. At the top of the sheet a note in ink reads '3:f. 3¼l. by: 1:f. 1.l.'. The illustration depicts an impressively carved Māori wall panel, or *poupou*, that was previously thought to be lost, but which is now known to be held at the Institute of Ethnology in the University of Tübingen. Deposited there in 1937 by the daughter of the Viennese museum director Ferdinand von Hochstetter, it was not until 1996 that the panel's first-voyage provenance was recognized using Miller's illustration, which thereafter aided the panel's full restoration.

Volker Harms writes of the wall panel: 'The carver, perhaps Paikuerangi, lived in the second half of the 18th century and belonged to the school of Te Raawheoro, to which the carving is to be attributed. The school was part of the Te Aitanga-a-Hauiti tribe, who occupied the area around Tolaga Bay' (Kaeppler *et al.*, *James Cook and the Exploration of the Pacific*, p. 177). Tolaga Bay, Uawa to its inhabitants, was an important Māori centre for carving. It seems likely that Banks obtained the panel in the bay, at Pourewa Island on 28 October 1769, for an unoccupied, partially built dwelling with such carvings inside was found when the voyagers explored this island. Local oral traditions have suggested that the panel may have been given to Banks as a present, while some scholars have speculated that it could have been given instead to his travelling companion, the Ra'iatean priest Tupaia.

In his journal Banks wrote: 'We saw also a house larger than any we had seen tho not more than 30 feet long, it seemd as if it had never been finishd being full of chipps. The woodwork of it was squard so even and smooth that we could not doubt of their having among them very sharp tools; all the side posts were carvd in a masterly stile of their whimsical taste which seems confind to the making of spirals and distorted human faces. All these had clearly been removd from some other place so probably such work bears a value among them' (*Banks's Journal*, vol. 1, p. 421).

Poupou carvings represented Māori ancestors and adorned the inside of houses belonging to high-ranking individuals. As such they were precious objects, full of significance for their owners, that could be, and were, moved from one dwelling to another. It appears that the example found by Banks had been moved in this way. The panel shown in the illustration is the only one of its kind from the first voyage, since opportunities to obtain other panels did not occur during the mission. The present illustration, like those below of canoe carvings also observed on Pourewa Island, record fine examples of local craftsmanship as witnessed by the voyagers.

88 'The Head of a Canoe'

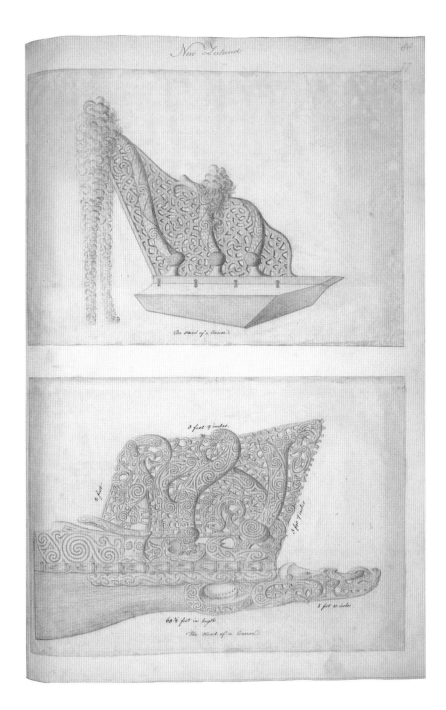

By Herman Diedrich Spöring.

The British Library, London.
BL Add. MS 23920, f. 77(a,b).
28.1 × 39.5 cm; 28.1 × 39.4 cm.

These are pencil illustrations by Herman Diedrich Spöring of the carvings on a Māori canoe prow seen on Pourewa Island in Tolaga Bay. Tolaga Bay was an area with a flourishing tradition of carving when the *Endeavour* expedition arrived there in October 1769. Both illustrations are annotated by Spöring on the recto in ink as above. The first, (a), is annotated on the verso in ink by the artist 'About 5 feet high. New Zeeland'. The second, (b), is annotated on the recto in ink with measurements for the canoe-prow carving, clockwise as follows '2 feet 3 feet 9 inches. 3 feet 7 inches 5 feet 10 inches. 68½ feet in length.'

Illustration (a) was apparently sketched in draft by Spöring at BL Add. MS 15508, f. 33(c). Two other drawings featuring the ornate canoe head shown in (b) exist, one at the Department of Prints and Drawings, the British Museum, folio 201.c.5, no. 271 (Spöring), and the other at BL Add. MS 7085, f. 32 (a copy of Spöring by Charles Praval).

Both Banks and Cook described the canoe for (b) in their journals, giving overall length measurements corresponding to those in Spöring's illustration of 68½ feet. Banks recorded in his journal entry for 28 October 1769: 'This morn we went ashore in an Island on the left hand as you come into the bay calld by the natives *Tubolai*. Here we saw the largest canoe we had met with: her leng[t]h was 68½ feet, her breadth 5, hight 3:6: she was built with a sharp bottom made in 3 peices of trunks of trees hollowd, the middlemost of which was much longer than either of the other two; Her gunnel planks were in one peice 62 f 2 in leng[t]h carvd prettily enough in bass releif, the head was also richly carvd in their fashion' (*Banks's Journal*, vol. 1, pp. 420–1). Of canoe prow (b), Dame Anne Salmond has observed: 'It is a detailed depiction of an elaborate, sinuous piece of carving, associated with the Te Raawheoro carving school' (*Two Worlds: First meetings between Maori and Europeans, 1642–1772*, p. 174).

While at Tolaga Bay, Cook named Spöring Island after Banks's draughtsman, but it has now reverted to its Māori name of Pourewa. This is the same bay rendered above by Banks as 'Tubolai'.

89 Treasure box *papahou*

The British Museum, London. Oc,NZ.109.a, 109.b. (Edge Partington and C.H. Read registration slips, designated 'Cook collection' by Read.) Length 65, width 16.5 and depth 7.5 cm. Wood and flax cord.

A treasure box is a finely carved container with a sliding lid, which was hung by cords from the rafters inside a house. It held the most valued possessions, or *taonga*, belonging to its owner. These might include personal items worn about the body as well as family or tribal heirlooms passed down the generations. Feathers, combs, neck pendants and other ornaments made of nephrite, stone, bone, shell or teeth were kept in treasure boxes. The carvings on these boxes varied in style, but would be ornate and well executed as befits a container of precious objects. Treasure boxes could also be undecorated, but few plain ones were collected, doubtless because a lack of carving made them less attractive to Europeans. Possessions held in treasure boxes were considered *tapu*.

This box is of the *papahou* type, being flat and rectangular in form as compared with the more common *wakahuia* treasure boxes, with their oval 'canoe-shaped' form. Another box type was the *powaka whakairo*, which was square in shape with sharp edges and a flat bottom, but few of these survive in modern collections. *Papahou* treasure boxes were an ancient regional type found in the northern and western areas of the North Island. The present example is complete, with a diagonal-plaited flax cord attached through the two lid corners to the box. At each end of the box there is a rounded tongue projecting from a low-relief *wheku* face, the larger of these two faces being adorned with *rauponga* carving. The lid is also carved with *rauponga* patterns, and the sides and bottom of the box display alternately three *wheku* and two *manaia* figures. These carvings are remarkably bold in both design and execution. There are traces of red ochre on the box surface.

The British Museum has a large collection of treasure boxes, one of the *wakahuia* kind perhaps also having been gathered on a Cook voyage and recorded by James Edge-Partington as 'Cook collection', Oc,NZ.113a–b. *Wakahuia* boxes were typical of central and eastern areas, but their use spread across the country and they continued to be made into the nineteenth century, unlike those of the *papahou* type, which died out possibly as a result of turmoil during colonization (see number 90 below).

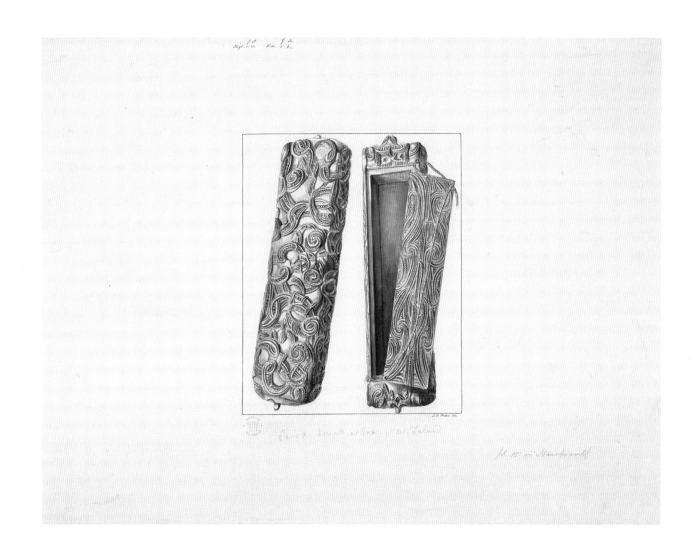

90 'Carvd trunk or box New Zeland'

By John Frederick Miller.

The British Library, London. BL Add. MS 15508, f. 22.

16.5 × 20.2 cm.

This signed pen and wash illustration by J.F. Miller of a Māori treasure box, drawn for Joseph Banks following his return from the Pacific, is annotated in pencil as given above. At the top of the sheet a note in ink reads 'High. 1:$^{f.}$ 11.^{I} Wide 0:$^{f.}$ 6.$^{In.}$'. This illustration shows a treasure box in a top view, sliding lid open, as well as a bottom view. Since this box would have been hung from the rafters of the house, its bottom was visible and hence decorated. A diagonal cord secures the lid to the box through two corner holes. The close agreement between this illustration and the treasure box at number 89 above confirms that object's first-voyage provenance. R. Ralph also produced a pen and wash drawing of this box, BL Add. MS 23920, f. 74(a), and it appears that this was then used for the engraving published in the official account of the *Endeavour* voyage, vol. 3, plate 15.

179

91 Paddle *hoe*

The Museum of Archaeology and Anthropology, University of Cambridge. D.1914.67. 180 × 14 cm. Wood.

This elegant paddle, or *hoe*, one of a pair from the *Endeavour* voyage held at Cambridge, is decorated with carving on its handle and with white-on-red ochre scroll work designs (*kowhaiwhai*) painted on the blade. There is a faded inscription in copperplate style visible on the blade, but it is illegible. The other paddle is D.1914.66. Each was fashioned from a single piece of wood. Their handles are carved at the terminals into open loops, and the raised grips at the junction with the blade are richly decorated with *taratara-a-kai* relief carving. It is possible that these paddles were gathered just south of Poverty Bay on 11 October 1769 (see number 78). Scholars regard them, and others like them held elsewhere, as displaying a distinctive style of painting and grip carving developed in the Poverty Bay area. The two Cambridge paddles represent the earliest known examples of their kind brought back to Europe.

There is another paddle of a similar type at the British Museum, Oc,NZ.150, 'Cook collection', which has been likened to one shown in a post-voyage illustration for Banks by J.F. Miller, BL Add. MS 15508, f. 29. James Edge-Partington, Charles Hercules Read and later Adrienne Kaeppler (see

Kaeppler, *Artificial Curiosities*, p. 203) all pointed out the resemblance, which is indeed very close both in terms of general form and blade decoration but not exact. A further paddle held at the Great North Museum: Hancock, NEWHM C589, has been attributed to the *Endeavour* voyage due to its likeness to the first of three painted *hoe* blades featured in Parkinson's splendid pen, wash and watercolour illustration at BL Add. MS 23920, f. 71(a). The match is, again, not an exact one, but the paddle concerned is certainly similar to the others described above and probably is that depicted by Parkinson. The second and third paddle blades in his illustration have not been located (see Jessop, *The Exotic Artefacts from George Allan's Museum*, pp. 109 and 133; see also Hooper, *Pacific Encounters*, p. 129).

At least 22 early paddles are known to exist in various collections worldwide, a number of which may have a Cook-voyage provenance. Of these, perhaps eight or more closely resemble in style the two Cambridge examples, among which are the paddles mentioned above at the British Museum and in Newcastle. Others in this particular group are to be found at the Museum of New Zealand, Wellington (one or possibly two); the Sunderland Museum; the Hunterian Museum in the University of Glasgow; the Institut für Ethnologie, Göttingen; the Museo di Storia Naturale at the University of Florence, and there is a second example at the British Museum. Research into the origins of paddles thought to derive from early-period contacts is ongoing and at present it would appear that some, if not all, of the eight mentioned above were gathered in the Poverty Bay area during the first voyage (see paper by Salmond, *Artefacts of Encounter*, 2011 version).

Benjamin West included a *hoe* in his portrait of Banks, which is situated behind Banks's right shoulder. A *hoe* formerly in the James Hooper collection, but now in the Museum of New Zealand, ME014921, has been identified with that painted by West, but the link is not proven despite the visual resemblance in shape, proportion and carving in both (see Phelps, *Art and Artefacts of the Pacific*, pp. 27 and 38–9). What does appear certain from the terminal carving featured by West is that neither of the Cambridge paddles, nor Oc,NZ.150 at the British Museum, can be the one seen alongside Banks in the portrait (see number 143).

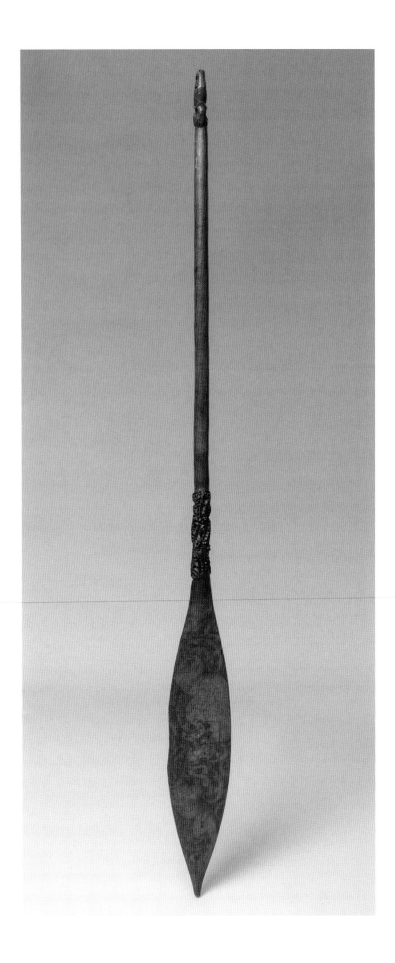

181

92 Canoe bailers *tiheru*

1. The Pitt Rivers Museum, University of Oxford. 1887.1.381. Length 50, width 32 and height 14.5 cm. Wood.

2. The British Museum, London. Oc,NZ.123. (Edge-Partington registration slip, 'Cook collection'.) Length 51, width 23 and height 16 cm. Wood.

Like so many Māori objects, these canoe bailers, or *tiheru*, are gracefully proportioned and artfully carved. Produced from a single piece of wood, they display classic *manaia*, human figure forms, on the handle end and posterior rim. The bowls are smoothly rounded in shape as are the curved openings at their ends. The first was given by Joseph Banks to Christ Church, Oxford, at an unknown date before mid-January 1773. Interestingly, prior to that it had been damaged and then carefully repaired using what appears to be New Zealand flax, evidently while still in Māori possession. The twine used was neatly inset to avoid drag or wear when the bailer was in use. Contemporary museum repairs with raffia have also been made. Moreover, it is one of a pair of bailers that Banks included in his donation to Christ Church, the other being from the Society Islands and, like other objects in his Oxford gift, these bailers show an attempt on his part to provide representative examples of Polynesian craftsmanship, in this case from more than one Pacific area (see number 65). The second bailer shown here is an undamaged *tiheru* from the British Museum, Oc,NZ.123. James Edge-Partington registered this bailer as a 'Cook collection' object and it exhibits all the fine qualities of its Oxford counterpart. Following the *Endeavour* voyage the same or a similar bailer was illustrated for Banks by J.F. Miller (see number 93 opposite).

92.1

92.2

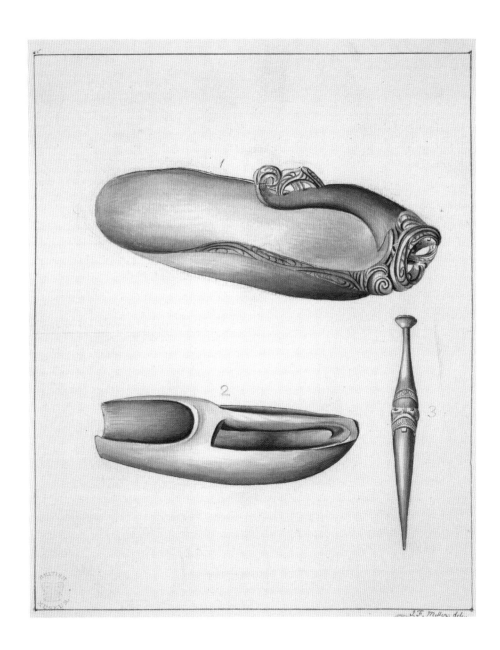

93 Artefacts from Tahiti and New Zealand

By John Frederick Miller.

The British Library, London. BL Add. MS 15508, f. 28.
20.3 × 16.5 cm.

This is a pen and wash illustration, probably dating to 1771–2, signed in ink 'J.F. Miller del'. A key on the folio in pencil in Banks's hand reads '1 Skid or Scoup from New Zeland 2 D⁰· from Otaheite 3 Planting stick from New Zeland'. Miller depicts a weaving peg and two canoe bailers. The lower bailer is from Tahiti and the upper bailer and weaving peg are both from New Zealand (see numbers 65, 83 and 92).

94 'Motuaro, Bay of Islands'

By Herman Diedrich Spöring.

The British Library, London. BL Add. MS 23920, f. 43(a). 26.7 × 41.6 cm.

This is a pencil drawing by Herman Diedrich Spöring, annotated anonymously in pencil on the folio as given above, and in ink on the verso by Spöring 'The Town on Motuaro-Island. New-Zeeland'. It shows a fortified headland *pā* on Motuarohia Island, in the Bay of Islands, which is featured in profiles and drawings made both by Spöring and Sydney Parkinson (see number 95 overleaf). Among these is a panoramic set of three coastal profiles of Motuarohia drawn in pencil by Spöring, with the *pā* appearing in profile 3, BL Add. MS, 15507, f. 35(a). The detailed study seen here was later copied by John James Barralet, seemingly with the intention that it should be engraved for the official *Endeavour* voyage account edited by Dr John Hawkesworth, but it was not included in the final publication; this is BL Add. MS 15508, f. 20.

A *pā*, or as it is often written in the voyage journals 'hippa' or 'heppah' from *he pā* meaning 'it is a pa', was a fortified Māori village or retreat, located where possible at an easily defended position to act as a stronghold if attacked by another tribe and protect tribal territory. The *pā* at Motuarohia was sighted by the voyagers on 4 December 1769. According to J.R.H. Spencer, who in 1974 surveyed the headland on which it stood, and returned to photograph the location in 1981: 'Spöring's pencil sketch of the *pā* and his coastal profile of the headland are, according to my reconnaissance of the area, highly accurate representations of the topography.' Spencer located the offshore position of *Endeavour* from which Spöring observed the headland, probably through a telescope. He noted that: 'The sandy bay which, on the left of the sketch, has canoes drawn up on the beach and structures on a small terrace on the hillside behind can still be seen today. The drawing shows tracks leading up from the beach on to the spurs and ridge line, and they are still clearly in evidence. The positions of terraces, which in Spöring's sketch bear houses or similar structures enclosed by palisades, agree well with those revealed by a recent survey.' He added: 'The foreshortening effect produced by using a telescope is seen in the presence of a large *whata* (foodstore) or a similar structure sketched in on the shores of Waipao Bay, Moturua Island, to the right of the *pā* headland' (Carr, *Sydney Parkinson*, pp. 265–7). The visual accuracy of Spöring's drawing was probably the reason why it was selected to be copied by Barralet after the voyage.

95 'View of the Hippa upon the Island of Motuaro in the Bay of Islands, New Zealand'

By Sydney Parkinson.

The British Library, London. BL Add. MS 23920, f. 43(b). 24.7 × 37.5 cm.

This unsigned pen and wash drawing by Sydney Parkinson is annotated by him in ink on the verso as given above. Another verso note in pencil by Parkinson, partly cut away from the lower edge, may read 'Koomatteewaraeea', apparently a transliteration or perhaps a mistransliteration by him of a Māori name for this bay. On the folio in a later hand is the comment in ink 'Fortified tops of Hills'. Parkinson here depicted a fortified headland settlement on Motuarohia Island that was also drawn by Herman Diedrich Spöring (see number 94 above). Parkinson prepared an initial outline sketch of the headland, BL Add. MS 9345, ff. 61v–62, but this and his finished drawing of the location are less true to its topography than Spöring's drawing of the same area.

In the present illustration Parkinson has shifted the sandy beach on which Spöring showed some canoes to a position beneath the *pā*, where in reality rocks ought to appear and no buildings. Parkinson added sea caves on the headland point where in fact there should be only crevices, and in place of the background hills he has introduced elevated romantic peaks. The *pā* is out of proportion, its palisades are altered into neatly arranged uprights, and there is a fighting platform from which defenders appear to be repelling attackers, an event not supported by journal descriptions made at the time of *Endeavour*'s visit. In short, it seems that Parkinson indulged a moment of fancy, treating this scene partly as an imaginative subject, rather than adhering to the strict observational accuracy apparent elsewhere in his mission artwork. It may even be that he produced this finished illustration later on, using Spöring's drawings and his own preliminary sketch, possibly while *Endeavour* was off the east coast of Australia, since the sketch version is bound among drawings made at that stage of the voyage and not with those of New Zealand. If he was occupied with other work at Motuarohia, this would have been a sensible thing to do, and it may help account for the lack of fidelity in the finished piece.

J.R.H. Spencer described this illustration as an attempt at 'an idealised composition', and further remarked of Parkinson that: 'From this analysis of Parkinson's sketches of known localities in New Zealand it appears that in his topographical drawings he often gave free rein to the imagination, to provide the elements of a pleasing composition, even when they were absent from the actual scene. Many of his finished wash compositions of the Society Islands to be found in Add. MSS 15508, 23920 and 23921 must be regarded as combining elements of the imagination with those of reality.' Parkinson often incorporated imaginative elements into his figure and landscape work to produce interesting composite scenes showing local people and their daily activities set against striking backdrops. When depicting landscapes alone, he could be still more creative, apparently with the main aim of satisfying his own taste and artistic sense. He was, of course, capable of reproducing scenes precisely as they appeared, but was evidently under no compulsion to do so every time. Otherwise his coastal views, and certainly his illustrations of plants and animals, are executed with accuracy in keeping with the practical and scientific requirements for such work (see numbers 5 and 122–6; see also Carr, *Sydney Parkinson*, pp. 270 and 274).

96 'New Zealanders fishing'

By Sydney Parkinson.

The British Library, London. BL Add. MS 23920, f. 44. 29.8 × 48.2 cm.

This pen and wash drawing by Sydney Parkinson is annotated anonymously on the folio in ink as given above, and on the verso by Banks 'New Zealand'. Preparatory sketches by Parkinson for three of the figures in the central canoe are at BL Add. MS 23920, f. 65. The finished drawing depicts Māori fishing in what Cook called Queen Charlotte Sound, or Totaranui to its inhabitants. *Endeavour* visited the sound from 15 January to 6 February, and it provided what became a favourite South Sea base for Cook at the small bay he called Ship Cove, or Meretoto. By this time Cook had completed his survey of the north part of the country, in fact a great island Te Ika-a-Māui, and was unaware that he was now actually on the South Island, Te Waipounamu. Queen Charlotte Sound had earlier been settled by a succession of tribal migrants from the north and was a key site for crossings between the two main islands, but inter-tribal warfare was not unusual and there had been recent local disturbances when Cook arrived, as evidenced by some of the more grizzly findings made by the voyagers (see A. Salmond, *Between Worlds: Early exchanges between Maori and Europeans, 1773–1815*, pp. 65–7).

In the foreground of the drawing Māori women are seen using hoop-nets in which bait was placed to catch fish. Some are wearing the mourning caps (*potae-taua*) of widows, woven of rushes dyed black and decorated with black feathers, and cloaks associated with the area are also featured. In the background are large carved canoes thought to be of a type originating from the North Island. These may have been introduced by Parkinson to embellish the scene, but impressively carved canoes were encountered on 8 February shortly after *Endeavour* exited Cook Strait and briefly turned northwards to confirm that North Island was indeed a separate landmass, so perhaps Parkinson incorporated the present ones to represent this fact. The people of the North Island were considered by the voyagers to be better dressed, taller and of a superior stature as against their southern neighbours, whose canoes were judged to be 'but mean' by comparison with those of the north (see *Journal of a Voyage*, p. 119; see also *Banks's Journal*, vol. I, p. 465). For Parkinson illustrations of similar carved canoes and their warlike occupants, see numbers 97 and 98 below.

Cook stayed at Ship Cove some three weeks, and in total he visited this location five times during his three Pacific voyages. It provided fresh water and food, and also furnished *Lepidium oleraceum*, an herbaceous plant from the cabbage family Brassicaceae endemic to New Zealand, which when infused in beverages helped to prevent scurvy, a disease now known to result from a deficiency of vitamin C. The name 'scurvy grass' has been applied to a variety of plants rich in vitamin C that have long been known to help ward off scurvy during long voyages. The English common name for *L. oleraceum* is Cook's Scurvy Grass. Importantly from Cook's point of view, the local people were friendly on this his first visit. He landed pigs, planted vegetables and distributed seeds among them. Their way of life and country were observed, and Parkinson produced studies of their appearance and activities as shown here. Māori nets, cloths and twine made from *Phormium tenax* greatly impressed the British, as did the area's dense woodland, each being of possible use to a maritime nation needing cordage and timber for shipping. It was here, too, that the voyagers found unequivocal evidence of cannibalism (previously suspected from reports obtained elsewhere but not confirmed), to the British a shocking practice, but clearly accepted among the Māori.

However, the most impressive discovery was Cook's when he ascended a hill on Arapawa Island and saw a strait leading out to the Pacific Ocean in the east, thereby establishing that he had surveyed one great island to the north. During his visit of 1642, Abel Tasman assumed that no passage existed here, but Cook now showed that the Dutch navigator was mistaken. Banks is usually credited with naming the exit Cook Strait (to the Māori, Raukawa Moana), and Cook duly sailed through it to commence his circumnavigation of the South Island, which lasted from February to the end of March with no landings. New Zealand, he had proved, was not part of a large southern landmass. Cook then anchored in Admiralty Bay having completed his running survey and, on 31 March 1770, he sailed westward into the Tasman Sea.

97 'New Zealand War Canoe'

By Sydney Parkinson.
The British Library, London. BL Add. MS 23920, f. 49.
29.8 × 48.2 cm.

This pen and wash drawing by Sydney Parkinson is annotated anonymously in ink on the folio as above. Note one of the canoe occupants in the centre holding up a severed head. On the reverse of the drawing is a sketch of the Māori pulling the flax sail cord and his nearby companion holding a paddle. The Māori tradition of preserving heads was first observed by the voyagers on 20 January 1770 in Queen Charlotte Sound, so it seems likely that the present drawing was completed at or shortly after that date (see number 96 immediately above).

Parkinson relates that: 'One day, in particular, they brought four skulls to sell; but they rated them very high. These skulls had their brains taken out, and some of them their eyes, but the scalp and hair was left upon them. They looked as if they had been dried by the fire, or by the heat of the Sun. We also found human bones in the woods, near the ovens, where they used to partake of their horrid midnight repasts: and we saw a canoe the baler of which was made of a man's skull' (*Journal of a Voyage*, pp. 115–16).

Note Parkinson's inclusion of three kinds of Māori quarterstaff, the *tewhatewha*, *taiaha* and *pouwhenua*, as well as a paddle, *patu* and various cloaks, including one with a *taniko* border. Also featured are personal ornaments such as a *kuru* earring, *rei puta* neck pendant, *heru* hair comb and assorted head feathers (see numbers 85, 91 and 99–102). The high prow and elaborately carved stern of the canoe resemble those of the canoe in the next illustration (see number 98).

98 'New Zealand War Canoe bidding defiance to the Ship'

By Sydney Parkinson.

The British Library, London. BL Add. MS 23920, f. 50.
29.9 × 48.3 cm.

This pen and wash drawing by Sydney Parkinson pen, annotated anonymously in ink on the folio, probably dates to the period of *Endeavour*'s visit to Queen Charlotte Sound. While in the sound Parkinson sketched a number of Māori figures, some of which are in similar if not exactly the same postures as individuals illustrated here (see BL Add. MS 23920, f. 63; see also numbers 96 and 97).

Here Parkinson is concerned to capture the threatening gestures of the Māori when challenging an opponent. Banks gives a vivid description of the Māori challenge in his journal essay account of New Zealand: 'The War Song and dance consists of Various contortions of the limbs during which the tongue is frequently thrust out incredibly far and the orbits of the eyes enlargd so much that a circle of white is distinctly seen round the Iris: in short nothing is omittd which can render a human shape frightful and deformd, which I suppose they think terrible. During this time they brandish their spears, hack the air with their patoo patoos and shake their darts as if they meant every moment to begin the attack, singing all the time in a wild but not disagreable manner and ending every strain with a loud and deep fetchd sigh in which they all join in concert. The whole is accompanied by strokes struck against the sides of the Boats &c with their feet, Paddles and arms, the whole in such excellent time that tho the crews of several Canoes join in concert you rarely or never hear a single stroke wrongly placd' (*Banks's Journal*, vol. 2, pp. 29–30).

As in his other composite drawings, Parkinson here brings together in a dramatic scene a number of activities and objects variously sketched by him in separate studies, or otherwise observed or collected during this stage of the voyage. Strictly speaking the challenge shown was not a *haka* because it was performed in a canoe, whereas the true *haka* is a dance and chant executed on land that may, as Banks recognized, be used in 'war or peace'. The war *haka* is more properly called the *peruperu*. There are various types of *haka* and their performance depends on the occasion concerned, whether it is that of welcome, challenge, celebration or part of funeral rites for the deceased. Each tribe has its own *haka* as passed down the generations. New Zealand fighting forces still use the *haka*, but its modern fame is based on its use by the country's sports teams.

99 Five Māori cleavers and a brass replica

The Pitt Rivers Museum, University of Oxford. 1887.1.387–9; 1887.1.393; 1887.1.714; 1932.86.1.

1. PATU PARAOA
47.8 × 9.4 cm. Whalebone.

2. PATU
36.5 × 11 cm. Wood.

3. KOTIATE
8 × 13 cm. Wood.

4. WAHAIKA
44 × 11 cm. Wood.

5. PATU ONEWA
34.3 × 10 cm. Basalt.

6. PATU ONEWA
BRASS REPLICA
36.5 × 10.2 cm.

Māori hand-cleavers like these would have been used in close combat and this selection displays the range of shape and material employed by their makers for such fearsome weapons. It also shows how Banks included representative examples of certain object types in his donation to Christ Church, Oxford, to try to illustrate the scope of indigenous design and production. As such this is an instance of his organized approach to collecting. In battle such weapons are understood to have been used for slicing, jabbing and striking. They were regarded by the Māori as precious heirlooms to be passed down the generations, and on the field of battle as prized trophies when taken from opponents. Pacific weapons of all sorts were deemed by Europeans to be highly collectable, and many were gathered during and after Cook's voyages. Early voyagers variously denominated Māori cleavers as 'Patoo-Patoos', 'Patta Pattows' and by other spellings similar to these.

In his journal essay account of New Zealand, Banks commented on these weapons: '*Patoo patoos* as they calld them, a kind of small hand bludgeon of stone, bone hard wood most admirably calculated for the cracking of sculls; they are of different shapes, some like an old fashiond chopping knife, others of this or always however having sharp edges and a sufficient weight to make a second blow unnecessary if the first takes place; in these they seemd to put their cheif dependance, fastning them by a strong strap to their wrists least they should be wrenchd from them. The principal people seldom stirrd out without one of them sticking in his girdle, generaly made of Bone (of Whales as they told us) or of coarse black Jasper very hard, insomuch that we were almost led to conclude that in peace as well as war they wore them as a warlike ornament in the same manner as we Europæans wear swords' (*Banks's Journal*, vol. 2, p. 27. See also vol. 1, p. 419).

The *patu paraoa* displayed here is of whalebone, and is the longest of the cleavers, with a hole in the handle for a strap and some lightly incised lines on its rounded end. Jeremy Coote of the Pitt Rivers Museum has noted discoloured patches on both faces that he suggests may be bloodstains (see Coote, *Curiosities from the Endeavour*, pp. 11–14). The wooden *patu* is quite roughly worked and has a strap hole in its handle. Also of wood, the distinctive *kotiate* is shaped like a figure eight, with a perforated handle that has a face carved on its end. The wooden *wahaika* is in the shape of a crescent, again with a perforated handle for a strap. The *patu onewa* is of basalt, but with an unfinished strap hole in its handle. Banks had 40 brass replicas of this type cast by Eleanor Gyles of Shoe Lane, London, and then engraved with his name, shield of arms and the year by Thomas Orpin of The Strand. These were produced in the period March–April 1772 with a view to taking them on the second Cook voyage, from which Banks withdrew. A number were taken on the third Cook voyage and distributed among some of the peoples that were encountered. Six are known to survive in modern collections.

A comparable collection of five Māori cleavers is held at the Museum of Archaeology and Anthropology, Cambridge, as part of the Sandwich collection: *patu onewa*, basalt, D.1914.56; *patu paraoa*, whalebone, D.1914.57; *wahaika*, wood, D.1914.58; *wahaika*, bone, D.1914.59; *kotiate*, wood, D.1914.60. Further cleavers reside at the British Museum in its 'Cook collection', and many more examples that were collected during this period and afterwards survive in collections around the world today. In the portrait of Joseph Banks by Benjamin West there is a Māori cleaver, perhaps a *patu onewa*, placed in the bottom right-hand corner of the painting as seen by the viewer.

99.6

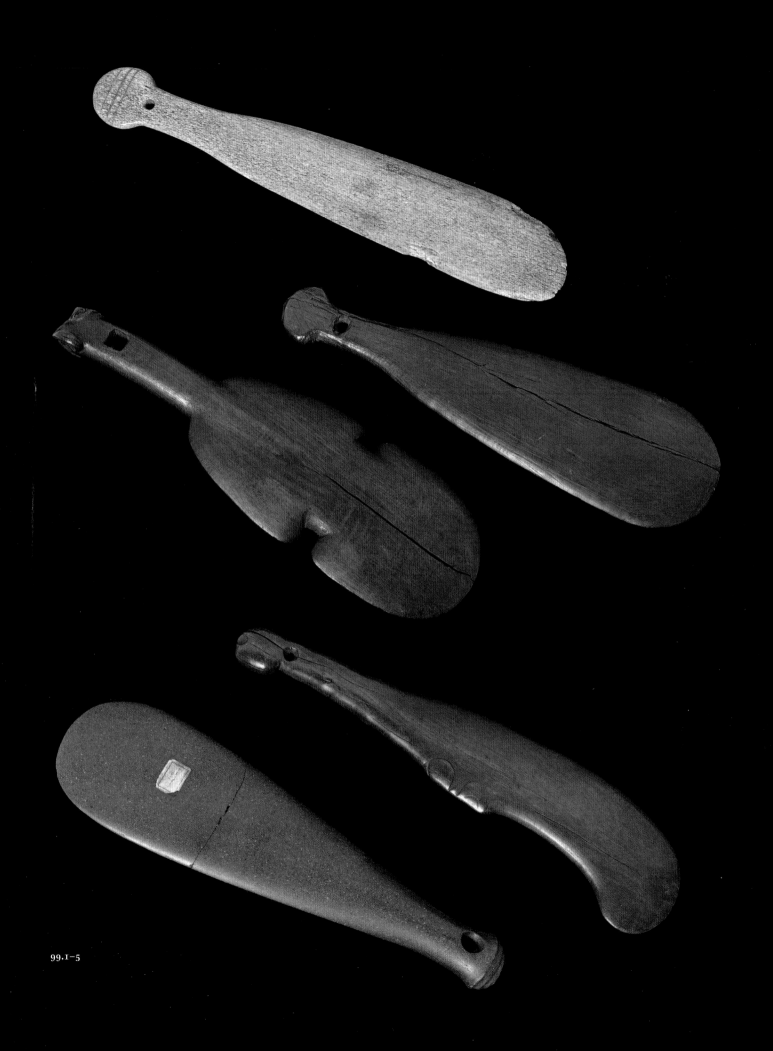

100–02 Three quarterstaves

100. TAIAHA
The Museum of Archaeology and Anthropology, University of Cambridge. D.1914.61. 184.5 × 8.5 cm. Wood and abalone shell (Haliotis iris).

101. POUWHENUA
The Museum of Archaeology and Anthropology, University of Cambridge. D.1914.62. 172 × 10.2 cm. Wood.

102. TWEHATEWHA
The Museum of Archaeology and Anthropology, University of Cambridge. D.1914.64. 137 × 22.5 cm. Wood.

TAIAHA

Quarterstaves were designed for parrying, stabbing and striking. This quarterstaff was fashioned from a single piece of wood, and is finely carved at its point but unembellished along its shaft. At one end the shaft forms a long flat spatulate blade designed for striking an enemy in battle. The pointed tip at the other end was used for stabbing, and is formed into a tongue protruding from the mouth of a face with inlaid eyes of haliotis shell. This device embodies a traditional Māori grimace displayed to opponents during challenges, a fitting touch for one of their weapons. The tongue is adorned with four double spirals. *Taiaha* were often decorated by attaching white dog hair and red feathers to the shaft beneath the face. On ceremonial occasions they served as a status symbol for Māori of rank. Although not used any more in combat, *taiaha* are still employed in the *wero*, the traditional Māori challenge given during the *pōwhiri*, a formal ceremony of welcome.

Number 100 would appear to be the earliest known example of its kind to have been collected by Europeans (see also numbers 101–2; see also the West portrait of Banks in which a *taiaha* features behind Banks's right shoulder, number 143). Known Māori quarter-staves from the *Endeavour* voyage are concentrated in the Sandwich collection at Cambridge. They were part of a collection of artefacts given by James Cook to his patron Lord Sandwich after the *Endeavour* voyage. Sandwich donated a selection of items from this material to Trinity College, Cambridge, from whence in 1914 and 1924 everything passed to Cambridge Museum of Archaeology and Anthropology. A group of weapons like this one displays a range of Māori design for its type, and was evidently included in the donation to Trinity with that aim in mind.

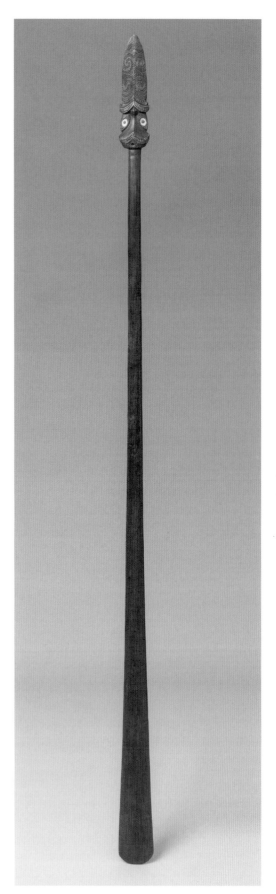

100

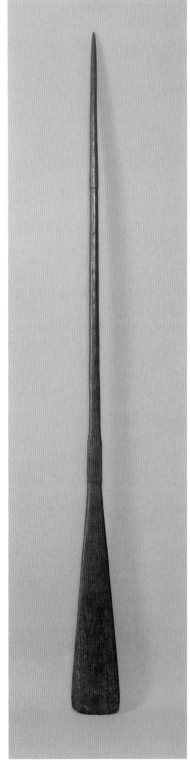
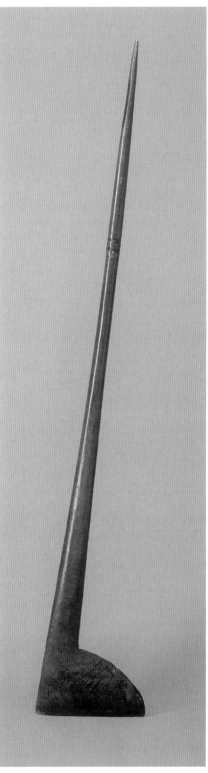

101

102

POUWHENUA

Like other quarterstaves, the *pouwhenua* was employed in close combat, the warlike Māori being highly skilled in hand-to-hand fighting. A fully trained warrior could wield such weapons with great speed and accuracy, using nimble footwork to maintain balance and outmanoeuvre an opponent, but it took years of practice to achieve this level of mastery. The present example was formed from a single piece of wood. At one end of there is a broad blade for striking, and the point at the other end was used for stabbing. Carving on the shaft is thought to indicate the hand positions required. This is one of a pair of *pouwhenua* in the Sandwich collection, Cambridge, the other being D.1914.63.

TWEHATEWHA

Made of wood, the *tewhatewha* is designed for striking at one end, where it has an axe-like blade, and for stabbing at the other, where it terminates in a sharp point. The present example displays some carved decoration on the lower end of the shaft. *Tewhatewha* were often adorned with a bunch of feathers hung from a drilled hole at the lower edge of the blade. A series of five Māori weapons, among them a *taiaha* and a *tewhatewha*, were illustrated for Banks by J.F. Miller following the *Endeavour* voyage, see BL Add. MS 23920, f. 70. Neither of the quarterstaves depicted by Miller appear to be held at Cambridge.

103   'Various kinds of Instruments Utensils &c, of the Inhabitants of New Zealand, with some Ornaments &c, of the People of Terra del Fuego & New Holland'

Stanfield Parkinson, ed., A Journal of a Voyage to the South Seas in his Majesty's Ship, The Endeavour (London, 1773; Fothergill reissue 1784). Plate 26. Drawn by S.H. Grimm. Engraved by T. Chambers.

The British Library, London. L.R.294.c.7. Rebound in half brown leather and brown cloth. Title, place and date of publication, gilt lettering on spine. Author given. 29 × 24.2 cm (closed). Plate 22.7 × 27.3 cm.

The key for this illustration from the posthumously published 1773 journal of Sydney Parkinson appears there on pages 128–31, as below.

Fig. 1. An Ornament for the Neck, made of three round pieces of Auris Marina, or ear-shell, the inside of which is a beautiful coloured pearl. These pieces are notched on the edges, and strung on a piece of plaited tape, made of white flax, and coloured red. It hangs loosely about the neck, and is two feet, eight inches and a half long.

2. One of their common Paddles; when used it is held by one hand at the top of the handle, in which there is a hole, and by the other at the bottom, where it is carved very neatly, being five feet, nine inches and a half long.

3. A Fish-hook, made of wood, and pointed with bone, which is tied on with twine; three inches and three quarters long.

4. A Fish-hook, made of two pieces of bone tied together; the line is fastened both at top and bottom; and, to the latter part, they tie some small feathers. The length of this hook is $4^{3}/_{8}$ inches.

5. A Fish-hook, made of wood, pointed with bone; about two inches and a half long.

6. A large Fish-hook, made of wood, and pointed with bone, having the end, to which the line is fastened, curiously carved; eight inches and a half long.

7. A Fish-hook, made of human bone; one inch and a quarter long.

8. A Fish-hook, made of wood, pointed with shell; five inches $^{5}/_{8}$ in length.

9. A Fish-hook, made of wood, and pointed with a substance that looked like one part of the beak of a small bird; two inches and a half long.

10. A Fish-hook, made of wood, and pointed with bone; three inches and a half long.

11. A Fish-hook, made of bone; one inch and a quarter long.

12. An Ornament made of bone, probably of some deceased relation, and worn in the ear; one inch and three quarters long.

13. and 14. are treated of in the accounts of Terra del Fuego and New-Holland. ['Their noses had holes bored in them, through which they drew a piece of white bone about three or five inches long, and two round.' p. 147; 'Both men and women wear necklaces, and other ornaments made of a small pearly perriwincle, very ingeniously plaited in rows with a kind of grass.', p. 7.]

15. A piece of Wood, part of the head of a canoe, singularly carved; nine inches and a quarter in length.

16. A favourite Ornament, which resembles a human face, made of wood, coloured red, and is much like some of the Roman masks. The eyes are made of the fine coloured ear-shell mentioned No. 1, laid into the wood. This was six inches long; but they have different sizes. Some of the smaller ones have handles carved very ingeniously; these they frequently held up when they approached the ship: perhaps it may be the figure of some idol which they worship.

17, 18, and 19, are Figures of Patta-pattoos, or War-bludgeons. They have holes in the handles of them, through which a string is passed and tied round the wrist when they make use of them. Numbers 17 and 19, are made of wood; the former is about fourteen inches long, and the latter twelve. Number 18 is about fourteen inches in length, made of a hard black stone, a kind of basaltes, and similar to the stone of which the Otaheiteans paste-beaters and hatchets are made.

20. Is a kind of Battle-axe, used either as a lance or as a patta-pattoo. The length of these is from five to six feet. The middle part of them is very ingeniously carved.

21. An Ivory Needle, made of the tooth of some large marine animal, with which they fasten on their cloaks. This is about six inches $^{3}/_{8}$ in length; but they have of various sizes; and some of them are made of the circular edge of the ear-shell mentioned in No. 1.

22. An Instrument made of the bone of some large animal, probably of a grampus, which is used sometimes as a paddle, and at others as a patta-pattoo, and is about five feet long.

23. A Wedge or Chisel, made of the green stone, or Poonammoo, as they call it, and sometimes of the Basaltes. These wedges they sometimes tie to a wooden handle, and then use them as hatchets and hoes. They are of various sizes, from one to eight inches in length.

24. A Whistle, made of wood, having the out-side curiously carved. Besides the mouth-hole they have several for the fingers to play upon. These, which

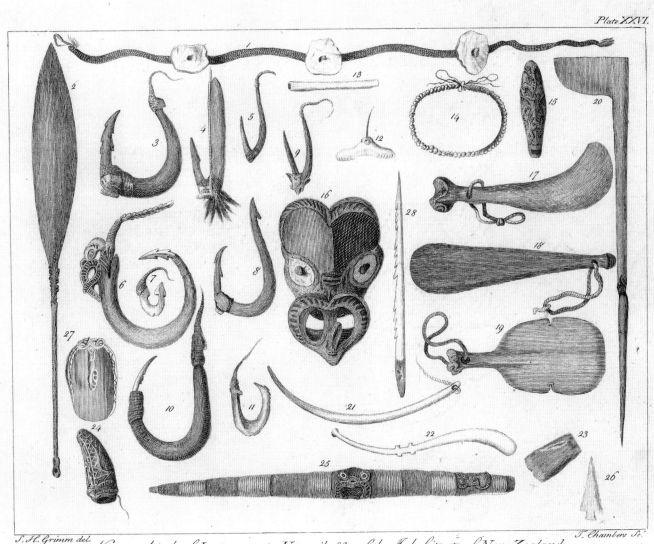

Various kinds of Instruments Utensils &c. of the Inhabitants of New Zealand, with some Ornaments &c. of the People of Terra del Fuego & New Holland.

are worn about the neck, are three inches and a half in length, and yield a shrill sound.

25. A Trumpet, nineteen inches and a half in length, made of a hard brown wood, which they split, and carefully hollow out each side so as to sit neatly again, leaving an edge on each side; and joining them together, they are bound tight with withes made of cane: it is broadest in the middle, which is rather flat, and gradually tapers to the ends that are open. In the middle of it there is a large hole which represents the mouth of a figure somewhat like a human one, having hands and feet, the parts of which are carved round the instrument: the head is not unlike the mask, No. 16. Another such like mask is also carved near one end of the trumpet. They produce a harsh shrill sound.

26. Is spoken of in the account of the people of Terra del Fuego. ['They use bows and arrows with great dexterity. The former are made of a species of wood somewhat like our beech; and the latter of a light yellow wood feathered at one end, and acuated at the other with pieces of clear white chrystal, chipped very ingeniously to a point.' p. 8.]

27. A singular kind of hand-scoop, or water-bailer, made of one piece of wood: the handle of it proceeds from the edge and hangs over the middle, and both it and the edge are very ingeniously carved. It is about eleven inches long, eight inches wide, and near six inches deep under the handle.

28. The head of a spear, made of bone, about six inches in length.

104 'CHART of NEW-ZEALAND, explored in 1769 and 1770, by Lieut: I: COOK, Commander of His MAJESTY'S Bark ENDEAVOUR.'

By James Cook and Isaac Smith. Engraved by John Bayly.

The National Maritime Museum, Greenwich, London. G263:1/2. 49 × 39cm. Scale approx. 1:3,600,000.

This is a print engraving of a classic chart by James Cook and his draughtsman and relation Isaac Smith. The chart shows *Endeavour*'s track from 6 October 1769 to 1 April 1770 around the two main islands of New Zealand. It was published in the official three-volume account of the *Endeavour* voyage, edited by Dr John Hawkesworth. Engraved by John Bayly (see number 134), it appeared in volume II of that work, facing page 281. A similar chart engraved by B. Longmate was included as plate 25 in Sydney Parkinson's posthumously published voyage account (for other manuscript and engraved prints of New Zealand's coastal outline arising from this voyage, see David, *The Charts & Coastal Views of Captain Cook's Voyages*, vol. 1, pp. 160–71).

Cook showed that New Zealand was not part of a larger landmass, as some had thought, but was actually comprised of two main islands. Note Banks Island on the east coast of South Island, a rare error by Cook since this is not an island but a peninsula. With *Endeavour* worn from her travels and the southern hemisphere winter fast approaching, Cook decided against sailing towards Cape Horn at high latitudes in search of any undiscovered landmass, preferring to make for the east coast New Holland (Australia as it became), and from thence home between New Holland and New Guinea via the Dutch East Indies. This offered the chance of new discoveries, in particular when charting northwards along the east coast itself and thereafter probing the strait previously found by Torres in 1606. A later voyage would have to resolve lingering doubts as to what might lie hidden deep in the southern oceans, and in his journal Banks speculated with considerable foresight on the course that such a voyage might take (see *Banks's Journal*, vol. 2, pp. 38–42).

On 19 April 1770 the coast of New Holland was sighted and over a week later a favourable bay was found into which *Endeavour* sailed. During the next four months Cook sailed northwards along Australia's east coast, charting its outline as he went and occasionally stopping for fresh water and other supplies. At these stops Banks and Solander, the first European naturalists to explore this coast, collected hundreds of specimens that Sydney Parkinson and Herman Diedrich Spöring drew. The Aboriginal inhabitants were observed and described, despite their tendency to retire from the strange new visitors. They showed little interest in gifts offered by the *Endeavour* crew, and this and the limited contacts that took place meant that relatively few man-made objects were gathered during this stage of the voyage.

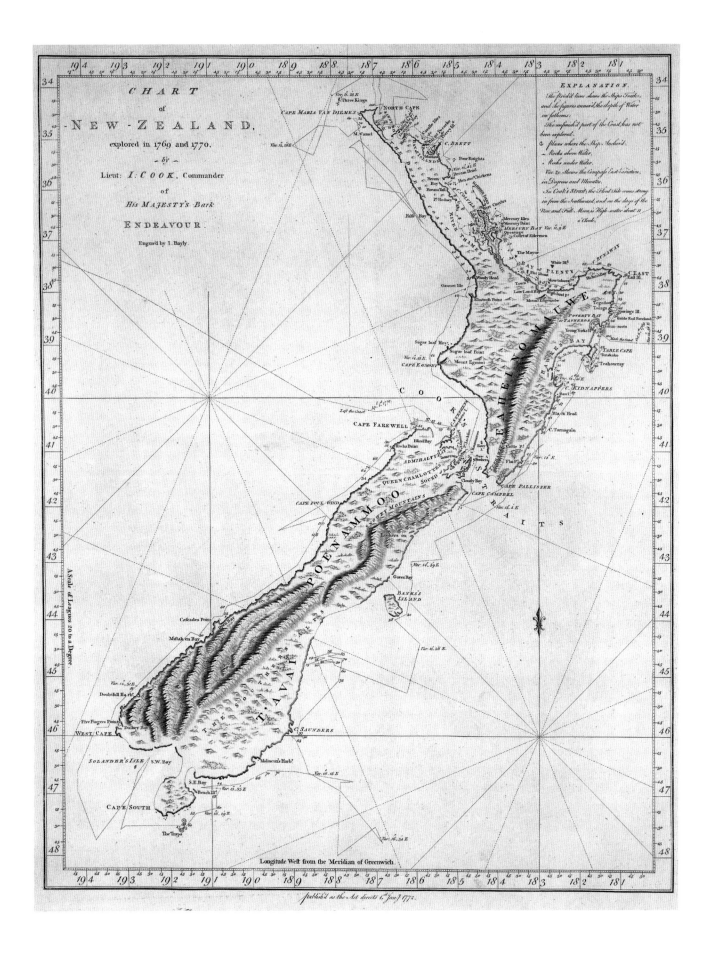

Exploring Collections from the Endeavour Voyage

NEIL CHAMBERS

BENJAMIN WEST'S MASSIVE PORTRAIT shows the youthful Joseph Banks as a collector of Pacific natural history and ethnographic objects (see frontispiece and number 143). Banks and his team expanded the scientific scope of the *Endeavour* mission to include these pursuits alongside its primary objectives in astronomy and geographical discovery. Banks's later career and reputation would be based on his achievements on the *Endeavour* voyage. After his final trip abroad to Holland in early 1773, Banks settled into the life of a metropolitan savant and country gentleman. His main collections were housed in London, not far from the capital's various clubs, learned societies and other institutions, and for the rest of his life they were made freely available for study by scholars from home and abroad. In November 1778 he was elected president of the Royal Society, which placed him at the head of the capital's senior intellectual body. The status that this position provided and his own experience as a voyager enabled Banks to fulfil the largely self-defined role of an unofficial advisor to government on such matters as exploration, trade and settlement at a time when the state lacked the sort of expertise that he and those in his circle were able to provide.

As such he was consulted on how to mount discovery expeditions as well as how to publish their results and distribute their collections at the end. Presidency of the Royal Society brought with it a place at the British Museum as an *ex officio* trustee, and using his many contacts Banks supplied the museum with all manner of 'natural and artificial curiosities' (as natural and man-made products were then known). The Royal Gardens at Kew were also enriched by collections of seeds and living plants flooding into the country during Banks's time as their unofficial director from 1773 onwards. In effect, Kew acted as a clearing house for botanic gardens located in various British possessions spanning the tropics, and as a repository for collections sent back by a series of exploring Kew gardeners. Incoming plants and seeds included numerous species valued not only because they were new to Western naturalists but also for their practical uses as food, in medicine or in manufacturing (see numbers 41 and 80). Throughout a period of rapid European and particularly British expansion in the Indo-Pacific region Banks was peculiarly well placed to draw on a network of navigators, settlers and colonial officials to secure natural history specimens and other objects from overseas.

Yet while his involvement in the administration of collections and collecting across the range of bodies with which he was involved remained fairly broad, in his private collecting Banks increasingly focused on developing his herbarium and alongside it his library, rich in travel and natural history literature. Material falling outside these concerns might be directed by him to a suitable body or just as likely be shared with a fellow collector taking more of an interest in it than he did. Similarly, that resulting from voyages launched under the Royal Navy that he helped to plan did not belong to him and would therefore generally be directed to the most appropriate institution with which he was connected. Living plants and seeds might, for example, be assigned to Kew while man-made objects might be given to the British Museum. Banks's own ethnographic and zoological collections from the three Cook voyages were divided in this sort of way at a fairly early stage. The artefacts were largely dispersed by the early 1780s, with many of them first being offered to the British Museum, and in 1792 the zoological specimens in alcohol were split between John Hunter and the British Museum.

Banks's herbarium and library were, by contrast, core lifetime collections that he steadily supplemented over the years and eventually bequeathed intact to the nation. Largely as a result of this approach, the scope and coherence of the surviving plant specimens, and the range of illustrative material connected with them, is far greater than it is for any other branch or discipline that Banks and his team pioneered in the Pacific during the *Endeavour* voyage.[1] Many plant specimens and drawings from that epoch-making mission are to be found today at the Natural History Museum, South Kensington (to which the British Museum's natural history collections were transferred when it was built in the 1880s), while Banks's books are now held in the British Library, St Pancras (having formerly also been accommodated at the British Museum, but latterly moved to the new building, opened in 1998).

Botany was, of course, Banks's particular interest and that of his travelling companion the Swedish naturalist Daniel Solander. Plants were also more easily collected, preserved and then transported than specimens from any other branch of natural history. As a result of this, plants formed the largest part of the collections that Banks and his team gathered and illustrated on the *Endeavour* voyage. It is estimated that some 30,400 individual plant specimens were collected by Banks and Solander during the mission, representing more than 3,600 described species. Of the latter total about 1,400 species were then new to Western science. Notwithstanding this botanic focus, more than 1,000 species of animals were collected, mainly comprising birds, fishes, arthropods and molluscs, but only five mammals were among them. With the exception of some arthropods and molluscs, and a small number of fish in London and Paris, few animal specimens now survive from the voyage. This is in no small part due to the difficulty in preserving such specimens, not least using eighteenth-century methods, as well as to the fact that after the voyage many were distributed by Banks to various collectors and details of their whereabouts eventually lost. Some were even destroyed in World War II by German bombing, the skull of a kangaroo thought to derive from the *Endeavour* voyage becoming a notable zoological casualty of that struggle when the museum of the Royal College of Surgeons was struck in May 1941 (see numbers 116–21).

Otherwise, under Banks's supervision the plant specimens were carefully maintained and increased by a series of outstanding librarian-curators, in succession Daniel Solander, Jonas Dryander and Robert Brown. Similarly, the graphic record of the voyage – plants, animals, figures and landscapes – as commissioned by Banks and undertaken by Sydney Parkinson, Herman Diedrich Spöring and Alexander Buchan, were safely housed by Banks first at New Burlington Street and afterwards at Soho Square, his London base from 1777 until his death in 1820. For Banks, art and collecting were always inextricably linked, and his strong tendency was always to favour observational accuracy in the works of the many artists that he patronized. All of Banks's voyage artists died during the mission, leaving behind many finished and still more draft illustrations. Numerous drawings of new plant species were therefore completed and engraved by others in London for Banks's unpublished *Endeavour Florilegium*, and the resulting illustrations and copper plates were retained by him and may be counted among the surviving voyage collections (see numbers 5 and 122–6).

The most prolific artistic fieldworker on *Endeavour* was certainly Sydney Parkinson, a young Scottish Quaker who originally came to Banks's attention when in the employ of the Hammersmith-based nurseryman James Lee, and who initially worked for Banks painting mainly zoological specimens brought back from his first expedition abroad to Labrador and Newfoundland in 1766. A number of these earlier works were completed at the request of the acquisitive Welsh naturalist Thomas Pennant for his own publications on animals. Parkinson acted as Banks's natural history artist during the *Endeavour* voyage, and his work was supervised by Banks and Solander, who predominantly but not exclusively selected plant specimens for illustration and advised on aspects of their structure and colour. More of Parkinson's illustrations were completed during the Atlantic stage of the mission than in the Pacific, and a wider range of natural history covered, because in the Atlantic the quantity of specimens being gathered was smaller. More mainly Oceanic animals were depicted at this stage, and Parkinson took his first, tentative steps in the field of landscape and figure art for the mission. In the Pacific, however, the number of plants coming aboard at each landfall grew enormously, and so Parkinson increasingly concentrated on these, resorting to sketches annotated with the specimen's colours in the vain expectation that he would later complete his drawings. While the specimens were fresh, their living form and colours were still apparent. This was the time to record essential information. Banks frequently added notes to Parkinson's drawings indicating the collection location of each specimen, and lists and draft descriptions of the plants were also made. Thereafter the plant specimens were pressed and stored.

Parkinson's artistic work in other branches of natural history did not altogether cease as the voyage progressed, but mostly it continued at sea when the influx of plants abated. Consequently, many marine species feature among the 268 animal drawings by him now held at the Natural History Museum, London. Moreover, what appears to have been an agreed policy was pursued with regard to zoological illustration, with those species whose form or colour could not easily be preserved – fish, delicate marine invertebrates, to a lesser extent birds, especially sea birds – being illustrated while more robust specimens – mammals, reptiles, molluscs, insects and other arthropods – were set aside for later attention. Parkinson's work on birds shows growing maturity. His illustration, for example, of the red-tailed tropic bird displays more competence and flare than his earlier, pre-voyage bird drawings for Banks (see number 36). Parkinson's illustrative work on invertebrates also shows his skill and meticulous accuracy, in stark contrast to the minimal effort devoted to mammals as a class. Even his two illustrations of the kangaroo are incomplete sketches owing, probably, to the many plants that had to be drawn on the East Coast of Australia, see numbers 5 and 116.

Parkinson's fellow Scot, Alexander Buchan, remains an obscure figure. It is not clear whether he had much if any training as an artist before departing on the voyage as a landscape and figure draughtsman, and it would seem to be for the simplicity and faithfulness of his work that he was chiefly valued by Banks. An epileptic, Buchan died at Tahiti in April 1769, and so his contribution to the mission's artwork was inevitably a small one. He nevertheless produced some good coastal profiles for Cook and his work at Tierra del Fuego is noteworthy. Here it contrasts with the efforts of Parkinson, himself already venturing into the scenic field, since Buchan's uncompromising renditions of people and places display none of the picturesque influences apparent in Parkinson's Fuegian landscape. Evidently Parkinson had absorbed more of the fashionable trends then current in British landscape painting than his countryman, and these are often apparent in his ensuing landscapes and his views of Pacific people in their natural settings (see numbers 30–2). But, it should be added, they are far less apparent when ethnography was in any significant way the focus of his attention, and are firmly excluded where natural history or coastal profiles were to be recorded. For work in these areas a more documentary approach was adopted by all concerned, in the case of the coastal profiles as a matter of navigational necessity, in that of natural history so as to satisfy the requirements of taxonomy, and in ethnography doubtless to accord with Banks's own desire for a reliable visual record of the peoples encountered and their way of life.

Like all the voyagers, including their employer, Buchan and Parkinson were subject to the preconceptions and indeed limitations of their own society in what they chose to observe and how they then represented it. Yet neither they nor Banks envisaged the sort of classical transformation that would be imposed on the original illustrations of Fuegians during later preparation of the official voyage account as edited by Dr John Hawkesworth. With all three mission artists dead, and Cook and Banks planning another Pacific venture, neither in any case harbouring literary ambitions at this time, work on the official account was left to artists, engravers and an editor in London who lacked direct experience of the Pacific. Moreover, Hawkesworth was given the formidable task of blending the journals of Cook and Banks so that they read as one. He and his artists rendered the images and text that were published according to the neo-classical and philosophical precepts with which they were most familiar, and as a result of this and various other errors of approach and of fact the resulting work was subjected to withering criticism upon publication. In it a plate of the Fuegians portrays them living in pastoral contentment by the sea rather than sheltering from the cold and rain in a hillside clearing as was actually the case. Cook rejected the book when he first saw a copy at the Cape during his return journey in HMS *Resolution* – he would head the official account of the second voyage.

Interestingly, there are a small number of natural history illustrations by Buchan from the Atlantic stage of the voyage, probably completed under Parkinson's instruction. Twenty-one of these have been identified at the Natural History Museum, mainly comprising insects and fish. It appears that early in the *Endeavour* voyage each of the Scots ventured into the others' designated field of expertise. This was just as well, for after Buchan's death Parkinson shouldered the responsibility of depicting landscapes and figures alongside his already enormous commitments in natural history. Adapting his plans and making the most of his team's versatility, Banks therefore called on his clerk, the Swede Herman Diedrich Spöring, to support Parkinson in his artistic duties.

Spöring was born and educated at Åbo (Turku), then in Sweden but now in Finland, where his father was a professor of medicine at the university. In 1755 he moved to Britain, where he worked in London first as a watchmaker and, from 1766, as a clerk to Daniel Solander at the British Museum. He assisted Solander in his work cataloguing the natural history collections at the museum, a favourite haunt of the youthful Banks and almost certainly where Banks first met Solander and, as a result, became aware of Spöring. Spöring's role on *Endeavour* was to fair copy the plant and animal descriptions of the naturalists and to label artwork, but at Tahiti he also started to produce landscape and ethnographic illustrations as well as coastal profiles and some natural history drawings. Spöring's artwork displays the sort of neatness and precision to be expected of a former watchmaker and a diligent amanuensis. Nine outstanding zoological drawings by him survive at the Natural History Museum, including two of rays caught at Sting Ray Bay, a name overturned by Cook in favour of Botany Bay, a change itself reflecting the emphasis given to that discipline on this mission. The life-like accuracy of this small group of drawings is what has most impressed modern scholars, and they are held to have been unsurpassed in that regard during the voyage (see numbers 111–12).

Parkinson and Spöring supplied most but not all of the ethnographic and scenic drawings made during the *Endeavour* voyage, which total over 200 surviving sketches and finished illustrations, some three-quarters of them attributable to Parkinson. At Tahiti and the Leeward Islands, Spöring drew views of and from Fort Venus. His technical eye was also turned on religious architecture at Tahiti, where he drew a *marae* and then one of the objects to be found inside such structures, a *fare atua*, or deity house, the latter being drawn to scale (another *fare atua* at Huahine was also depicted to scale; see BL Add. MS 23921, ff. 27(c), 29(a) and 27(a) respectively). The *pahi* boathouse at BL Add. MS 23921, f. 10(a), might in places almost have been drawn with a ruler, as also parts of Oborea's famous canoe (see number 68). Their stilted manner, however, only serves to reinforce a sense of visual accuracy in Spöring's art. His creative faculties, such as they were, were evidently not stimulated by sights like these, although his eye for detail certainly was.

At New Zealand Spöring's output increased to include more landscapes, artefacts and a picturesque arched rock at Tolaga Bay that drew raptures from Banks. Spöring might even have taken Banks's journal hint that the Tolaga arch could be used to frame a distant view of a boat, for *Endeavour*'s pinnace is visible through it in Spöring's illustration of this feature. His grotto-like composition was duly selected for inclusion in the Hawkesworth edition, although there it was further embellished to suit artistic taste. Similarly, Spöring's depiction of the defensive *pā* situated on a coastal arch at Mercury Bay was altered for public consumption when it appeared in the official account, being modified with various details and increased in scale to create a grandeur that Spöring's original hardly possessed. Later, after withdrawing from the *Resolution* mission and sailing instead to Iceland, Banks would again write rhapsodically about geological formations, this time those found at Staffa in the Hebrides as he made his way northwards. The Hebridean islands that he visited on this trip might just as well have been located on the far side of the world for all that was known of them in the late eighteenth century, and judging from Banks's journal the simple living and honest hospitality of their inhabitants might also stand comparison with that of the 'noble savages' previously encountered in the Pacific (Johnson was far less kind about the Scots during his tour of the country). On Staffa, though, Banks overcame his initial amazement to produce a detailed description of the island's impressive basalt structures, along with measurements and carefully executed illustrations by his accompanying artist John Frederick Miller. This was more typically Banks's mode of observation and writing, and was more in keeping with his vision for an analytical approach to the wonders of nature. Thomas Pennant subsequently included Banks's description and illustrations in his published tour of Scotland (he did not visit Staffa), and the Gothic novelist and critic Horace Walpole thought them the best thing in the whole book. Thus through Banks's explorations were remote islands on either side of the Earth revealed to eighteenth-century perception.[2]

Unpublished at the time, though admirable, are Spöring's intricately rendered Tolaga Bay canoe-head illustrations complete with size measurements. Tolaga was a great centre for Māori carving, and the voyagers were suitably impressed by the size and decoration of the canoes encountered there (see numbers 88–9, included in the present work alongside other illustrated Māori carvings depicted for Banks by J.F. Miller after the voyage). Had Spöring survived, he clearly could have undertaken ethnographic drawings of the sort given by Banks to Miller. Spöring like Parkinson was doomed, however, to die of fevers contracted in the Dutch East Indies.

Parkinson, too, is impressive by the New Zealand stage of the voyage, having gained in confidence with his figure work and finding much in Māori behaviour, dress and possessions to occupy his pen and brush. He continued to produce a range of work, as he had in Tahiti, including draft sketches for finished illustrations, full landscapes, landscapes with figures and objects, studies of individuals and their skin decoration and personal ornaments and, in particular, striking depictions of Māori canoes and their occupants. Some of his illustrations of Māori life suggest, as do similar illustrations in the Society Islands, works that bring together scenic and human elements in one carefully observed set-piece composition (see, for example, numbers 59, 70–1 and 96–8). Such drawings may not depict events in every respect as actually witnessed, but they capture in convincing detail items of dress, objects such as weapons and tools, attitudes and facial expressions, as well as the elaborate carving of Māori canoes. Parkinson might prepare for these finished pieces with sketches of certain details in them, and base the resulting illustration broadly on what he had seen. One benefit of such an approach was to allow him to record a large amount of information in a single combined image rather than by doing so in separate studies, an important consideration given the weight of botanical illustration that he also carried. In this way he conveyed a vivid sense of the new places and peoples that were encountered. Some of these scenes with figures were selected for inclusion in the official voyage account, although for that they were altered in various ways by artists and engravers in London who were quite unfamiliar with the Pacific world revealed in Parkinson's illustrations.

Parkinson's wash-and-ink landscapes are more picturesque and imaginative than any other artwork that he, or indeed his colleagues, produced on this mission. Where figures or man-made objects are lacking, his landscapes are more obviously pleasing compositions, including elements almost certainly not present in the scenes themselves. As backdrops to people and their possessions they also convey this impression. The ethnographic information and plants assembled in the foreground of such illustrations appear to be fairly typical of that witnessed by the voyagers. They bring before the viewer aspects of island life in greater detail for closer scrutiny, and were probably based on things that Parkinson saw or collected. The background is exotic and attractive but may be generalized. Parkinson's interest in tropical weather, light and atmosphere as seen in a number of his offshore landscapes, not least those featuring habitations and canoes of various kinds, has been noted by scholars, but, it should be added, he did not apply colour to any of these illustrations. That was almost exclusively reserved for his botanical art (see number 70). It is almost as if landscapes were for Parkinson a creative outlet from the observational rigours of his strictly scientific, ethnographic and coastal illustrations. This is not to say, however, that his views of such places are unreliable or even fanciful. Parkinson's landscapes and figure illustrations are a rich and informative record of the people and places visited during the voyage.

A diligent and likeable character, Parkinson was able to win the confidence of local people sufficiently to record their dress, posture, various ornaments, skin markings and other characteristic features. He produced a number of studies of tattoos at Tahiti and of Māori *moko* in New Zealand. His brother, Stanfield, commented in the posthumously published edition of Sydney's surviving voyage papers (a fair copy voyage account by Sydney was lost) that 'Sydney Parkinson had made, at his leisure hours, a great many drawings of the people at Otaheite and the neighbouring islands, as also of the New-Zealanders, particularly of some who were curiously marked in the face; and that he frequently sat up all night, drawing for himself or writing his journal'.³ Indelible skin markings were in no sense new to Europeans, but Parkinson's drawings of them in the Pacific provide some of the most iconic human images of the whole voyage (see numbers 78 and 84). He also drew the facial expressions of different peoples, such as those of the dancers seen in the Society Islands and those observed in Māori *haka* performances. He did not produce many personal portraits as such, but instead tried to show characteristic differences between island peoples. His work in these areas was more accomplished than that of Spöring or Buchan, but is nevertheless regarded as technically limited in its handling of the human form and somewhat stereotypical due to the generalized nature of what he was trying to represent.

Uniquely important is the small collection of artwork by Tupaia, a Ra'iatean priest and navigator who sailed on *Endeavour* when it departed the Society Islands. Formerly thought to be by an anonymous 'Artist of the Chief Mourner', believed by some to be Banks himself, Tupaia's illustrations were preserved by Banks and have descended to the British Library, BL Add. MS 15508. A remarkable man, Tupaia was the first Polynesian to draw on paper, to sail on a Western vessel for England (although he died on the way) and to produce Western-style charts, displaying his considerable geographical knowledge. His illustrations are especially significant for recording an indigenous view of Tahitian culture as well as those locations visited during the ensuing voyage across and then out of the Pacific, namely New Zealand and Australia (see numbers 50, 58, 67, 69, 77 and 109).

Banks's failure to publish the new plant species gathered on the voyage has been much discussed by scholars. It appears that work on a multi-volume illustrated botanical work was firmly under way from 1773, and that this effort lasted until at least 1784, by which time five London artists and eighteen engravers had been employed by Banks to work up Parkinson's drawings and sketches from the voyage. Not every detail of the original sketch was exactly copied through to the final copper plate, but a close similarity was maintained and the essential botanical facts preserved. Banks and Daniel Solander selected the plants for inclusion and they oversaw the work, with Solander drafting the many systematic descriptions. Banks said that both his and Solander's names would have appeared together on the title page had the work ever emerged, and he expended a vast amount of money in the abortive attempt to see that it did. Disbound in the 1980s for better storage, the former contents of eighteen volumes held in the Natural History Museum, London, attest to the industry of those involved in preparing the botanical illustrations from Parkinson onwards. The contents of a further three disbound volumes containing the zoological illustrations from the voyage also reside at the museum. It would seem, however, that in the end Banks's deteriorating finances during the period of the American Revolution, his commitments to other bodies such as the Royal Society, the British Museum and the Royal Gardens at Kew, and the premature death in 1782 of Solander all combined to thwart final publication.

Despite this, Banks's experiences during and after the *Endeavour* voyage provided a pattern and experience for ensuing publications that he oversaw from that of Cook's third-voyage account through to the organization and preparation of the official account of the voyage of HMS *Investigator*. In fact, Banks was a key figure as an instigator, contributor and background editor of numerous travel and botanical publications in this period, and through his intervention the Admiralty, and especially its long-serving secretary John Barrow, achieved much greater proficiency in producing such works than initially it had for the bungled *Endeavour* voyage account. Eventually, in the 1980s, Alecto Historical Editions printed one hundred sets of the 738 extant engraved copper plates of Parkinson's illustrations, as prepared by Banks for his planned *Florilegium*. The plates are held at the Natural History Museum, London, along with Parkinson's botanical sketches and completed drawings as well as the zoological drawings and paintings by Parkinson, Buchan and Spöring. The landscapes and figure drawings now reside at the British Library, St Pancras, where Banks's library is also held.

Banks's herbarium and that of Sir Hans Sloane form the basis of the General Herbarium of the Natural History Museum. The Banks herbarium includes a considerable number of type specimens, the original specimens on which species descriptions and names are based, making them not only historically interesting but also scientifically important to taxonomists. Some of the Banks-Solander specimens may be closely matched to voyage or *Florilegium* illustrations, and they are therefore indisputable voucher specimens for these illustrations. The vast majority of specimens, however, coincide with but are only partially represented in the drawings, and the remainder have no obvious visual connection with the artwork, although their collection details suggest that they are the obvious voucher specimens in terms of existing drawings and manuscripts. During Banks's lifetime some duplicate specimens were distributed to fellow naturalists, and this practice continued well into the modern period, only ceasing at the Natural History Museum in the 1930s. Indeed, the museum distributed a large number of Banks duplicates to at least fourteen other major institutions, in Wellington, Auckland, Sydney, Edinburgh, Berlin, Halle, Copenhagen, Stockholm, Paris, Vienna, Washington, St Louis, New York and Calcutta. A number of *Endeavour* specimens fell victim to fire when the General Herbarium was damaged by bombing on 10 September 1940 during the first blitz on London, but duplicate material still in the museum was available to make good the losses.

Probably the first Cook-voyage zoological specimen to be incorporated in the British Museum collections was a New Holland parrot from the physician and naturalist Dr William Watson in early April 1772. How Watson obtained the parrot is not clear, but since he sponsored Banks's election to the Royal Society in 1766 and was an active museum trustee it seems likely that the bird came from the collections then housed at Banks's London residence in New Burlington Street. The history of the zoological collections from the *Endeavour* voyage is complex, owing to their wide and largely undocumented dispersal, but it shows that Banks usually directed such material to established collectors in relevant fields or to fellow naturalists and institutions with a particular interest in it. High among these was the British Museum, but zoological specimens from the *Endeavour* voyage circulated far more widely than that, and as a recipient of both natural and artificial material from Cook's next two voyages Banks also had to decide what of this he wanted to keep and where to assign the rest.

Banks estimated that he and his party returned from the Pacific with some 500 birds and 500 fishes as well as many new invertebrates. The former estimate is perhaps too high, and the number of identifiable *Endeavour* bird specimens that now survive is limited. The latter estimate is probably a fairer reflection of the fish that came back, for which a number of specimens are still to be found. Latham, Pennant and Kuhl were among the naturalists that studied Banks's birds, and Banks presented specimens to Marmaduke Tunstall, whose collections in due course passed to the Great North Museum: Hancock, although a rainbow lorikeet given by Banks to Tunstall cannot now be located there, see number 127. It was in January 1773 that the British Museum received some New Zealand birds from Banks, a forerunner of many subsequent donations not only of zoology but also of Cook-voyage artefacts that he would consign to Bloomsbury. The later boast, however, of the museum proprietor and showman William Bullock that he obtained most of the bird collections from the *Endeavour* voyage via the Royal College of Surgeons is almost certainly an exaggeration. The college did indeed possess an important collection of animals in spirit given by Banks to the surgeon John Hunter in 1792 (see below), but this contained more than just birds and had accrued from various Cook voyages, and perhaps from other sources too, so the number of first-voyage birds that Bullock could have obtained in this way was surely limited. If the college's curator William Clift is to be believed, some birds that were passed to Bullock by the college actually fell apart when he removed them from their jars for mounting. Any first-voyage birds that Bullock did receive by this route in good enough condition to survive would have been sold with the rest of his collection in 1819.

The Natural History Museum, London, holds about 50 fish specimens in spirit and a few additional dry ones from the Cook voyages, of which about 25 derive from the *Endeavour* voyage. These are dispersed through the museum's spirit collection. A valuable though slightly smaller collection of 44 fish is at the Muséum national d'Histoire Naturelle, having been given by Banks to the French ichthyologist Pierre Marie Auguste Broussonet, probably in 1780–2 and 1786 when Broussonet was in England pursuing his research. Formerly held in the Faculty of Medicine at Montpellier, Broussonet's home town, these were later transferred to Paris by Cuvier. As with many Banks gifts of this sort, the material concerned may contain specimens from a number of sources, including Cook's final two voyages, but it would appear that as many as ten of the fishes in Paris could be *Endeavour* specimens.

Delicate marine invertebrates such as jellyfish have, by contrast, not survived. Parkinson drew a number in loving detail during the *Endeavour* voyage and descriptions were made too. Some preserved in alcohol might afterwards have been passed by Banks to the British Museum and to John Hunter, but they have since been lost. Time and poor curation tend to claim their heaviest toll among such specimens. The Quaker naturalist John Fothergill obtained shells from the first voyage through his connection with Sydney Parkinson and Banks, and these as well as Fothergill's corals and insects were purchased in 1780 by the anatomist William Hunter, brother of John. Thus a number of first-voyage shells and almost certainly other Cook-voyage natural history specimens are to be found today among William Hunter's collections in the University of Glasgow. There too are Pacific artefacts of very early date, some of which could certainly derive from the Cook voyages, perhaps even that of *Endeavour*. Corals were not illustrated during the *Endeavour* voyage, doubtless because their forms are permanent. None of the salps described or figured during this voyage is known to have survived.

Shells were extremely popular in this period and a number of Cook-voyage examples were passed by Banks to eminent collectors such as the Duchess of Portland. The duchess probably also obtained various shells from Solander, who arranged and listed her collection between January 1778 and June 1779. It is known, too, that she obtained them through the dealer George Humphrey, who simply went aboard the second-voyage vessels when they returned and purchased what he could from the crew, including the bulk of the shells. He later did a similar thing when the third Cook mission returned to England. Apparently the Duchess gave him £15 for some of the shells that he obtained in this way. Johan Alströmer, an old student acquaintance from Solander's days at the University of Uppsala, obtained valuable first-voyage shells from both Banks and Solander as well as many other shells from the duchess during a visit that he made to England from 1777 to 1778. These he returned to the family collections in Sweden along with equally precious insects and artefacts from Banks. After William Hunter's death in 1783 his nephew Mathew Baillie had the use of his uncle's collections and, following the death of the duchess in 1785, Baillie acquired shells at the Portland sale that are now held in the Hunterian collections, Glasgow. John Hunter made purchases too as did Humphrey, perhaps buying back some of the shells he had previously sold the duchess (he had been forced to sell his own collection in 1779 to pay his creditors, but afterwards started to build it up again, including by re-acquiring items in this way). Shells from the sale also went to collections on the Continent, but relatively few of these can now be traced. Tunstall certainly received shells from Banks since he sent around 200 first-voyage specimens to Linneaus in Sweden, some of which probably still reside in the Linnean Society of London as a result of its founder James Edward Smith having acquired the Linnean collections in 1784.

Perhaps most significant of all in this respect, Banks shells may be found in the Natural History Museum among a collection numbering some 1,120 specimens, representing 392 species.[4] These were part of the shell and arthropod collections that he gave to the Linnean Society in 1805, which were passed to the British Museum in 1863 when the society could no longer maintain collections other than those of Linnaeus on which it was based. On arrival at the museum the shell collection was still housed in Banks's original cabinet along with many contemporary labels. These collections were transferred to the Natural History Museum when that was later established. Decorated with white-on-green neo-classical plaster mouldings in the Adam style, Banks's splendid cabinet was loaned to the Victoria and Albert Museum in 1925 and after a period in more than one museum in Cambridge it eventually returned home to the Natural History Museum. It was clearly not only an elegant but also a convenient store place for such specimens until Banks at last felt they could safely be passed to a body likely to preserve them. Banks had been a founder member of the Linnean Society back in 1788 and he evidently made this choice because it possessed its own museum and pursued a broad range of natural history. Perhaps, too, the standard of accommodation and curation available at the overcrowded British Museum made that a less inviting option. Banks's surviving shells definitely include material from a variety of sources, among them Solander and each of Cook's Pacific voyages (see numbers 26, 35 and 115).

The cabinet also contained Banks's insects and crustaceans. The Danish entomologist Johann Christian Fabricius, a frequent visitor to London and its collections, described a number of insect type species using Banks's material. He researched, too, other collections in London and elsewhere that included insects from the Cook voyages, and about a third of the 1,500 or so new species described in his *Systema Entomologiae* of 1775 were based on Banksian specimens. These would have included first-voyage material and that from various other sources. Fabricius's ensuing works also drew on Banks's insect collections, which are rich with type specimens, although all too frequently these are now hard to locate owing to the loss over time of labels and identifying data. The condition of some of the specimens has also deteriorated due in no small part to poor curation during the nineteenth century, and it would appear that there are large gaps in certain genera, possibly as a result of sales by the Linnean Society when it gave up this material in 1863. Estimates of the size of the Banks collection of insects at the Natural History Museum vary, but the total may amount to nearly 4,700 specimens judging from old accession registers.[5] Banks's surviving crustaceans at the museum are all dry pinned specimens, and these too were worked on by Fabricius, although they like the insects are often imperfectly labelled and as a result it is now hard to identify Fabrician types among them. The Banks shells, crustaceans and insects are today housed separately, the former two being in the Department of Zoology and the latter in the Department of Entomology. William Hunter also obtained Banks insects, some possibly with Fabricius's assistance, which are now in Glasgow.

Among the few mammals brought back or drawn during the *Endeavour* mission, the most notable was certainly the kangaroo, no specimens of which survive from the voyage. This paucity is probably explained by the fact that the places visited were not especially rich in mammals, and that when ashore the naturalists devoted most of their attention to the enormous number of new plants that were encountered. At sea, where more time might be available for other things, marine biology tended to predominate in zoological observation and collecting. Modern estimates suggest that Banks returned from the Pacific with 5 mammals; 107+ birds; 248+ fishes; 370+ arthropods; 206+ molluscs; 6 echinoderms; 9 salps; 30 medusae and some other animals.[6]

Following the return of the first *Resolution* voyage in August 1775, Cook's collections as well as duplicate insects and other material belonging to mission naturalist Johann Reinhold Forster were sent to the British Museum to be distributed by Daniel Solander. Those things intended by Cook and Forster for Banks, then out of town, would have been added to his existing collections or divided up probably much as above. According to Solander's letters, Cook sent to the museum for Banks four casks of birds, fish and other items. Cook also forwarded a box of Cape plants for Banks, but his shells were all for Lord Bristol. Solander informs us in September that Forster wanted one of each insect species to be allotted to Banks and the museum, with Banks sharing in the rest of the Forster material in order of priority thus: first, the British Museum; second, the Royal Society; third, Banks; fourth, Marmaduke Tunstall; fifth, Ashton Lever. It was a pretty much typical series of Cook-voyage recipients. The museum's Book of Presents for this month indicates that in addition to some more insects Forster also donated fish, birds and mammals. In a separate note Banks recorded that he obtained many natural history specimens from the Forsters, although in September 1778 he refused a collection of shells offered by Forster senior. Significantly, Banks purchased the second-voyage natural history illustrations by Johann Georg Adam Forster in four volumes, two of plants and two of animals. Lieutenant Charles Clerke intended for Banks some bird drawings by a midshipman, and surgeon's mate William Anderson had apparently assembled a good botanical collection. Both these men would sail on the next Cook mission never to return and both would bequeath their collections from it to Banks. Solander further noted that various officers and crew offered 'curiosities' to Banks from the voyage. Tobias Furneaux, commander of *Adventure*, sister ship to Cook's *Resolution*, may well have been among them.

Following the return of the ill-fated final Cook voyage, Banks was again a major recipient of the collections that were made. There was no designated naturalist on this mission. Instead Banks's prime sources of natural history and also of artefacts were his paid collector, David Nelson, the by then surgeon William Anderson and Commander Charles Clerke. The latter two both succumbed to tuberculosis while at sea. In a moving deathbed letter Clerke commended surgeon's second mate William Ellis to Banks, whose mainly bird drawings Banks subsequently bought. Banks therefore acquired a great mass of the natural history illustrations made on all three Cook voyages, undoubtedly the largest concentration of such material to be possessed by any single individual connected with these missions. This he would bequeath to the British Museum, and the vast majority of it is now to be found in the Natural History Museum, London. He also obtained many natural and man-made specimens from each of the voyages.

The fate of Banks's animals in spirit was largely decided in 1792 when he split them between the anatomist John Hunter and the British Museum. The year before the rear premises of Banks's Soho Square residence fronting Dean Street had been extensively altered the better to accommodate his library and herbarium, these by then being his main focus in collecting, and so it seems likely that as a result of this he chose to shed his animal collection, containing specimens from more than one Cook voyage as well as from other sources. The 344 items that Banks sent to Hunter were numbered and labelled 'J.B.' by Hunter's assistant William Clift. It may well be that some of these specimens were afterwards dissected or boiled and then stripped of their flesh to show their anatomy. Those that remained intact were purchased by the nation along with Hunter's other collections after his death in 1793 and used to found the museum of the Royal College of Surgeons. There they were included in a catalogue of natural history specimens compiled in 1806 by George Shaw, who designated the Banks material somewhat anomalously as the 'New Holland Division' (a copy of this catalogue was made by Clift six years later, and both versions are in the college library). Banks probably also made later donations of Pacific material to the college. As a British Museum trustee he was certainly involved in the controversial sale to the college in 1809 of a large number of unwanted oesteological and natural history specimens held in the museum's basement, which may well have included items from his earlier 1792 donation. This was a loss to the museum, but also a cause of complaint in the college, since a lot of the material was in very poor condition on arrival there (doubtless the reason that the trustees thought it suitable for purposes like dissection). Much later, in 1845, the college donated some 348 wet specimens to the British Museum and a small number from the 'New Holland Division' were probably among them. It is quite possible too that some of the 1809 material returned to the museum on this occasion, thus having once been sold by it only to be returned to its collections years afterwards. Few, however, of the specimens from Banks's 1792 clear-out can now be identified, most being lost, decayed over time or even destroyed by bombing in World War II.

Typically man-made objects were obtained by gift or exchange during the *Endeavour* voyage. There was brisk trade with islanders for water, food, 'natural and artificial curiosities' and sex, in return for which metallic objects such as iron nails were eagerly sought by the Tahitians since in Polynesia the making of metal was unknown. On all his Pacific voyages Cook struggled to regulate the exchange of the latter two commodities, which led to the spread of disease on the one hand and on the other caused the stripping by sailor and Polynesian alike of metal from his vessels. Today the surviving artefacts from the 2,000 or more that were gathered during Cook's three Pacific voyages, and the illustrations and written records that were also made during them, add significantly to what we know of Polynesian culture prior to and at the time of the first major European contacts. This is especially important since the customs and traditional crafts of the Pacific were permanently altered or altogether lost as a result of European penetration. Surviving collections demonstrate, too, the workmanship and brilliance of Polynesian culture as Europeans encountered it in this period, and have latterly attained high status among present-day members of that culture as forming an important part of their heritage. This has raised questions of ownership but it has also encouraged the exploration of new and creative ways of engaging with such objects through the direct involvement of contemporary Polynesian thinkers and artists in their study and exhibition.

But whereas Carl Linnaeus had supplied a system for arranging natural products (see numbers 15–16), at this time there was no equivalent method for artefacts. Indeed, ethnography as a scientific discipline did not exist, and what was observed of or gathered from indigenous societies depended on the various choices and opportunities available to navigators, naturalists or missionaries, with Europe's philosophers, historians and writers afterwards applying their own interpretations to voyage results (see the Admiralty instructions and Lord Morton's 'Hints' to the voyagers, numbers 6 and 10). If objects were formally arranged at all on arrival in Europe, it was usually according to their type or place of origin. Thus various artefacts were exhibited geographically in the British Museum's South Sea Room to show the customs, religions, trades and products of the different societies from which they came. Such objects together with the detailed accounts of indigenous cultures that were brought back caused a storm of interest in Europe, where people found it no less difficult to fit reports of the newly discovered cultures into their existing world views than did the peoples of the Pacific their first experience of the strangely dressed travellers in their extraordinary vessels. Patrons, collectors and prize hunters all sought objects for their cabinets or for sale, not infrequently for both. Such artefacts were often delicate, hard to preserve or bulky in size, making their transport, storage and display a perennial problem. Many would subsequently deteriorate beyond recall, be sold, exchanged or given away by the individual collectors and institutions that held them, commonly without documentation, and their identity or existence lost. As a result, locating and verifying those that survive today is notoriously difficult, but in a number of cases not without hope. Journals, letters, illustrations, museum labels, contemporary lists or guides and sale catalogues can all assist in the attempt, and, fortuitously, some objects fell to individuals or institutions that safeguarded them and information about their past.

Certainly, the most important collections brought back on the *Endeavour* in July 1771 were those gathered by Banks and his party and those belonging to the commander. Banks's were housed in his New Burlington Street residence where visitors could see them. In November 1772, William Sheffield, keeper of the Ashmolean Museum, Oxford, described Banks's home as 'a perfect museum; every room contains an inestimable treasure', adding that 'I passed almost a whole day here in the utmost astonishment, could scarce credit my senses'.[7] What Banks had established in the capital was, in effect, an early South Sea museum, anticipating the opening of similar displays at the British Museum in 1775. The British Museum's South Sea Room contained collections from the Admiralty that were gathered during the Pacific voyages of Samuel Wallis and George Carteret and, especially, those from the first and second voyages of Cook. Admiralty papers include a summary list of *Endeavour* objects that in August 1771 were passed by Cook to his patron the first lord, Lord Sandwich. Sandwich divided these objects between the British Museum and Trinity College, Cambridge, where he had studied as an undergraduate (keeping a few items for himself).[8] Both donations were made in October 1771, and that given to the British Museum set an important precedent, for henceforward the museum would be the recipient of various collections from missions launched under the Admiralty. The Cambridge portion now forms the largest concentration of known first-voyage material for which contemporary lists survive. Cook passed much of his collections from this and his ensuing voyages to Sandwich (who may have given more material to Trinity in 1775), the king, the British Museum and to that great museum showman, Sir Ashton Lever, although there were other recipients and a considerable amount naturally went to Cook's immediate family.

Lord Sandwich's donation of Cook material to Trinity College, Cambridge, provided an example that Banks apparently followed. At an unknown date before mid-January 1773, he donated an important collection of *Endeavour* artefacts probably directly to his old college Christ Church, Oxford, although the gift may have been to a college fellow and friend, the anatomist John Parsons. Banks received an honorary degree from the university in November 1771, which might also have encouraged him to make his donation, and we know that his collection was present in the university by 16 January 1773 since on that date his friend Thomas Falconer wrote to him mentioning sight of it there. Banks's collection remained in the care of the college until it passed in part to the Ashmolean Museum in 1860, and was deposited in its entirety at the newly opened Pitt Rivers Museum in 1886. Similarly, in 1776 the second-voyage naturalist Johann Reinhold Forster donated to the University of Oxford a carefully listed collection of man-made objects that he and his son had made during the mission. He too had received an honorary degree from the university shortly beforehand. The collection included examples of nearly every ethnographic object collected by the Forsters during their voyage, but the conscientious organization and coverage applied to this gift was in stark contrast to the unsystematic way in which they subseqently disposed of their remaining artefacts for money or to secure favour and influence. The Forster donation was held at the Ashmolean Museum until 1886, when it was also transferred to the Pitt Rivers Museum as part of a move to situate the Ashmolean's ethnographic material in a dedicated university institution.

Banks continued to shed material mainly in fields that he did not wish to pursue as a private collector and, as explained, he favoured the British Museum with various gifts from Cook's voyages. In late October 1778 he sent to the museum a 'collection of artificial curiosities from the South Sea Islands'.[9] This gift probably included not only *Endeavour* voyage objects but also second-voyage material acquired from the likes of Cook and the Forsters. Royal Society fellows were well represented among museum trustees and staff, and the donation was astutely made just prior to Banks's almost unanimous election to the society's presidency that November. This in its turn brought a place at the museum as an *ex officio* trustee, a duty he evidently took seriously judging from his regular attendance at board and committee meetings over the years. So circumstanced, Banks was thereafter able to use his position as a senior figure within London's learned institutions and as an unofficial advisor to government on voyages of discovery to channel large quantities of material to (and in some notable cases away from) the museum, much of it coming from future voyages and settlements in the Indo-Pacific region. In this way the early Pacific collections of the museum were greatly enhanced, but, popular though it was with the public when exhibited, such material was not always highly regarded by staff who lacked a system for organizing it and found its storage and preservation difficult. Lodged in the Department of Natural and Artificial Curiosities because they had mostly been gathered in the field by naturalists or navigators, and lacking a dedicated department of their own until as late as 1946, the ethnographic collections were subject to sales, or worse to destructive clearances of the overcrowded basement area where many were kept when not on display.

The 1777 move to his London base at 32 Soho Square also implied new working arrangements, for this was where Banks concentrated his herbarium and library collections, and where until his death he welcomed many scholars wishing to consult them. Banks obtained much of the natural history material from Cook's final voyage, and he doubtless also acquired various artefacts since so many were gathered during that mission. Most of this haul came from David Nelson, William Anderson and Charles Clerke, although gifts from other individuals were forthcoming. The majority of Cook's material went to Lever following two requests by the antiquary Daines Barrington to Lord Sandwich in letters of 16 June and 3 October 1780. Barrington also indicated that Clerke's collections should be directed to Lever,

this despite the dying Clerke's written wish that they go to Banks, one still contained in Banks's surviving correspondence.[10] Banks, however, did not want to retain such artefacts and on 10 November 1780, just over a month after the return of the mission, he donated to the British Museum 'A collection of artificial curiosities from the South Sea Islands, the West Coast of North America and Kamchatka; lately brought home in His Majesty's ships Resolution and Discovery'. His example was rapidly followed, for on the same day various officers and men from this mission also donated natural and artificial collections to the museum, to be followed towards the end of the month by further contributions from yet more of the voyagers.[11]

These gifts included many objects from the Northwest Coast of America, some of which survive at the museum and now form part of an important early series relating to that region as catalogued by James Edge-Partington towards the end of the nineteenth century, namely NWC.1–117, and eighteen additional items, Q78.Am5–21 and Q79.Am1. Often rather confusingly referred to as the Cook-Banks collection, this series is a mixed one which also includes objects that definitely post-date Cook's voyages, in some cases by many years, as well as others that have come from places he never visited or that are simply of unclear provenance. A detailed modern catalogue of this series has been published based on considerable scholarly detective work.[12] In the series a number of Northwest Coast objects, probably from Cook's third voyage and hence donated in 1780, have been located using two sorts of contemporary label still found or recorded as having been present on the objects concerned. One set of these labels indicating the name and area of origin of the object is thought to be by Solander, who worked at the museum from 1763 and in 1773 became under-librarian responsible for the natural and ethnographic collections, a post he held until his untimely death in 1782. Solander's labels may also be found on some objects from the central Pacific, not included in the NWC series, but almost certainly coming from one of the Cook voyages if not definitely the last. The second set of labels, albeit giving item numbers only, may possibly derive from Banks.

The NWC series also contains objects acquired during later voyages to the Northwest Coast that Banks obtained and gave to the museum. In 1789 the fur trader George Dixon provided material that Banks conveyed to the museum. In 1790 the Scottish surgeon and botanist Archibald Menzies brought back further material from a fur trading voyage under James Colnett that Banks also passed to the museum. Menzies supplied yet another consignment in 1796 after an important surveying voyage under George Vancouver. Registration at the museum was still primitive at this time and individual items were rarely recorded, something that considerably increases the difficulty inherent in locating surviving objects today. So the itemized list of Dixon material that was entered in the minutes is unusual, and stands in these years as an isolated but not unique attempt to record a voyage donation more fully. An illustrated account of Vancouver's voyage was also published, and using these sources a small number of Northwest Coast objects from Dixon have been identified in the museum. No such list exists for the material gathered by Menzies during his voyage under Colnett, but a small number of object labels with signed descriptions by Menzies have survived, and a comparison of these with entries in the Book of Presents as well as checks among the collections themselves have enabled scholars to locate a few objects thought to be from this voyage. It was Banks who distributed the collections made by Menzies after he returned from a sometimes troubled mission under Vancouver. Banks rapidly sent the surviving plants and seeds to Kew Gardens, and he forwarded to the British Museum various artefacts and natural productions, among which were objects gathered on the Northwest Coast of America. Beforehand he had a numbered list compiled of everything, possibly because this collection had, in effect, the status of public property having been obtained by Menzies under orders during an official mission, and perhaps also because Banks by now suspected what might happen to some of it after entering the museum. Several Northwest Coast objects included in the NWC series, and a smaller number of Oceanic artefacts separate from it, are today identifiable from this voyage using Banks's list and matching numbered labels currently or formerly attached to the objects.[13] Objects from the Leverian museum have been identified within the series, some of which may derive from Cook's final voyage, and these appear to have arrived at the museum at dates during and even after the second half of the nineteenth century. Other NWC objects apparently of eighteenth-century origin have at different times come from a variety of collections by gift or purchase. The NWC series is held at the British Museum along with a range of other material relating to Cook, and doubtless there are objects in the collections still to be identified as resulting from the historic voyages of his period.

As president of the Royal Society Banks also oversaw the transfer to the museum in 1781 of the Royal Society's important collections, although these may have contained less Cook-voyage material than is sometimes supposed given that the museum was already viewed as a more fitting destination for such donations. J.R. Forster, it is true, presented the society with material from the second Cook voyage, but this would also have made its way to Bloomsbury in 1781. The decision to relinquish the society's collections was probably prompted by the move in 1780 to new quarters in Somerset House, but it followed too Banks's own previous conduct in directing valuable material of various kinds from a range of sources to the national repository. As he put it to the Dutch botanist Jan Ingenhousz in a letter of May 1782: 'I am Sorry that Mr Jacquin is so angrey that I have not yet fulfilld my Promise of sending him Arms & curiosities from the South Sea the reason I have not yet done it is that in order to give a preference to the British Museum who Engagd to fit up a room for the sole purpose of receiving such things I long ago sent all mine down there consisting of several Cart Loads'.[14] Banks doubted that his donation had been dealt with by staff, and he even thought that some might be returned to him in due course. Whether any did come back is unclear, possibly it did, but what is certain is that by the early 1780s he had divested himself of the bulk of his Pacific artefacts, ensuring that the British Museum received a generous share.

Likewise, as a guide but not necessarily an invariable rule, from at least 1770 onwards collections acquired by the Admiralty on missions launched under the Royal Navy were regarded as belonging to the nation and so might be sent to its main museum. The actions of both Sandwich and Banks had helped to encourage this practice within official circles and among returning navigators. Nevertheless, officers and men from each Cook mission gave away, traded or sold to other recipients many of the objects that they brought back, and there was a lively international market for their 'curiosities'. Dealers like George Humphrey obtained large collections from the Cook voyages. Even Banks himself gave away material of various sorts to friends and associates, among those receiving artefacts from him being Sir William Hamilton, Charles Francis Greville, Thomas Pennant, John Lloyd, William Bullock and the Swedish merchant and collector Johan Alströmer. So distribution was wide but not, in Banks's case at least, altogether aimless for he tended to give material to established collectors or specialists likely to make use of it in their own work.

The collection given by Banks to the Alströmers, a family of Swedish landowners with extensive commercial concerns, offers interesting possibilities since it may well have been handed on by Banks late in 1778 and must therefore have included material from the first and also the second Cook voyages. Given that it was described at the time as nothing less than a set of duplicate artefacts from Banks's collection as a whole this seems all the more probable. Some objects do indeed appear to derive from locations visited on the second mission. Others cannot have come from a Cook voyage since they are from areas not visited by him and were probably acquired by the Alströmer family through their trading activities and contacts. Banks also gave shells, plants and insects to the Alströmers, but these are seemingly not now extant. The Alströmer artefacts were later given by the family to the Royal Swedish Academy of Sciences and thereafter were passed by the academy to the Etnografiska Museet, Stockholm, where they now reside. Stig Rydén published the collection in the 1960s in a catalogue of 67 listed objects (see numbers 54, 81, 82 and 107).[15]

During the late eighteenth and the nineteenth centuries learned private societies, like learned private individuals, increasingly viewed the British Museum as the most fitting repository for collections of national scope and importance. Both the Royal Society and later the Linnean Society followed this pattern when they felt unable to provide sufficient funds or space to maintain their own museums. The British Museum and other public institutions of suitable standing were seen as offering the continuity as well as the accommodation, curation and access that individuals or small bodies could not. It was a trend repeated on a local scale with philosophical societies in Newcastle, Manchester, Liverpool, Leeds, Sheffield and Birmingham, all building up collections, often including various artefacts, much of which in due course passed to city and regional museums. The early gifts made to the universities of Cambridge and Oxford might, partly, be understood in this broad context. The 'ancient universities' were regarded by first Sandwich and then Banks as appropriate repositories for important gifts of Cook-voyage material that subsequently went to specialist university museums created later in the nineteenth century to support teaching and research. These donations have led to the successful preservation of known first-voyage material. Yet it should be stressed that the existence largely intact of concentrations like these is in many ways exceptional. Much that came back on Cook's vessels was dispersed and did not survive. Moreover, where it did occur the gradual transfer of private collections into public bodies of various kinds did not mean that the collections themselves always received better treatment than had formerly been the case. Nor did it mean that they remained whole, as much was incorporated into existing holdings, or later sold, exchanged or again transferred not infrequently with some loss of material and accompanying information.

The gift of *Endeavour* artefacts from James Cook to Lord Sandwich and from him to his old college, Trinity College, Cambridge, was accommodated at the Wren Library alongside other object collections, making the library an important museum until the twentieth century, and a popular public attraction. In 1914 and 1924 the Sandwich collection, as it is here called, was deposited at what is now the university's Museum of Archaeology and Anthropology. Surviving contemporary lists verify the provenance of this important collection and it stands as a rare example of a known *Endeavour* voyage concentration. Similarly, at an unknown date before mid-January 1773, Banks gave a collection of *Endeavour* objects probably directly to his old college, Christ Church, Oxford. The college passed part of the collection to the Ashmolean Museum in 1860, and then everything was deposited at the newly opened Pitt Rivers Museum in 1886. Interestingly, in the case of Banks's gift to Christ Church an attempt seems to have been made to provide a representative sample of the material culture of the different societies encountered during the *Endeavour* voyage, or at least of the sorts of objects that had been gathered from them. Banks's donation, numbering 27 surviving items, furnishes examples of objects connected with activities such as fishing, war, music, working tools, domestic life, items of dress and their manufacture, and in consequence it displays the materials and design used in their production and sometimes even a range of types. The donation by Lord Sandwich to Trinity College also incorporated a range of objects selected from those collected at the main locations visited on *Endeavour*.

The more common alternative to donations of the kind that Sandwich and Banks made was for private collections to be broken up and sold when an individual lost interest in them or died. Parts of these might be purchased by or eventually find their way into public respositories, but the integrity of the original collections was forever lost. Much material disappeared altogether and what remained often continued to change hands privately, some of it coming to rest among later generations of owners who had all but forgotten from whence it originally came or how it was acquired. The historical associations of such material once being rediscovered, however, it is not unusual to find objects of great worth again appearing for sale or circulation. Lever and Bullock, the foremost commercial museum proprietors of Banks's day, both gave up their collections, which finally went under the hammer in 1806 and 1819 respectively, with the British Museum bidding for parts of the latter but, according to some sources, not the former on Banks's advice. Lever had, in fact, already given up his museum in 1786, using a lottery in which it was won by a law-stationer and agent called James Parkinson for a ticket costing a guinea.[16] Only 8,000 of the 36,000 tickets on offer were taken up by the public, and when Parkinson eventually sold the museum in 1806 it raised the low sum of £6,642. 13s. 6d, of which a number of lots were actually bought back by the owner at a total cost of £1,600. Lever struggled to maintain his museum from about the time that he received Cook's third-voyage collections, and an offer to sell it whole to the nation in 1783 was declined following the turmoil of the American Revolution, as also two offers to the empress of Russia in 1786. Banks was apparently consulted in 1806 regarding purchasing the entire museum for incorporation in the British Museum at the price of £20,000 and advised against this, but the allegation that he also blocked auction purchases simply because he disliked Lever is based chiefly on hearsay. Limited space and poor finances were just as likely to have weighed heavily with trustees and ministers when considering the acquisition of such a large collection, one that had already outstripped the resources of both its previous owners, and such concerns would become ever more pressing in the lean years after the end of the Napoleonic Wars. The failure to bring Lever's museum into public ownership nevertheless still draws adverse comment from scholars. Clearly petty differences, obsessions of various kinds, personal rivalries and even outright hostility played their parts in shaping the colourful and dynamic world of collecting that emerged in this period.

Lever's and Bullock's museums differed from Banks's collections in the sense that they incorporated a much broader range of material and were publicly displayed for a fee, something necessitated by the enormous cost of running them. Both were established outside the capital but moved into it, presumably to increase their audience and to draw on the many avenues for acquiring new material that London above all cities offered in the late eighteenth and early nineteenth centuries. Lever's Holophusicon, first situated in London at Leicester House and then moved by Parkinson to the Rotunda at Blackfriars, was in effect succeeded by Bullock's dazzling displays at the Egyptian Hall in Piccadilly. Indeed, Bullock purchased various lots at the 1806 Leverian sale and, like Lever before him, was an active player in the market for curiosities and natural history of all kinds. It would appear that he was on friendly terms with Banks, from whom he may well have received Cook-voyage items, among them a Tahitian mourning dress in 1810 and later a Hawaiian helmet and cloak reputedly given to Cook by the chief Kalani'ōpu'u (Banks himself did not collect these items, now in Wellington; see number 58). Since Bullock's museum was started in the late 1780s most Cook-voyage objects that he obtained must have come from post-Cook voyage sources and he also collected Pacific material from subsequent explorers of the region. This often makes it difficult to determine what came from particular expeditions. The wide and poorly documented dispersal of material from these museums during and after their final break-up notwithstanding, it is still possible to trace specimens from Lever and Bullock using contemporary illustrations, original display labels, sale catalogues and lists of later collections (see number 127). Banks is said to have obtained some material from the 1806 Leverian sale (probably through an agent, the dealer Thomas Atkinson) and Bullock certainly did, in particular a number of valuable bird specimens from Cook's voyages, birds being strongly represented in Lever's collections since his career as a collector in Manchester started with an aviary of nearly 4,000 live specimens. It is known that Bullock purchased four ethnographic lots at the Leverian sale.

Leverian sale material was scattered to places near and far, including the capital (for example Edward Donovan's museum of British natural history); Walworth in Southwark (Richard Cuming, who pursued Leverian objects throughout his life); Romsey in Hampshire (Latham's collection, so mainly birds); Knowsley Hall in Merseyside (Earl of Derby, whose collections now form the basis of Liverpool's World Museum); Widdicombe House in Devon (Cook artefacts now held at the Museum of Archaeology and Anthropology, Cambridge, many from the third voyage); Glasgow (William Hunter's collections, passed in 1807 to the university by his nephew) and further afield to collections in Paris, Berlin and especially Vienna (the imperial cabinet of Franz I of Austria, to which went many Cook-voyage natural history specimens and artefacts, the largest concentration of the latter today held outside Britain being at the Museum für Völkerkunde). Material from the Bullock auction went to, among others, collections at Knowsley, Edinburgh, Paris, Berlin, Leiden and Vienna. The work of scholars in various fields has helped outline the broad distribution of the Lever and Bullock collections, but there is much detailed work still to be done in tracing what survives.[17] Cook material also circulated among many smaller private collectors, and as their collections lapsed into obscurity so too, in many cases, did the treasures that they contained. Unrecognized Cook-voyage objects must exist in various locations around the world today.

The Museum of Archaeology and Anthropology at Cambridge is especially fortunate to hold more than one major Cook-voyage collection of artefacts. Not only does it possess the Sandwich material from Cook, amounting to about 100 known first-voyage objects, but it also holds a valuable collection of objects formerly belonging to the voracious Welsh naturalist and travel writer Thomas Pennant. The Pennant collection comprises objects that were passed to the museum by the 9th Earl of Denbigh and his wife in 1912 and 1913, the earl's father having married Pennant's great granddaughter and thereby obtained this material. It contains a mix of objects evidently from more than one Cook voyage and perhaps also from later voyages. There are items from Tonga, Hawai'i and the Northwest Coast of America, none of which were visited by Cook in *Endeavour*, while Tahiti and New Zealand are also represented, both having been explored on all the voyages. Pennant was granted extensive access to Banks's collections until at least 1783 when the two men fell out, so it is likely that he acquired first-voyage objects from Banks, but material from other Cook voyages could also have come via Banks or indeed contacts such as the Forsters, Cook or Lever. Recent suggestions for possible Pennant collection objects from Banks include a Tahitian noseflute, chisel and barkcloth beater, although the last differs in terms of length and groove numbers from those indicated on surviving *Endeavour*-voyage beater illustrations.[18] Even so, it is still conceivable that it and the other two objects were acquired through Banks, but firm documentary evidence to confirm their provenance has yet to be found. To these two premier collections was added Cook material from the Holdsworth family of Widdicombe House, Devon, obtained in 1921–2. Much of this was originally acquired by the family at the sale of the Leverian museum in 1806, the Cook items in it therefore deriving from the second and third voyages. The Cook-voyage collections held at the Museum of Archaeology and Anthropology come mainly from these three sources, an achievement in no small part due to the contacts and efforts of its first curator, Baron Anatole von Hügel, with help from his wife Eliza, and those of his successor Louis Clarke.

Thomas Pennant was an important figure in Banks's early career and the publication of material from his collections. At that time Pennant was the more senior figure within British travel and natural history circles, and from his first trip abroad to the coasts of Newfoundland and Labrador in 1776 Banks granted the Welshman's numerous requests for information and specimens, while to an extent Pennant reciprocated with useful introductions and information of his own. Their relationship was in this sense a mutually beneficial one, but there was a significant imbalance in it since Pennant's desire to publish from Banks's collections was always far greater than that of Banks himself. For example, during his 1772 trip to Iceland, Banks produced a pioneering survey of the basalt formations on the Hebridean Island of Staffa, and Pennant later included this material in his own published tour of Scotland. Yet in 1783, when Pennant purchased from J.F. Miller some drawings he completed for Banks during the Iceland expedition, there was a falling out. Banks had already severed relations with Miller in November 1776 after he exhibited without permission some plant drawings made for Banks. Now the illicit material acquired by Pennant was incorporated in his forthcoming *Arctic Zoology* (London, 1784–7), and he defended his actions to Banks by asserting a tenuous claim to it on behalf of the public. He further requested some views of the Northwest Coast of America from James Cook's final voyage, which Banks was then helping to prepare for publication. It was a good example of the high-handed manner in which Pennant was accustomed to treat Banks, but he was by this time addressing the president of the Royal Society with a much more powerful network and higher station than his own. Banks sharply rejected any claim to work that was rightly his, and referred Pennant to the Admiralty for Cook's views, offering to defend himself should he be attacked. Despite a hasty apology from Pennant, relations between the two men remained cool thereafter, and Pennant was never again able to plague Banks in the way he once had done (see numbers 38, 47 and 128).

Such an episode illustrates tensions in the natural history network from which Banks no doubt derived useful experience as to how to conduct himself and, as his standing in it increased, how to police the behaviour of others too. Banks developed a strict but generous code for the use of his own collections that, on the whole, does him credit. Access was readily granted to scholars wishing to use his library and herbarium for research purposes with, over time, those collections relating to ethnography or to branches of natural history other than botany being passed to colleagues or institutions that took a greater interest in them. One consequence of such an approach was actually to enhance his standing within learned circles rather than diminish it. This, the quality and organization of his personal collections, as well as his control over collecting and collections more widely all made Banks an especially valuable contact for numerous naturalists and navigators seeking to develop their own research and careers. And although neither Pennant nor Miller emerge untarnished from the contretemps with Banks, Miller must nevertheless be regarded as an important figure in the history of Banks's *Endeavour*-voyage collections and their illustration. He did not sail on the voyage of course, but all of Banks's mission artists having died during it, John Frederick and his brother James were among those afterwards employed by Banks to illustrate the findings. Miller almost certainly came to Banks's notice through his father, Johann Sebastian Müller, a Nuremberg-born artist and engraver who emigrated to England in 1744. Müller's most significant work in botanical illustration was his *Illustratio Systematis Sexualis Linnæi* (London, 1770–7), including 108 hand-coloured plates and a duplicate uncoloured set, which itself provided a useful point of comparison for Banks as he toiled away on his *Florilegium* with two of Müller's sons assisting as artists. Banks's library copy of Müller's work is at the British Library, pressmarks 74/453.i.15 and 74/460.h.18.

J.F. Miller produced most of the *Endeavour* artefacts drawn for Banks after the voyage. Mainly in pen and wash, Miller's illustrations, like his plant illustrations for Banks, are clearly and convincingly delineated. They include studies of fishing hooks and tackle, bailers, adzes, chisels, weapons, deity figures, wood carvings and so on. Some of these are commonplace objects not of high symbolic or religious status, while others are elaborately decorated rarities, sometimes even being of unexplained original meaning and purpose (see number 86). A range of material is featured and sometimes this includes more than one example of an artefact type in the same illustration, presumably for comparison. The importance of Miller's illustrations lies in the fact that he was commissioned to record accurately a number of specific objects from the voyage and so what he shows us are all items definitely brought back to Europe. However, the identification in modern collections of these objects remains problematic, many having been dispersed without documentation to explain their history or altogether lost. That said, comparison with existing objects is by no means fruitless and some interesting parallels are perhaps apparent (see numbers 56, 57, 87, 89, 90, 108, 110 and, worth pondering also, numbers 39, 43, 64, 65, 78, 79, 83, 84, 85, 91, 92 and 93). There are 31 folios attributed to Miller in British Library Additional Manuscripts 15508, 23920 and 23921, showing in part or in full 130 depictions of objects gathered by Banks and others during the voyage. Miller also produced 99 finished drawings of plants from the sketches of Sydney Parkinson for inclusion in Banks's *Florilegium* (one for the Society Islands; 26 for New

Zealand; 61 for Australia; 11 for Java), while his brother James produced 85 drawings for that work (27 for New Zealand; 52 for Australia; 6 for Java). These are held at the Natural History Museum, London. After withdrawing from the second Cook mission to the Pacific in 1772, Banks took the team that he had assembled for that mission to Iceland, including as draughtsmen the two Miller brothers and John Clevely. J.F. Miller's work during this trip is outstanding, not least his illustration of Fingal's Cave at BL Add. MS. 15510, f. 42 (published in 1774 by Pennant in *Tour in Scotland and Voyage to the Hebrides 1772*, vol. I, facing page 263). Somewhat overshadowed by unfortunate lapses of conduct, Miller's work on Banks's collections nevertheless deserves greater recognition.

Over nearly half a century Banks channelled an impressive array of 'natural and artificial curiosities' to the British Museum and to other institutions or individuals with whom he was connected. This material belonged to him from his own travels or, increasingly, it arrived from contacts and collectors in his far-reaching network. In the end he bequeathed his herbarium and library for the lifetime use of his librarian Robert Brown, both to be passed to the British Museum on Brown's death or beforehand if he agreed to do so. Thus, in 1827 these collections were transferred to the museum, where Brown took up the post of underlibrarian with the title Keeper of the Sir Joseph Banks Botanical Collections, numbering by this time some 23,400 specimens. Shortly thereafter Brown was made responsible for the Botanical Branch, a new specialized division of the Natural History Department, which in the 1830s was reorganized into a Botanical Branch, a Zoological Branch and a Mineralogical Branch the better to reflect the disciplinary divisions in its growing collections. The arrival of Banks's herbarium with Brown prompted these changes.[19] Banks's library, the herbarium's working partner during his lifetime, went to the Department of Printed Books under Henry Hervey Baber. A library devoted in the main to natural history and travel literature, it amounted to about 7,900 books and 6,100 unbound tracts, in total about 14,000 items. Banks's long-serving librarian Jonas Dryander catalogued the library in five exemplary volumes, published from 1796 to 1800, an updated version with annotated additions thereafter being kept at Soho Square. This was all made possible because, if nothing else, Banks concentrated on his collecting. In so doing, he held back the most prized of his collections until last – and may even have delayed their transfer to the museum following his death to be sure that arrangements there would be adequate to ensure their proper care.

The increasing concentration of such material in a public institution like the British Museum was to an extent foreshadowed by donations made soon after the *Endeavour* voyage. But such an influx placed a strain on already limited space and resources so that when Banks's main collections arrived at the museum it was being rebuilt to the designs of Robert Smirke. Its departmental structures would also need to be refashioned to suit its burgeoning collections and the evolving disciplines on which they were based. As a result, in the final quarter of the nineteenth century a new building was constructed at South Kensington to house the natural history collections, while at Bloomsbury the British Museum would henceforward concentrate on its collections of books, papers, vases, statues, coins and medals as well as ethnography. Opened in April 1881, the new museum was the location to which Banks's herbarium and his *Florilegium* drawings and plates were assigned. These are therefore now to be found there, along with the zoological drawings from the *Endeavour* voyage and some specimens of insects and shells, plus various systematic descriptions and lists made during or soon after the mission. With them went some duplicate library books belonging to Banks, and many other natural history illustrations and manuscripts that he had collected over the years.[20] Banks's library was, however, kept at the British Museum because it was there that the national library collections were held. Many of the *Endeavour* landscapes and figure drawings were also retained at Bloomsbury since these were not regarded as part of Banks's *Florilegium* illustrations or his various natural history papers. The books were later transferred to the British Library along with the landscape and figure drawings and a number of maps that Banks also collected. So were the illustrations from Banks's trip to Iceland in 1772. But at Bloomsbury, in the Department of Prints and Drawings, there are still to be found over 500 drawings of animals by a number of artists – Sydney Parkinson, Francis Masson, J.F. Miller, Peter Mazell, Peter Paillou, John Webber, Frederick Nodder, George Stubbs and others – who were employed by Banks at one time or another to work on various zoological subjects. Specimens and drawings associated with or derived from Banks's activities, including during the *Endeavour* voyage, are to be found in smaller numbers in repositories around the globe.

Research to establish what voyage objects survive from the *Endeavour* and other voyages is ongoing. Documentary evidence to link material, not least man-made material, to a particular voyage or collector is often lacking, so attributions are frequently at best circumstantial. Illustrations like those by Miller can help, but connections between drawings and particular objects in present-day collections are not in many cases now certain or possible. Benjamin West's splendid portrait depicts objects similar to if not actually the same as known examples from the *Endeavour* mission (see numbers 37, 45, 54, 74, 80, 81, 91, 99 and 100). Note in its bottom right-hand corner as viewed what appears to be a drawing in an open folio of the New Zealand flax plant. This symbolizes not only Banks's interest in botany but also its practical applications, for this plant was one that he saw widely employed in New Zealand for making cloth and cordage, both of potential use to a maritime nation like Great Britain. Thus when Banks is shown pointing at his Māori cloak or *kaitaka*, the possible original of which is held at the Pitt Rivers Museum (number 74), he is represented as drawing attention not only to his role as a Pacific collector but also to the utility to be derived from indigenous manufactures and natural products of this sort. In that sense the West portrait captures a number of themes bound up in such voyages and the collecting to which they gave rise. It indicates, too, major themes in Banks's own future career as a collector generally, and as a naturalist and economic botanist in particular.

THE EAST COAST OF AUSTRALIA

At the time of Cook's arrival on the east coast there were over 300,000 and perhaps as many as 750,000 indigenous people living in Australia in some 400 distinct tribes or language groups. Contacts were limited since local people tended to avoid the men of *Endeavour*. The two groups most closely encountered during the voyage were the Gweagal clan at Botany Bay and the Guugu Yimithirr tribe at Endeavour River. Both Cook and Banks noted the apparent simplicity of Aboriginal life and the contentment that it seemed to provide in contrast to the relentless quest for material gain that, in its various forms, characterized European society. However, the limited nature of contacts meant that the voyagers gained little real understanding of indigenous society and culture. When they landed Aboriginal peoples were primarily hunter-gatherers with a complex oral culture and spiritual values based on a reverence for the land and a belief in the Dreamtime. The Dreamtime explains the creation of the world by totemic spirit beings and also the reality of present-day Dreaming. It establishes the structures and rules of society. Rituals, songs and dance refer to the Dreamtime, in which tribal ancestors and spirit creatures do not die but merge with the natural world and so remain part of it. As a consequence there is a profound reverence for the land and what it represents that is reflected in Aboriginal art, music and carvings.

Before *Endeavour* arrived on the east coast in April 1770, Malay and Chinese traders had visited the north and west coasts of the continent. The Dutch had visited all of the coasts apart from that on the east side, calling these lands New Holland. Moreover, it has even been suggested that the Spanish and Portuguese may have reached the east coast at an earlier period. The coast that Cook would call New South Wales was sighted on 19 April 1770, and then for over four months he sailed northwards along it, charting and periodically making stops to obtain supplies. The longest stop was at Endeavour River for repairs following the near wreck of the ship on a reef on 10 June 1770. Having narrowly escaped disaster Cook continued northwards, rounded Cape York, and then sailed for Batavia (Jakarta), there to suffer the loss of many crew members as a result of diseases contracted at this notoriously unhealthy Dutch outpost. He never returned to the east coast mainland after his pioneering 2,175 mile running survey of it. His orders did not require him to do so and perhaps he remembered the traumatic events of June 1770.

During his third Pacific voyage Cook briefly visited Tasmania (known as Van Diemen's Land) in January 1777. He suspected it might be a large island separated from the mainland by a strait, but did not establish the fact. George Bass and Matthew Flinders would do that in 1798. Flinders, a Lincolnshire protégé of Banks, later circumnavigated Australia in HMS *Investigator*, proving it to be a single landmass. Flinders favoured the name of Australia for the continent rather than the then current Terra Australis, and his preference was eventually adopted.

105 'Two of the Natives of New Holland, Advancing to Combat.'

Stanfield Parkinson, ed., A Journal of a Voyage to the South Seas in his Majesty's Ship, The Endeavour (London, 1773; Fothergill reissue 1784). Plate 27. Drawn by Sydney Parkinson. Engraved by T. Chambers.

The British Library, London. L.R.294.c.7. Rebound in half brown leather and brown cloth. Title, place and date of publication, gilt lettering on spine. Author given. 29 × 4.2 cm (closed). Plate 27.3 × 22.6 cm.

This impressive engraving by Thomas Chambers of two Aborigine warriors encountered at Botany Bay was published in 1773 in the posthumous edition of Sydney Parkinson's journal. Any original illustration by Parkinson of the two figures is now lost and it may be that the plate was produced using more than one source. Number 106 overleaf is a field sketch by Parkinson of Aborigine warriors that may have been undertaken in preparation for this or another lost illustration, and Chambers might have incorporated details from this material. No doubt Chambers had Parkinson's written description of events at Botany Bay on which to rely, as quoted later in this entry. Note that in Chambers's engraving the first warrior's shield has two eye holes in it, as described by Parkinson in his journal, but these do not appear in number 106, where the shields are nevertheless shown as circular. Note as well that this warrior is shown holding a sword in accordance with various journal comments to that effect, the voyagers being unfamiliar with the different forms of Aboriginal weaponry. In his journal essay account of New South Wales (August 1770) Banks supposes that what the voyagers mistakenly described as swords were actually *woomeras* (see *Banks's Journal*, vol. 2, p. 133), but in fact the object concerned was probably a club, throwing stick or (non-returning) boomerang. Val Attenbrow takes this view in *Sydney's Aboriginal Past*, pp. 88 and 96. A small number of boomerangs held in different collections have been associated with the *Endeavour* voyage but not convincingly. Other details of interest include the body markings on both warriors, which were painted on using a fine white clay. These markings accord generally with those observed in the bay. Banks and Parkinson later noted nose-bones being worn at Endeavour River, but it is known that they were also common in the Sydney area. The second warrior wields a spear with three barbs cut out of each side of the point (see, again, Attenbrow, *Sydney's Aboriginal Past*, pp. 88–9).

The coast of New Holland (Australia) was sighted on 19 April 1770 and by 28 April a favourable bay was found which *Endeavour* entered. Cook would later give this bay the most famous name of any that he recorded on this voyage, Botany Bay. The bay area was inhabited by Aboriginal people, some of whom made signals and apparently brandished weapons as *Endeavour* came in. The ship was anchored at a southern cove just inside the entrance to the bay (Kurnell), where the people carried on fishing in their canoes or with their shore activities, seemingly unperturbed by the new arrival. However, when Cook landed with a party of some 30 or more men including Banks and Solander, two men of the Gweagal clan, which lived in that part of the bay, offered brave resistance to the British group. Tupaia could not converse with the local people and from this it was clear that their cultures were different. The warriors shouted defiantly and shook lances despite attempts by the explorers to signal their friendly intentions, and were eventually driven away by the use of small shot, but not before throwing lances in return. On inspecting nearby bark huts the British found some children left in one of them, and so decided not to enter it, but instead to leave behind some beads and ribbons (these were later found abandoned). They removed some weapons, for which see numbers 107–8 below.

Parkinson provided a vivid description of the historic landing: 'On our approaching the shore, two men, with different kinds of weapons, came out and made toward us. Their countenance bespoke displeasure; they threatened us, and discovered hostile intentions, often crying to us, Warra warra wai. We made signs to them to be peaceable, and threw them some trinkets; but they kept aloof, and dared us to come on shore. We attempted to frighten them by firing off a gun loaded with small shot; but attempted it in vain. One of them repaired to a house immediately, and brought out a shield, of an oval figure, painted white in the middle, with two holes in it to see through, and also a wooden sword, and then they advanced boldly, gathering up stones as they came along, which they threw at us. After we had landed, they threw two of their lances at us; one of which fell between my feet. Our people fired again, and wounded one of them; at which they took the alarm and were very frantic and furious, shouting for assistance, calling Hala, hala, mae; that is, (as we afterwards learned,) Come hither; while their wives and children set up a most horrid howl. We endeavoured to pacify them, but to no purpose, for they seemed implacable, and, at length, ran howling away, leaving their wives and children, who hid themselves in one of the huts behind a piece

Two of the Natives of New-Holland, Advancing to Combat.

of bark. After looking about us a little while, we left some nails upon the spot and embarked, taking with us their weapons; and then proceeded to the other side of the bay, where we had seen a number of people, as we came in, round a fire, some of whom were painted white, having a streak round their thighs, two below their knees, one like a sash over their shoulders, which ran diagonally downwards, and another across their foreheads. Both men and women were quite naked, very lean and raw-boned; their complexion was dark, their hair black and frizzled, their heads unadorned, and the beards of the men bushy. Their canoes were made of one piece of bark, gathered at the two ends, and extended in the middle by two sticks. Their paddles were very small, two of which they used at one time; and we found a large lump of yellow gum in their gigs [spears] which seemed to be for striking fish. Some of their weapons had a kind of chisel at their ends, but of what substance they were formed we could not learn' (*Journal of a Voyage*, pp. 134–5).

106 Two Aboriginal Australians and other sketch drawings

By Sydney Parkinson.

The British Library, London. BL Add. MS 9345, f. 14 verso.
18.4 × 23.5 cm.

This is a series of ten verso pencil sketches by Sydney Parkinson. In them he depicts things almost certainly seen at Botany Bay during the period that *Endeavour* was there, 28 April–6 May 1770, and, along with available journal accounts and perhaps a lost finished drawing based on the sketches, they may have supplied details for the print engraving in number 105 immediately above. The sketches are of two Aboriginal figures (their body markings and weapons), some bark canoes (a kneeling figure is shown paddling in one), two shields (circular in shape and painted in their middles) and a bark hut. The small outline, top left, would appear to be of an Aboriginal paddle, which the voyagers noted were small enough for one to be held in each hand. One of the central figures is preparing to launch what appears to be a multi-pronged fishing spear using a *woomera*, a hand-held wooden implement usually two to three feet long with a hook at one end into which the spear butt is placed, thereby operating as an extension to the throwing arm in order to provide considerable additional force to the throwing action. He holds before his body a circular shield, but it has no eye holes to match those described by Parkinson in his journal, although two such holes do appear in the preceding published engraving. On the recto of the present illustration is a sketch by Parkinson of two birds derived from George Edwards's *A Natural History of Uncommon Birds* (London, 1743–51), plate 295. This sketch is annotated 'Curasso & Cushew Birds Ed. 295'. The original Banks *Endeavour* copy of this publication is at the British Library, pressmark 435.g.3–4.

Joseph Banks recorded that: 'Our boat proceeded along shore and the Indians followd her at a distance. When she came back the officer who was in her told me that in a cove a little within the harbour they came down to the beach and invited our people to land by many signs and word[s] which he did not at all understand; all however were armd with long pikes and a wooden weapon made something like a short scymetar. During this time a few of the Indians who had not followd the boat remaind on the rocks opposite the ship, threatning and menacing with their pikes and swords – two in particular who were painted with white, their faces seemingly only dusted over with it, their bodies painted with broad strokes drawn over their breasts and backs resembling much a soldiers cross belts, and their legs and thighs also with such like broad strokes drawn round them which imitated broad garters or bracelets. Each of these held in his hand a wooden weapon about 2½ feet long, in shape much resembling a scymeter; the blades of these lookd whitish and some though[t] shining insomuch that they were almost of opinion that they were made of some kind of metal, but myself thought they were no more than wood smeard over with the same white pigment with which they paint their bodies. These two seemd to talk earnestly together, at times brandishing their crooked weapons at us in token of defiance' (*Banks's Journal*, vol. 2, p. 53).

107 Three fish spears and a lance

1. The Museum of Archaeology and Anthropology, University of Cambridge. D.1914.1. 138.4 × 2 cm. Wood, bone, resin and plant fibre. Fishing spear. 3 prongs, with 3 bone tips intact.

2. The Museum of Archaeology and Anthropology, University of Cambridge. D.1914.2. 132 × 1.5 cm. Wood, bone, resin and plant fibre. Fishing Spear. 3 prongs, with 1 bone tip intact.

3. The Museum of Archaeology and Anthropology, University of Cambridge. D.1914.3. 138.4 × 2.2 cm. Wood, bone, resin and plant fibre. Fishing Spear. 4 prongs, with 1 bone tip intact.

4. The Museum of Archaeology and Anthropology, University of Cambridge. D.1914.4. 153 × 1.6 cm. Wood, resin and plant fibre. Lance.

A relatively small number of objects were obtained from Aboriginal peoples during the time that Banks and Cook were on the east coast of Australia because contacts with local tribes were limited and European gifts were generally shunned. Aborigines may have attached little value to such offerings or perhaps avoided them for cultural reasons because to accept entailed an unwanted obligation. The Museum of Archaeology and Anthropology, Cambridge, holds an extremely valuable set of four spears known to have been gathered from New South Wales during the *Endeavour* voyage. This set comprises three fish spears at D.1914.1–3 and a lance at D.1914.4. They probably derive from encounters at Botany Bay, when two Gweagal warriors opposed the landing voyagers, throwing spears at them before being driven away. There was an initial fear that the spears were tipped with poison, probably due to the resin used to make them, but this was not the case. Some 40 to 50 spears were taken from nearby huts and others collected in the days immediately afterwards (see number 105; see also *Banks's Journal*, vol. 2, pp. 54–5, and *Cook's Journals*, vol. 1, p. 305).

Aboriginal fish spears or gigs were employed when fishing in shallows or from canoes for fish, turtle and crab. The first two included here have three wooden prongs. Only the first spear, D.1914.1, has all its tips in place, the tips of the spears not yet having been clearly identified as fish or mammal in origin, although Banks thought that they came from fish bones. In each case the prongs are attached to the wooden shaft by a plant cord binding and resin, and the surviving tips are attached to their prongs in the same way. The straight upright flowering stem (scape) of the grass tree *Xanthorrhoea*, probably *X. resinosa* Pers., was used to make these spears, with the resin employed in their construction also being gathered from the leaf bases of this plant. All three fish spears have had their shafts cut back or broken, presumably after being collected, so their original lengths are unknown. Such spears often incorporated two-piece shafts, which allowed their length to be adjusted according to the depth of water being fished. In their original state an upper measurement of more than 15 and perhaps nearly 20 feet is quite possible. In his journal Banks commented on the fishing spears seen at Botany Bay, which were predominantly multi-pronged as compared with spears later observed northwards along the coast, where he noted that single prongs, made of wood or of the spine of a stingray, sometimes with barbs fitted on them, were in general observed. Banks's limited contacts with people at these locations meant that his comments were not wholly accurate since both spear types would have been in use at each of them. He also commented more than once on the impressive accuracy of Aboriginal men seen fishing with their spears (see *Banks's Journal*, vol. 2, pp. 126–7; see also vol. 2, pp. 93, 95 and 132–3).

The lance has a single prong of hardwood tapering to a point, and this is secured to the shaft by strips of plant fibre and resin. Scholars believe it to be a fighting or a land hunting spear. Doubtless both uses were possible. Most Aboriginal tools and weapons were multi-purpose, their uses varying according to need or circumstance. In the Sydney area spears of this kind typically ranged from about 6 to 16 feet. A visual check of the Cambridge lance reveals a shaft end lacking obvious evidence of cutting back or breaking. It appears that only one single-pronged spear was collected at Botany Bay. After the initial clash with two Aboriginal warriors, Banks recorded that: 'We however thought it no improper measure to take away with us all the lances which we could find about the houses, amounting in number to forty or fifty. They were of various leng[t]hs, from 15 to 6 feet in leng[t]h; both those which were thrown at us and all we found except one had 4 prongs headed with very sharp fish bones'. It might be added that more than one opportunity to collect spears took place in the Botany Bay area. Three days later both Banks and Cook were ashore when some crew members were followed by tribesmen, who, according to Banks, threw at the crewmen four spears, which were then collected. Cook records three spears thrown, but he also noted that two days afterwards another was hurled at William Monkhouse by a man hiding up a tree who ran away, and this might also have been collected (see *Banks's Journal*, vol. 2, pp. 54–8; see also *Cook's Journal*, vol. 1, pp. 308 and 310).

Some Aboriginal spears were certainly in Cook's possession, since after the *Endeavour* voyage he included the ones seen here in a collection of artefacts given to Lord Sandwich, who passed

a selection of this material to Trinity College, Cambridge, where he had formerly been an undergraduate. In 1914 and 1924 the college deposited the Sandwich material at the Museum of Archaeology and Anthropology, including this remarkable sample of the few arms gathered by the voyagers on the east coast. A putative 'New Holland' spear also survives in an important collection of South Sea artefacts that Banks passed to the Alströmer family in Sweden probably in late 1778, see 1848.01.0060. Accompanying this gift, most likely made via Johan Alströmer, who visited London from 1777–8, was a collection of Pacific natural history specimens. In 1848 the Alströmer family donated these artefact and natural history collections to the Royal Swedish Academy of Sciences and the former are now housed in the Etnografiska Museet, Stockholm. Coming from Banks such a spear would be a likely candidate for first-voyage status, although by 1777 he possessed material from Cook's second Pacific voyage, which must also have formed part of his gift. Moreover, further similar gifts by Banks from other sources might have been made to the Alströmers in later years. The contentious 'New Holland' attribution derives from a catalogue of the collection as donated to the academy, but the original source of this catalogue is not known and later alterations and various errors mean that it is not entirely reliable. Stig Rydén thought the spear was Polynesian, though from which location he knew not, stating that the catalogue designation was 'surely erroneous' (see *The Banks Collection*, pp. 70 and 94).

The Museum of Archaeology and Anthropology in Cambridge also holds a *woomera* and hafted stone axe, both considered to be from the Sydney area and occasionally suggested as possible first-voyage items, although scholars more convincingly argue that these objects resulted from contacts at a period after the arrival of the First Fleet. Part of the Widdicombe collection, much of which was originally obtained from the 1806 sale of the Leverian Museum, which certainly contained Cook material but mainly from the second and especially the last voyage, these two items cannot be matched to any of the Leverian lots known to have been purchased at that time. Their source or sources therefore remain unclear (see, respectively, 1922.994 and 1922.995). The great expert on the Leverian museum and its surviving contents is Adrienne Kaeppler (see her *Holophusicon. The Leverian Museum ...*, pp. 17, 41, 90 and 188).

108 Bark shield

The British Museum, London. Oc,1978,Q.839. Length 97, width 29 and depth 12 cm. Bark and wood.

The *Endeavour* landing at Botany Bay on 28 April 1770 was opposed by two Aboriginal warriors, who were eventually driven off after throwing spears. They fled leaving everything behind, including their weapons, and these the voyagers felt it prudent to remove, including 40 to 50 spears (see numbers 105–7). In his voyage journal Banks wrote: 'Defensive weapons we saw only in Sting-Rays bay [Botany Bay] and there only a single instance – a man who attempted to oppose our Landing came down to the Beach with a shield of an oblong shape about 3 feet long and 1½ broad made of the bark of a tree; this he left behind when he ran away and we found upon taking it up that it plainly had been pierced through with a single pointed lance near the centre' (*Banks's Journal*, vol. 2, p. 133; see *Cook's Journals*, vol. 1, p. 396).

Aboriginal shields made of bark are known as elemong shields. It is generally agreed that the one seen here was collected at Botany Bay, since Banks's journal description above broadly agrees with its appearance, and a post-voyage illustration by J.F. Miller for Banks records a visually similar shield from New Holland with a distinctive hole located near its handle (see number 110). The equivalent hole in the museum shield was probably caused by an impact from a pointed object, perhaps a spear or adze, presumably during clashes between rival Aboriginal groups. Recent radiographic images of the shield show that the hole has a rough edge consistent with it being the result of a blow, and have revealed more marks on the surface suggesting other strikes from a sharp implement, evidently a weapon or weapons of some kind. Moreover, traces of white kaolin clay have been found on the shield face, suggesting that it was painted, perhaps at its centre with a mark or symbol as noted by Sydney Parkinson in his journal comments in number 105. The Aboriginal warriors observed at Botany Bay wore painted white patterns on their skin almost certainly of the same clay. An additional small circular hole at one end of the shield is not explained by voyage records or museum staff. It is neatly rounded and seems to have been deliberately created by tools of a later date. The shield is made of the Red Mangrove *Rhizophora mangle*, and its handle is of flexible green mangrove wood. The Red Mangrove has a northerly distribution starting some 400 miles above Sydney, so it would appear that travel and trade reaching at least that far was undertaken among the tribes on this part of the coast.

Charles Praval produced an illustration of an Aboriginal man holding a similar shield, perhaps copied from a lost Parkinson drawing, and Endeavour River has been suggested by Joppien and Smith as the location for that illustration (see BL Add. MS 15508, f. 13, and Joppien and Smith, *The Art of Captain Cook's Voyages*, vol. 1, pp. 56–8; for Praval, see number 73). This lost illustration, Joppien and Smith further suggest, may exceptionally have resulted from a pose contrived for the purpose of the drawing, with the subject wearing bits of what appear to be a European shirt, possibly from Cook, and covering his nakedness with a shield. Joppien and Smith also suggest that a sketch by Parkinson of a man holding what appears to be a spear may have been in preparation for the supposed lost drawing, BL Add. MS 9345, f. 20. Parkinson rarely attempted completely nude figures, possibly because of his Quaker roots, but the artificial nature of the copy image is still difficult to construe given his usual concern to record faithfully details of indigenous life and objects. Whether the shield featured by Praval is the same as the museum one is not clear, as it lacks identifying marks, in particular the central hole, but this must remain a possibility.

109 Aboriginal Australians in two canoes

By Tupaia.

The British Library, London. BL Add. MS 15508, f. 10(a). 26.3 × 36.2 cm.

This is a pencil and watercolour illustration by Tupaia, the Ra'iatean priest and navigator travelling with Banks to England on *Endeavour*. It was painted on the east coast of Australia, probably at Botany Bay. It shows Tupaia's interest in the detail of two Aboriginal bark canoes, vessels naturally of concern to him as an expert navigator, but different in design to those that he knew in the Society Islands or had encountered elsewhere on the voyage. Aboriginal canoes were employed for transport and for fishing, and most early records of people seen on the coast, in bays, estuaries and on rivers include frequent mention of them.

Here one of the canoe occupants is shown fishing with a four-pronged fishing spear (see number 107). Banks noted four Aboriginal Australians fishing in separate canoes as *Endeavour* entered Botany Bay on 28 April 1770. In each canoe the occupant 'held in his hand a long pole with which he struck fish, venturing with his little imbarkation almost into the surf'. He also described the construction of the canoes in Botany Bay, this being from a single piece of bark tied together at both ends, and supported in the middle by small bows of wood. Paddles of about 18 inches in length were, he wrote, held in each hand – just as shown by Tupaia. Such craft were estimated to be about 12 feet, might carry up to three people and be moved in shallow water by the use of a pole. Moreover, 'In the middle of these Canoes was generaly a small fire upon a heap of sea weed, for what purpose intended we did not learn except perhaps to give the fisherman an opportunity of Eating fish in perfection by broiling it the moment it is taken'. Others thought that such fires were carried in canoes in order to stay warm and start fires ashore more easily. Probably they served more than one purpose (see *Banks's Journal*, vol. 2, pp. 53, 126 and 134–5; see also number 106 above).

110 Artefacts from Australia, New Zealand and New Guinea

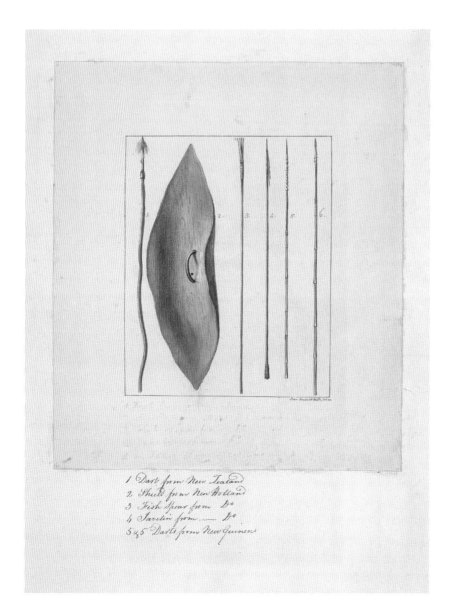

By John Frederick Miller.

The British Library, London. BL Add MS 23920, f. 35.
20.6 × 16.5 cm.

This is a pen and wash drawing by J.F. Miller for Joseph Banks, signed in ink 'John Frederick Miller del: 1771', and annotated in ink on the folio with a key '1 Dart from New Zealand 2 Shield from New Holland 3 Fish Spear from D$^{o.}$ 4 Javelin from D$^{o.}$ 5–6 Darts from New Guinea'. A similar key by Banks in pencil on the lower recto margin has partly been erased. Items 2, 3 and 4 are objects gathered on the east coast of Australia during the *Endeavour* voyage, very probably in the case of 2 and 3 if not also 4 at Botany Bay. There is an impressive Aboriginal bark shield held at the British Museum, London, which has been attributed to the *Endeavour* voyage using the evidence provided by this illustration and Banks's journal comments (see number 108). There are three Aboriginal fishing spears in the Sandwich collection, Cambridge, one of which has four prongs, although scholars do not think that it is an identical match for spear 3 in the present illustration (see number 107; see also, in respect of 4, a possible close study by Parkinson of the same spear's head at BL Add. MS 9345, f. 20). The rear-facing points on spear 4 are probably fish bones, perhaps from the spine of a stingray.

111 *Trygonorhina fasciata* Müller & Henle, 1841

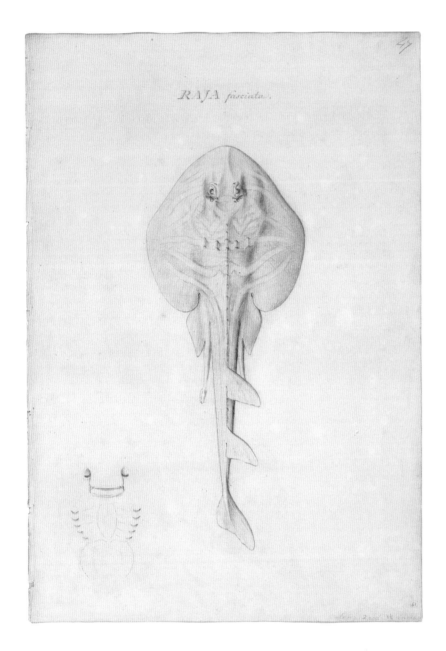

By Herman Diedrich Spöring.

The Natural History Museum, London.
Wheeler, Catalogue, Part 3 Zoology (1986), 51(1·47).
52.5 × 36 cm.

From 29 April to 6 May *Endeavour* was anchored in a bay that gave Banks and Solander their first opportunity to investigate the flora and fauna of a coast hitherto unknown to European naturalists. Due to the haul of plants gathered there Cook changed the name that he initially chose for this location, Sting Ray Bay, to the most famous of those recorded anywhere on this voyage, Botany Bay. Cook's original name reflected the number of fish seen in the bay and here is a finished drawing by Herman Diedrich Spöring of a stingray caught there on 29 April 1770. Held at the Natural History Museum, London, the drawing shows a dorsal view of the specimen as well as details of the ventral side of the head. The illustration is entitled 'RAJA fasciata.' At the foot of it there is a pencil annotation 'Long. 2 ped: 1½ unicas', and on the verso another pencil note reads 'Trygonorhina fasciata Müller und Henle'. Stingrays are cartilaginous fishes related to sharks, and are common in coastal tropical and subtropical marine environments throughout the world. Most stringrays, including this species, have a barbed sting on the tail that is used in self defence. This handsomely marked species is commonly known as the Fiddler Stingray.

112 *Aptychotrema banksii* (Müller & Henle, 1841)

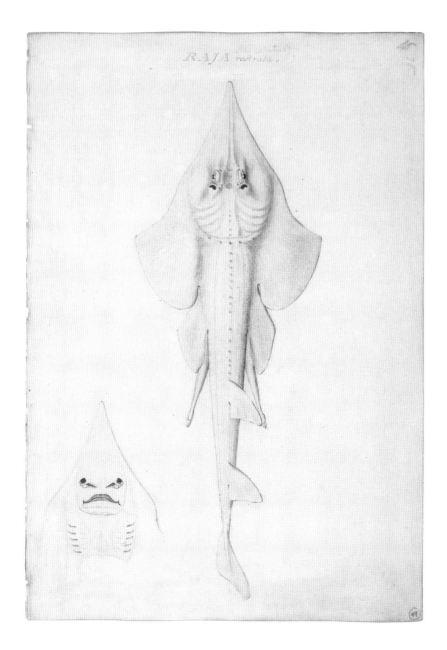

By Herman Diedrich Spöring.

The Natural History Museum, London.
Wheeler, Catalogue, *Part 3 Zoology (1986), 49(1:45).*
42.6 × 35.8 cm.

This finished pencil drawing of a stingray by Herman Diedrich Spöring, showing a dorsal view and detail of the ventral side of the head, is entitled 'RAJA rostrata', with a pencil annotation above the title 'Rhinobatos L?' Pencil annotation also appears on the verso, 'Rhinobates (Rhinobates) Banksii Müller und Henle'. According to written records by Daniel Solander the specimen itself was collected on 29 April 1770 and therefore in Botany Bay. This species of ray is endemic to the east coast of Australia. It grows to 1.2 m in length and is recognized by its long triangular snout. Commonly known as the Eastern Shovelnose Ray, it is currently classified as *Aptychotrema rostrata* (Shaw and Nodder, 1794). A specimen from the voyage that was given by Banks to the French naturalist Pierre Marie Auguste Broussonet is now held in the Muséum national d'Histoire Naturelle, Paris, but this does not appear to be the one illustrated by Spöring (see Wheeler, *Catalogue*, Part 3 Zoology, pp. 50–1).

Since Sydney Parkinson was inundated by botanical material at this historic location, Spöring drew examples of the marine life found there. Just nine zoological drawings by Spöring, all finished, exist in the Natural History Museum's collection of *Endeavour*-voyage animal illustrations. All are of sea creatures and they are especially fine works. Indeed, Wheeler considered that Spöring's 'drawings of rays, sharks, a bony fish, and crustaceans were not excelled in the zoological material', and further assessed them as 'more life-like and accurate than drawings of comparable subjects by either of the other artists [Parkinson and Buchan]' (Wheeler, *Catalogue*, Part 3 Zoology, pp. 7–8; see also *Banks's Journal*, vol. 2, pp. 56 and 60–1).

113   *Banksia serrata* Linnaeus f., *Suppl. pl.*: 126 (1782)

Engraving by Gabriel Smith after watercolour by John Frederick Miller after sketch by Sydney Parkinson.

The Natural History Museum, London. Diment et al., Catalogue, Part 1 Botany (1984), A 7/326, col. engraving 1983 BF: pl. 285. Plate height 46 × width 29.5 cm; plant ill. height 42 cm.

This is a coloured print from a copper plate engraved by Gabriel Smith, based on a finished 1773 watercolour drawing by J.F. Miller. Miller's watercolour was worked up from an annotated outline drawing by Sydney Parkinson, and the drawings, plate and print, as well as a lectotype specimen gathered at Botany Bay, are all held at the Natural History Museum, London.

Banksia is a large genus of trees or woody shrubs containing over 170 species, all but one of the extant species being endemic to Australia. Over 90 per cent of all *Bansksia* species occur only in south-west Western Australia. Most banksias have spectacular elongate inflorescences, each flower spike consisting of hundreds or thousands of individual flowers. The genus *Banksia* was created to honour Banks by Carl Linnaeus the younger in his *Supplementum Plantarum* (Brunswick, 1782).

114   Four-pounder gun

The National Maritime Museum, Greenwich, London. KTP0071. Overall length, 181.5 cm; diameter at breech, 33 cm; diameter at the muzzle, 25.5 cm; diameter of the bore, 8 cm; width across trunnions, 46 cm. Weight of gun, approximately 607 kg. Iron. Mounted on a replica sea carriage.

On 6 May *Endeavour* continued on the journey northwards along the east coast of Australia. James Cook soon sighted and named Port Jackson, but he did not enter it and so missed discovering one of the finest natural harbours in the southern hemisphere. Governor Arthur Phillip, commanding the First Fleet, would later be the first European to enter Port Jackson and anchor at what he called Sydney Cove in January 1788. Cook sailed onwards naming features and charting the coast, keeping close to it but quite unaware that he was travelling towards some of the most difficult waters to navigate of the entire mission, those in which lay the Great Barrier Reef. There were few landfalls for Banks, in fact landings on seven days only, along with three outings afloat in his dingy, and there was no contact at all with any indigenous people during this period.

Then, on the night of 10 June, having travelled some 1,500 miles from Botany Bay, *Endeavour* struck a reef lying on the inner side of the Great Barrier Reef off the coast of present-day North Queensland. Cook had decided to proceed at night under double-reefed topsails where shoals and various islands had been noticed because he believed he was moving out into deeper and hence safer waters, all the while having a man taking the depth by heaving the lead. Yet after an apparently safe sounding (17 fathoms) the ship ran onto an unseen reef, it being dark and the tide high. This was duly named Endeavour Reef. Twenty-four hours of frantic effort freed the vessel from the coral outcrop on which she sat and she limped to the mouth of what Cook called Endeavour River, the site of present-day Cooktown. In order to save *Endeavour* many objects were cast overboard, among them six of the ship's guns and their carriages. This enabled *Endeavour* to sit higher in the water, limiting

damage from the reef, and eventually allowing her to be floated off it. The guns, however, were left behind. There they lay for two hundred years, undisturbed on the reef until 1969, when all six were recovered by archaeologists. The present gun is one of the six, the others being held in Botany Bay, Cooktown, Canberra, Philadelphia and Wellington. It is a cast-iron four-pounder gun weighing just over 600 kg, and was made in about 1725–50.

The ship was careened and repaired at Endeavour River during an enforced stay from 17 June to 4 August 1770, the longest period spent at one place on the east coast. This protracted stay was due, in no small part, to unfavourable monsoon weather conditions that delayed Cook's departure. A plate showing the beached ship under repair was published in the official voyage account edited by Dr John Hawkesworth, vol. 3, plate 19. During their time ashore Banks and his companions studied the local flora and fauna and they also met the local people. Kangaroos were first captured at Endeavour River, and the Guugu Yimithirr word *gangaru* was also first recorded there, being variously spelled by Cook and Banks as 'kangooroo' or 'kanguru' (see numbers 116–21 and 129). Cook eventually departed in early August, but was again almost wrecked when he passed through the Great Barrier Reef, only to be becalmed and almost carried back on to it by the incoming tide. Luckily he found what he called Providential Channel and sailed safely inside the reef again. It was a tense time, and a comparison of the accounts of Banks and Cook at this point shows how Cook drew on Banks's lucid expression in order to convey the drama and emotion of the crisis: it was not only for observations of natural history and ethnography that Cook turned to Banks when recording the *Endeavour* voyage (see *Banks's Journal*, vol. 2, pp. 104–8, and *Cook's Journals*, vol. 1, pp. 377–81). As a developing journalist he profited in various ways during his close working association with the younger man.

Having rounded York Cape, Cook took possession of the east coast for Britain, naming it New South Wales. He then made for Batavia (Jakarta), where disease killed many of his crew and much of Banks's remaining party. Tragically, Tupaia and his boy servant Taiato also died at this, the first major European settlement that they visited.

115 *Pyrazus ebeninus* (Bruguière) and *Ovula ovum* (L.) as *Amphiperas ovum* (L.) in G.L. Wilkins, *Bulletin of the British Museum (Natural History)*, vol. 1 (3) 1955, p. 102, and plate 16, fig. 11, and for the second specimen p. 103

The Natural History Museum, London.

1. Height 8, width 2.7 and length 2.7 cm and height 7.1, width 2.6 and length 3.2 cm, for two specimens.

2. Height 8, width 4.3 and length 5.2 cm and height 8.5, width 2.2 and length 5.2 cm, for two specimens.

These shells are part of a collection held at the Natural History Museum, London, which originally belonged to Banks and passed from him to the Linnean Society and afterwards in 1863 to the British Museum. They were very probably, but not certainly, gathered by Banks or a member of his party during the *Endeavour* voyage (see numbers 26 and 35). Banks and Solander collected shells during their long stay at Tahiti and also around New Zealand, although in the latter case shellfish were of interest mainly as a source of extra food. In Australia shells were collected at Botany Bay and during the prolonged stop at Endeavour River, most of which were recognized by the voyagers as previously undescribed. The natural distribution of both these specimens is known to include present-day New South Wales and Queensland. Most of the specimens in the Banks shell collection at the Natural History Museum now known to be Australian in origin are found on the east coast, thus making them likely first-voyage candidates. *Pyrazus ebeninus*, originally classified as *Murex aluco nigra*, is very common in the mud flats around Sydney, Botany Bay and the upper reaches of Port Jackson. *Ovula ovum*, at first classified as *Bulla ovum*, is commonly known as the White Egg Cowry, and has a wide distribution in the Pacific, where it was traditionally used to ornament canoes, in items of dress and to adorn the body.

115.1

115.2

116   *Macropus* sp. Unfinished pencil outline

By Sydney Parkinson.
The Natural History Museum, London.
Wheeler, Catalogue, *Part 3 Zoology (1986), 4.(1:4). 35.8 × 52.5 cm.*

This is an unfinished pencil outline by Sydney Parkinson of a kangaroo specimen encountered at Endeavour River, with annotated comments on the verso '[pencil, Parkinson] the whole body pale ash colour the ears excepting the base fine spec[k]led gray iris of the eye Chesnut. [pencil, Banks] Kanguru [ink, Banks] Endeavours River'. Sporadic sightings of macropods had been communicated to Europe by Dutch navigators as far back as the mid-seventeenth century, and sketchy descriptions by English and Dutch travellers appeared in print early in the next century, but it was the accounts, visual records and specimens from *Endeavour* that really brought these animals to the notice of Western society. While the vessel was being repaired at Endeavour River after its near shipwreck on a reef, Banks and his party had much longer to collect and observe the local flora and fauna than had been the case at Botany Bay. One of the most famous discoveries made at this stage of the voyage was certainly that of the kangaroo, three specimens of which were caught. Some flesh of these animals was eaten by the voyagers and declared by Banks on his first tasting it to be 'excellent meat', which it is. Consumption of various plants and animals in the places visited was necessary for survival, but such behaviour was also consistent with a central mission aim to assess the natural resources of each area, including local sources of food (see *Banks's Journal*, vol. 2, p. 94; for earlier possible sightings of a small kangaroo and some droppings at Botany Bay (1 May) see p. 57, and of possible kangaroo tracks at Thirsty Sound (30 May) see p. 73).

The first reports of a mysterious, fast-moving animal glimpsed by crew members at Endeavour River were recorded by Banks in his journal for 22-4 June. Banks did not himself see any such creature until the 25th, when out gathering plants: 'what to liken him to I could not tell, nothing certainly that I have seen at all resembles him'. On 6 July Banks led a party on a collecting expedition up the river and, after an uncomfortable night spent in a river-side camp plagued by mosquitoes, four animals were the next day observed that outpaced Banks's greyhound by 'making vast bounds' through some long grass. To Banks these animals resembled the small African rodent jerboa, by which he probably meant the lesser Egyptian jerboa *Jaculus jaculus*, today known not to be a related species, although similar to the kangaroo in having a long tail, short forelegs and long hind legs with which it hops at speed. Banks's was an early, tentative step in what was to be a long-running debate among Europe's naturalists as to how best to classify the animal now being witnessed for the first time by the explorers. A specimen had yet, however, to be obtained. One was eventually taken when, on 14 July, Third-Lieutenant John Gore shot a male of 38 lbs dead weight according to Banks, or 24 lbs clean. This was followed on 27 July by a large male of 84 lbs dead weight, again shot by Gore. Finally, on 29 July a small female of 8½ lbs was caught by Banks's greyhound. Thus, more than a month after the first sightings along the river, Banks at last obtained his own specimen of this remarkable animal. Solander produced a draft and a fair copy description of the specimens under the name 'Kanguru saliens' (see number 117 immediately below). Parkinson made two sketches, that seen here and another of a kangaroo crouching (see Wheeler, *Catalogue*, Part 3 Zoology, 3.(1:3), pp. 33-5). The specimen shot by John Gore on 27 July may well have been the source of Parkinson's drawings (for Banks's journal comments on these events, see *Banks's Journal*, vol. 2, pp. 84-100 and 116-17; see also *Cook's Journals*, vol. 1, pp. 359 and 363, though Cook omits mention of Banks's small specimen).

Note Parkinson's verso annotations on the present drawing regarding the colouring of the animal. He frequently recorded colours on his natural history sketches before they faded or were lost in the specimens themselves, with the intention of finishing the illustrations later. However, he died on the journey home and so many of his drawings from the busy Pacific stage of the voyage remained in sketch form only. What is interesting in this case is the fact that no finished illustration of such an extraordinary creature as the kangaroo was produced, and further that there are so few mammals in Parkinson's zoological work from the voyage as a whole – only four species in fact. Banks confessed to the paucity in this area immediately on his return home when he wrote to Thomas Pennant that 'Our Collections will I hope satisfy you, very few quadrupeds, one mouse however (Gerbua) weighing 80 lb weight' (see Chambers, ed., *Indian and Pacific Correspondence*, vol. 1, letter 22). Oliver Goldsmith

stated in his *History of the Earth, and Animated Nature* (London, 1774), vol. 4, pp. 351–3, that Banks also brought home a small specimen from the voyage in the form of a stuffed skin 'not much above the size of a hare'. This presumably was Banks's small female. The dearth in Parkinson's animal illustrations was almost certainly the result of his concentration on botanical material along the east coast of Australia and at other landfalls, and on marine life in his zoological drawings when at sea (see number 21). It must be remembered, too, that most of the places visited by Cook on this voyage are not richly endowed with mammals, which also helps explain the lack of illustrations of them. After the voyage many of Banks's zoological specimens were given away, and if published at all they appeared later under the names of other naturalists. He may have thought about eventually publishing this material himself, but to begin with he devoted his main effort to publication of the many new plant species obtained on the mission, a massive work that was itself never completed.

It remains uncertain precisely what kangaroo parts were brought back on *Endeavour*, but any that did return can only have come from three animals at most. Banks's remark to Pennant, Goldsmith's comment, Banks's and Cook's journals, as well as the illustration by Nathaniel Dance in number 118, a slide formerly held at the Royal College of Surgeons in 119 and George Stubbs's famous painting of a kangaroo in number 120 together indicate that at the least Banks may have returned with the skin and skull of one large male specimen, the skull of a smaller male and the skin of another still smaller specimen, namely that of the female. None of these kangaroo parts appear to have survived and the eventual fate of any kangaroo material that the voyagers did bring back remains unclear.

117 'Kanguru saliens', Daniel Solander draft description, Slip Catalogue, Mammalia, f. 90

The Natural History Museum, London. Wheeler, Catalogue, Part 3 Zoology (1986), 3.(1:3), S.C. Mammalia f. 90 (draft), ff. 91–5 (fair copy). Double foolscap folio 90, 32.5 × 20.3 cm unfolded; 10 × 16.3 cm folded.

Daniel Solander was a favourite pupil of the great Swedish naturalist Carl Linnaeus, under whom he studied at the University of Uppsala. Linnaeus sent Solander to England to conduct research and to promote his system of classification among its naturalists. Solander decided to remain in England and to work at the British Museum, where he was employed to catalogue the natural history collections according to his master's system. It was almost certainly at the British Museum that Banks and Solander first met in 1764. A likeable gossip and an extremely capable naturalist, Solander became firm friends with Banks, who arranged for him to travel on the *Endeavour* voyage as part of his team of illustrators and collectors. The Solander manuscript slip catalogue is held in the relevant scientific departments of the Natural History Museum, London – in the case of the present slip, that of Zoology. The catalogue contains Solander's systematic descriptions of specimens collected on the voyage, written in Latin on small sheets of paper that were kept in specially designed cases during the circumnavigation (see number 4). One description, in draft and fair copy, is of three kangaroo specimens gathered along the Endeavour River under the name 'Kanguru saliens'. The draft version is published here.

Alwyne Wheeler in his *Catalogue* described these important documents as follows: 'the account of Kanguru saliens by Solander exists in two forms, written on three sides of a double foolscap folded to a slip size (f. 90), and on five successive slips (f. 91–95). The foolscap draft (f. 90) is the earlier, containing many corrections, the slips (f. 91–95) are fair copies from the earlier draft, but are unique in that they give the measurements and weights of the three specimens examined [in fact the fair copy is the same as the draft copy in terms of content with only slight variations in order, and although the weights of all three animals are given in both slips just the measurements of the first male are recorded in them, for which see Wheeler below] ... while the foolscap draft (f. 90) may be the earlier it is a composite description which was based on at least two specimens (male and female reproductive organs are described). From Solander's manuscript (f. 95) the three specimens can be distinguished as follows: a young female of eight pounds weight, a male, two to three years of age, of twenty-four pounds weight, and an adult male of eighty pounds weight. The only length given refers to a specimen 28 inches body length and 26 inches tail length. (This would appear to be in keeping with the smaller male animal.) These weights differ from those given by other sources on the voyage, e.g. Banks's Journal for 14 July 1770, which records the weight as 38 lb. This source records the other specimens as 84 lb, obtained on 27 July 1770, and a third of 8½ lb on 29 July. Parkinson's drawing of a leaping kangaroo suggests he drew the adult male of 84 lbs' (pp. 33–4). Solander's two slip descriptions of the specimens are detailed, including comments on the genitalia of both the smaller male and the female specimens. He explicitly refers to the mammae of the female, so it is certain that although he did not record a pouch for this animal he must have seen one and therefore been aware of its marsupial nature. Solander based much of his description on the first immature male, and did not specifically mention the second adult male except when giving its weight, but he paid rather more attention to Banks's young female probably because of its gender. It is even possible that at this early stage he thought the specimens were all of the same species. Yet notwithstanding the fullness of Solander's descriptions, the information supplied in the journals of the other voyagers and the illustrations that were made, the identity of what was popularly termed 'Captain Cook's kangaroo' remained a contentious issue well into the twentieth century.

In Banks's day naturalists struggled to classify kangaroos based on the specimens brought back on *Endeavour* because no satisfactory category yet existed in Western science for such animals, and so to begin with naturalists related these animals to existing species. The first attempt to classify the animal according to the existing binomial classification system for animals established by Carl Linnaeus in 1758 was that of Statius Müller in 1776. Based on the first published image and description of a kangaroo that appeared in the official account of the *Endeavour* voyage as edited and published in 1773 by Dr John Hawkesworth (see number 121), Müller related it to the rodent genus *Mus* naming it *Mus canguru*. This is therefore the name which is established for the holotype of the species, a holotype being the single type specimen upon which the description and name of a new species is based.

Subsequent to Müller's *Mus canguru* the kangaroo was named as a jerboa (the *Yerboa gigantea* of Zimmerman, 1777, and the *Jaculus gigantea* of Erxsleben, 1777) and also as an opossum (the *Didelphis gigantea* of Schreber, 1777). The uncertainty surrounding classification of kangaroos was in due course clarified by the zoologist George Shaw, a specialist in Australian fauna who went on to lead a distinguished career in charge of the natural history collections at the British Museum. On the basis of differences in dentition between that of opossums and jerboas on the one hand and of the kangaroo on the other, as well as the anatomical structure of the hind feet of kangaroos, Shaw in 1790 described a separate new genus *Macropus* in combination with the name *gigantea* and then, in 1800, in combination with the name *major*. The distinctive nature of these animals was further emphasized when in 1811 naturalists assigned kangaroos to a new mammal infraclass Marsupalia (marsupials) based on their mode of reproduction and nurture, the new higher taxon being introduced by naturalists to cope with the various and increasing number of recently discovered animals that carry their young in pouches as does the kangaroo.

Taxonomists subsequently agreed that the holotype of Müller's *Mus canguru* could be identified as a young Eastern grey kangaroo, which was classified in Shaw's genus *Macropus* as *M. giganteus*. This was generally accepted until 1925, when Tom Iredale and Ellis Troughton presented evidence that this animal was not the Eastern grey, but in their opinion a smaller form, probably of the *robustus* series, basing their interpretation on analysis of the characters in Solander's description, including primarily the lack of hair between the nostrils and details of dentition, pelt colour and habitat. For their 1925 paper they acquired and published the text of Solander's draft slip. Then, in 1937, they changed their minds after obtaining by purchase the skins and skulls of two species local to the Cooktown area that had been captured by hunters. After comparing these with Solander's description, which they strenuously maintained was based almost exclusively on the first male of 38 lbs, they identified this particular specimen as a whiptail wallaby *Wallabia canguru* (now named *Macropus parryi*). The specimen of 84 lbs they treated as merely being appended by Solander to his slip description, and since it would exceed the mean weight for whiptails they considered that this specimen belonged to a different species and could be disregarded. To Banks's small female little regard was paid and these authors mistakenly recorded it as the second not the third specimen taken by the voyagers.

These conclusions were the start of an often heated controversy over precisely what species had actually been gathered and described during the voyage, and they were first challenged by Henry Raven in 1939. On the basis of his interpretation of Solander's data and a comparison between the engraving published by Hawkesworth in 1773 and modern photographs of a whiptail wallaby and Eastern grey kangaroo, particularly the characteristic markings on the head and hip of the whiptail, Raven concluded that the correct identification of the animal described by Solander was indeed *Macropus giganteus*. Terence Morrison-Scott and Frederick Sawyer concurred with this identification in 1950 and, contrary to Iredale and Troughton, pointed out that Solander's was a composite description based on both male and female specimens and on animals of different ages, and furthermore they suggested that as he wrote Solander may even have had three separate species as well as three separate specimens in front of him. Others have speculated about the number of species that were caught, and it seems almost certain that at least two were collected by the voyagers. Morrison-Scott and Sawyer argued that Solander's slip description could not therefore be decisive in arriving at the identity of the *Endeavour* kangaroo. They regarded characters such as dentition and the presence or absence of hair on the muzzle as ultimately inconclusive. Morrison-Scott and Sawyer did, however, draw attention to a drawing of a skull by Nathaniel Dance, which they thought likely to be that of the 84 lb animal shot on 27 July 1779, and this they identified as belonging to a young *Macropus robustus*, a common wallaroo. Their identification has, however, been challenged (see 118 immediately below).

The controversy was continued in a further paper by Iredale and Troughton in 1962 in what was a detailed and acrimonious rebuttal of Raven and of Morrison-Scott and Sawyer, particularly with regard to the interpretation of Solander's Latin

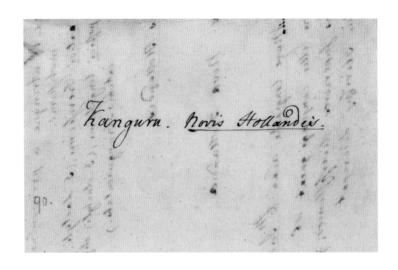

description of the animal's dentition. Iredale and Troughton also considered that Raven's comparison of Hawkesworth's figure with photographs of a whiptail and Eastern grey kangaroo disproved in several ways his claim that the Eastern grey is identical with Solander's description, and instead identified the description as of a northern form of the whiptail wallaby that lacks the striking cheek and body marks of the southern form. This distinction is certainly open to dispute since current experts do not recognize significant variations in pelage for this species, although there is some seasonal and individual variation. Iredale and Troughton again emphasized their view that Solander's description focused primarily on the specimen of 38 lbs, basing their arguments on this assumption and on Hawkesworth's text and engraved reproduction of the Stubbs portrait of an *Endeavour* kangaroo specimen (bizarrely they asserted that the engraving must have preceded the painting) that, crucially, was used by Statius Müller in the first truly scientific attempt at classifying these animals. It is not clear, however, that Solander's slips were ever seen or used by Hawkesworth or Stubbs, the former of whom mentions in the official mission account only two of the three specimens collected by the voyagers, placing the Stubbs plate with the 38 lb specimen and omitting altogether Banks's small female that may actually have been the source of that depiction (see numbers 120–1). Hawkesworth's description of the *Endeavour* kangaroos appears to have been largely based on the manuscript journals of Cook and Banks, except that at one point Hawkesworth stated that the 84 lb animal 'was not at its full growth, the innermost grinders not being yet formed', that is with unerupted inner molar teeth, whereas in his slips Solander described this specimen as an adult male, further implying that Hawkesworth had not seen the slips. The first obvious use in print of Solander's slip description of the kangaroo that the present author can find is that by Thomas Pennant in his 1781 edition of *History of Quadrupeds*, vol. 2, pp. 306–7 (Banks's library copy of this work in two volumes is at the British Library, pressmarks 461.i.15 and 461.i.16.)

In a postscript to their 1962 paper, Iredale and Troughton commented extensively on a paper published in the same year by John Calaby, George Mack and William Ride that fundamentally disagreed with their own findings. In an attempt to determine the specific identity of Müller's holotype, Calaby, Mack and Ride examined a wide range of specimens collected in 1960 from the type locality by a party from the Queensland Museum. They gave the specimens obtained as being five grey kangaroos, seven grey wallaroos, six antilopine wallaroos, seven whiptail wallabies and ten agile wallabies. Included among these was a young male Eastern grey kangaroo, Queensland Museum number J.10749, approximating in weight and with the same dental age as the 38 lb animal collected during the *Endeavour* voyage, a specimen that would later assume central importance when this debate was brought before the International Commission on Zoological Nomenclature (ICZN). The dentition of this specimen was found to be the same as that of the photograph of a skull in the Hunterian collection of the Royal College of Surgeons in London that is believed to have been given by Banks to the anatomist John Hunter (see number 119). Calaby *et al*. concluded that the specimen described by Hawkesworth and then designated *Mus canguru* by Müller was an Eastern grey kangaroo weighing 38 lbs since it was a young animal like the modern specimen, which also possessed similar limb measurements to that appearing in Hawkesworth. They further argued that Solander's slip description was indeed a composite one that included details applicable to more than one species, and they pointed out that not only the whiptail wallaby but also the grey wallaroo has an unfurred muzzle.

Calaby *et al*. agreed with Morrison-Scott and Sawyer that Solander's description of the incisors accords neither with the grey kangaroo nor the whiptail wallaby, but conforms with the condition in the wallaroo. They said that although Solander thought the 84 lb male was 'adultus', his description implies that the two larger specimens before him had unerupted last molar teeth, that is were not at their full growth. However, all of the male whiptail wallabies between 30 and 50 lbs in weight examined by Calaby *et al*. had erupted last molars, thus ruling out the possibility that either of the larger specimens described by Solander was a whiptail wallaby. Finally, as suggested, the most obvious character in whiptail wallabies is a prominent facial stripe and another on each thigh, a character that surely could not have been missed if a whiptail was one of the animals described by Solander. Calaby *et al*. thought that the 8½ lb specimen was probably a grey kangaroo. Notwithstanding all these arguments, in their postscript Iredale and Troughton persisted in the view that Solander's description mainly referred to the 38 lb animal, still for them a northern form of a whiptail wallaby. In addition, they threw doubt on the authenticity of the Hunterian skull, questioning whether it really came from the *Endeavour* voyage at all and suggesting that the slide image of it showed a mismatched cranium and mandible.

The identification of the holotype specimen of the kangaroo has important implications for kangaroo nomenclature and in the mid-1960s a number of zoologists, including protagonists from both sides of this lengthy debate, submitted a series of complex taxonomic and nomenclatural proposals to the ICZN in order to bring stability to the naming of the taxa involved, in particular to conserve the name *Macropus giganteus* for the grey kangaroo (see Calaby, Mack and Ride (1963) and 'Comments on the Proposed Stablilization of the Macropus Shaw 1790 Z.N.(S.) 1584'). As part of these it was proposed by Calaby, Mack and Ride in 1963 that the skull possibly given by Banks to John Hunter be declared the holotype of *Mus canguru*, and that, since this specimen had been destroyed, it be replaced by a neotype, namely the male grey kangaroo previously collected on 24 November 1960 at Kings Plains, some 20 miles south of the Endeavour River, and since then preserved as a male skin and skull in Queensland Museum, J.10749. The ICZN finally ruled in 'Opinion 760' (1966) that the names *Macropus* and *giganteus* be placed on the official list of generic and specific names in zoology, making *Macropus giganteus* the scientific name for the Eastern grey kangaroo, in effect a decision some 250 years in the making and one in agreement with Shaw's initial choice of that name. The commission also ruled that the neotype specimen for the Eastern grey kangaroo should indeed be that designated by Calaby, Mack and Ride. The commission decided that the name *Macropus major* would be available for the Sydney grey kangaroo if this is considered to be a separate specific or sub-specific taxon from *Macropus giganteus*, and they rejected the proposal by Troughton and McMichael (1964a) to apply the name *Mus kanguru* to the whiptail wallaby. The outcome of these rulings was to arrive at nomenclatural stability while leaving everyone free to take their own view of the identity of 'Captain Cook's Kangaroo'. Although the identities of the species actually gathered at Endeavour River remain debatable, one thing is clear, which is that such a term cannot be correct given that Cook never caught a specimen, and it ought more properly to refer to Gore and Banks as the only voyagers who apprehended them on this voyage (see number 116). No specimen was taken in the Botany Bay area and this commonly held misconception is also definitely wrong.

118 *Macropus robustus* Gould, 1841

By Nathaniel Dance (later Sir Nathaniel Holland).

The Natural History Museum, London. Wheeler, Catalogue, Part 3 Zoology (1986), 5.(1:5). 48 × 30 cm.

This is an undated watercolour illustration of a complete skull and below it the jaw of a kangaroo specimen, annotated in ink on the recto 'N. Dance'. Nathaniel Dance was a celebrated portrait painter and founder member of the Royal Academy. He eventually gave up portraiture for the life of an MP and propertied gentleman, although he continued to produce some landscapes and political caricatures. Shortly before Cook departed on his final voyage, Dance completed for Banks the famous portrait of the great navigator sitting in his naval uniform pointing at the east coast of Australia on his own second-voyage polar map of the southern hemisphere (see number 134). Dance's portrait was for many years hung over the fireplace in Banks's library at 32 Soho Square and is now held at the National Maritime Museum, Greenwich.

The present illustration shows a kangaroo specimen possibly brought back to England on *Endeavour*, but the specimen's subsequent fate is not clear and Morrison-Scott and Sawyer (1950) were unable to trace it. These authors identified the drawing as a subspecies of *Macropus robustus*, the common wallaroo. Dance's drawing, they suggest, may well show the skull of the 84 lb male specimen shot by John Gore at Endeavour River on 27 July 1770 (for which see number 116).

Calaby, Mack and Ride (1962) compared the skulls of animals that were collected in the Endeavour River area in 1960 with the Dance illustration reproduced in Morrison-Scott and Sawyer, and on the basis of careful scrutiny of dentition they concluded that the drawing is of a specimen of the grey wallaroo, *Macropus robustus robustus*, most probably derived from the 84 lb specimen shot on 27 July. However, Troughton and McMichael (1964b) disputed these findings, arguing that since Dance's drawing is undated it might have been completed much later than the *Endeavour* voyage from other specimens supplied once a settlement had been established at Sydney Cove. They further suggested that the drawing is imperfect in a number of respects and could not therefore be relied upon for an identification. These arguments were used by Troughton and McMichael to support their contention that the 38 lb animal shot on 14 July 1770 was a whiptail wallaby (for the controversy surrounding identification of the species gathered by the voyagers at Endeavour River, see number 117 immediately above).

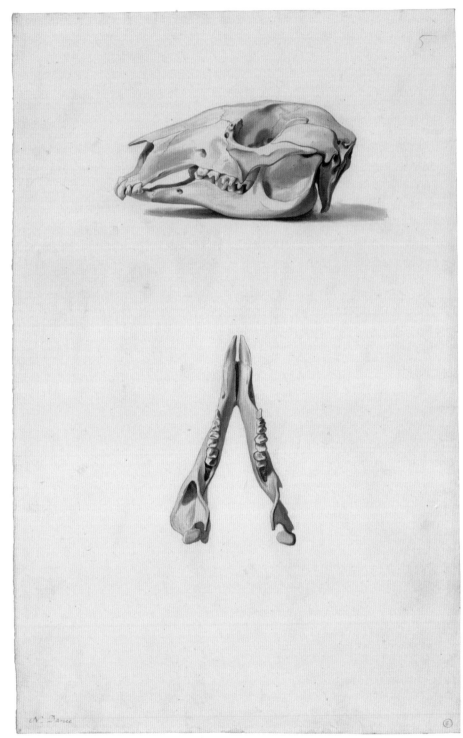

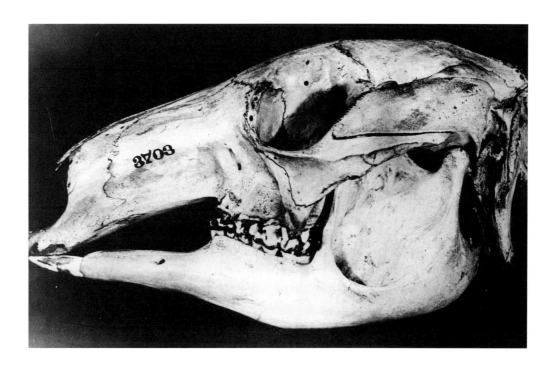

119 Black and white photographic print of the skull of a kangaroo

Museum of the Royal College of Surgeons (Hunterian Museum).
15.25 × 22.9 cm.

This is a photograph made from a lost lantern slide of kangaroo parts thought to belong to a single specimen brought back from the *Endeavour* voyage. The skull concerned was in the Hunterian collection of the museum of the Royal College of Surgeons in London until it was destroyed by German bombing in May 1941, but fortunately it had been photographed beforehand to make a slide to illustrate the George Adlington Syme Oration given in 1939 by Sir Alfred Webb-Johnson. The slide image was then published by Webb-Johnson in a resulting paper, 'Surgery in England in the Making', p. 23.

The museum of the Royal College of Surgeons contains the collections of Banks's contemporary and friend, the distinguished Scottish surgeon and anatomist John Hunter. Hunter referred to the fact that he was given a kangaroo skull by Banks in published observations made in 1790 in the journal of the colonial surgeon John White, and Morrison-Scott and Sawyer (1950) took the view that the slide was of the skull given by Banks to Hunter, numbered 1732 in a museum catalogue by Richard Owen in 1853 and then 3703 in a subsequent museum catalogue by William Flower in 1884. Morrison-Scott and Sawyer identified the specimen as the skull of a young Eastern grey kangaroo (to them a great grey kangaroo), and they designated the slide of it as the photo-lectotype of *Macropus canguru* Müller. If it did originate from one of the three specimens collected at Endeavour River during the enforced stay there, then these authors believed that it was probably the skull of the 38 lb animal shot on 14 July 1770 by John Gore. Calaby, Mack and Ride (1962) identified this skull as that of a young grey kangaroo which they also believed came from the 38 lb specimen collected near the Endeavour River. They noted that there can be no question of a lectotype since technically speaking only one specimen of this taxon was collected and a lectotype is a specimen later selected to serve as a type specimen for a species originally described from a set of syntypes. None of these authors indicated that the skull and jaw might not be from the same animal.

However, this identification was challenged by Troughton and McMichael (1964a), who considered that the slide represented a mismatched cranium and mandible, showing that at least two skulls of similar size were available at the time they were catalogued, thus casting doubt on the authenticity of both parts. They also pointed out that there is no extant statement as to the actual origin of the Hunterian skull in Banks's or Hunter's handwriting, and that there is therefore no positive evidence regarding its origin. They further argued that, even if this were the skull to which Hunter referred, there is no proof that it came from Cook's voyages since Hunter simply stated respecting the kangaroo that 'the only parts at first brought home were some skins and skulls' (see White, *Journal of a voyage to New South Wales*, p. 272). This could be interpreted, according to Troughton and McMichael, as meaning that the specimens were brought to England after the settlement of Port Jackson because it is known that various specimens, alive and dead, were sent home prior to 1790. These views, particularly as regards the mismatched parts of the skull, have not found support among other zoologists.

The original lecture slide showing the skull illustrates the tangled and often perplexing history of objects brought back from the *Endeavour* voyage, not least those relating to the kangaroo. It has not yet been traced among the collections held at the Royal College of Surgeons, and the version included here is a photograph made from the slide, possibly in the 1970s but for an unknown purpose. On the back of the photograph are stamped the name, address and telephone details of a now defunct London-based photographic company Fox Photos Ltd. The photographic image was apparently later reproduced, ironically enough, for a 2003 college exhibition entitled *What is missing*, looking at specimens destroyed by World War II bombing. It would seem that if the slide was sent to an outside firm at an earlier date to make a modern photograph then it was for some reason not returned to its original place, or that it was put back but subsequently lost.

120 'The Kongouro from New Holland.'

Commissioned by Joseph Banks and painted by George Stubbs, signed and dated 1772. Exhibited at the Society of Artists, London, 1773 (318, entitled 'A Portrait of Kongouro from New Holland, 1770').

The National Maritime Museum, Greenwich, London, ZBA5754. 60.5 × 71.5 cm. Oil on mahogany panel.

This is an oil painting by the animal painter and engraver George Stubbs, which was commissioned by Joseph Banks. It is one of a pair of paintings for Banks by Stubbs arising from the *Endeavour* voyage, the other being of a dingo (entitled 'Portrait of a Large Dog', and first exhibited alongside the kangaroo painting, 319). Stubbs was England's foremost illustrator of animals in this period, renowned for the accuracy of his depictions of horses, dogs and sporting subjects, frequently in paintings commissioned by the nobility to show them and their favourite steeds or pets. Largely self-taught, Stubbs applied himself to the study of anatomy in order to improve his art. The accuracy achieved in his horse paintings was chiefly the result of many dissections of these animals that he undertook in Horkstow in northern Lincolnshire during the mid-1750s, leading to his masterly *The Anatomy of the Horse*, an illustrated work that he engraved and had published in London in 1766. This work gained Stubbs the recognition not only of the gentry but also of leading anatomists and naturalists across Europe. By the time of Banks's *Endeavour* commissions Stubbs was well known in Royal Society circles for his work on exotic animal species, having produced various illustrations of these for the anatomist brothers William and John Hunter, who rated his work exceptionally highly. What the Hunters valued in Stubbs's work was its use as a reliable visual aid and source of scientific information for the study and teaching of anatomy. Banks's own focus in the artistic representation of natural history reflected similar priorities, making Stubbs a logical choice for him too. His extensive patronage of artists and engravers over many years demonstrates his central concern with the degree of precision achievable in art when recording natural history or antiquarian subjects. As Farington observed of Banks's taste: 'accuracy of drawing seems to be a principal recommendation to Sir Joseph' (*The Farington Diary*, vol. I, p. 27).

It is noteworthy, therefore, that Stubbs's version of a kangaroo is presented in such a statuesque manner looking backwards over its shoulder, and with significant deficiencies of detail, such as the hind feet where the middle digit is shown clubbed and almost without its prominent claw. As with other artists working on the voyage collections in London, Stubbs lacked first-hand experience of the Pacific world, and he had to rely on incomplete material supplied by others. In the present case he probably worked from a stuffed or inflated kangaroo skin, which would explain the oddly swollen form of his depiction. Moreover, the skin given to him might have been partly damaged. Which specimen skin it was remains unclear, but it would seem that it was a small one, and so possibly the young female caught by Banks's greyhound on 29 July 1770 (see Carter, *Sir Joseph Banks*, pp. 90–1, and number 116). Calaby, Mack and Ride (1962) examined an image of Stubbs's painting but were unable to identify it specifically with any species occurring at Cooktown.

Scholars have debated whether Stubbs had access to Sydney Parkinson's sketches of a kangaroo, showing more clearly the natural form of the animal and its posture, with the Natural History Museum zoologist P.J.P. Whitehead plausibly suggesting that this might not have been the case owing to the legal dispute between Banks and Stanfied Parkinson over the deceased artist's effects (see Whitehead, 'Zoological Specimens from Captain Cook's Voyages', pp. 182–4, and number 5). Whether Banks or Solander advised Stubbs also remains an enigma, but it is possible that they were able to provide only limited help if preoccupied with other ventures, such as preparing for a second voyage under Cook. Reproductions of the Stubbs image were steadily corrected once further information and specimens arrived in England, and for many years it remained the basis for numerous printed kangaroo illustrations. Indeed, more than a hundred different engraved versions of it have been identified in a range of literature up to the mid-nineteenth century, and its likeness has been adapted for flags, stamps, coins and medals. By contrast Stubbs's dingo never achieved the iconic status of its counterpart. It was painted from eye-witness reports alone, no specimen having been brought back, and so fell short of the true structure and character of that animal.

Stubbs's kangaroo painting was largely forgotten by natural historians from the late eighteenth century until 1957, when Averil Lysaght brought it back to their notice as the original for the engraved version that was first published in 1773 by John Hawkesworth in the official *Endeavour* voyage account (see the 1957 article by A.M. Lysaght, 'Captain Cook's kangaroo'). The Hawkesworth version is shown in number 121 immediately below. Stubbs's kangaroo and dingo paintings descended through the family of Banks's widow Dorothea, and until recently were kept at Parham Park in Sussex. Both have now been acquired by the National Maritime Museum, Greenwich, using major grants from various donors after government, drawing on expert advice, intervened to prevent their sale and export to Australia.

Interestingly, a young spirit-preserved specimen exists at the Natural History Museum, London, that was catalogued by the zoologist J.E. Gray in his *List of the Specimens of Mammalia in the Collection of the British Museum* (1843), p. 87, under *Macropus major* (=*M. giganteus*) and there recorded as 'From Capt. Cook's Voyages' (Whitehead, 1969). Whitehead was unsure why Gray suspected this wet specimen of being a Cook-voyage animal. Gray does not mention a specific voyage by Cook, but the *Endeavour* expedition would be the most obvious candidate, although here again no particular specimen from that expedition is stipulated in Gray's catalogue. Whitehead also observed that on page 145 of an earlier manuscript catalogue by Gray (the same number having been scratched, presumably by Gray, near the base of the jar containing the surviving specimen) there is the comment that it was described by Shaw, but Shaw in his 1800 description of *Macropus giganteus* believed that the first described specimens of the kangaroo from the *Endeavour* voyage were males. Whitehead noted that the wet specimen is a female, and so Shaw had evidently not seen it, nor indeed Solander's slip description in which both male and female gentialia as well as the mammae are described. In fact, it can here be confirmed that the wet specimen in question, GERM 145b, only weighs just over 3 lb, and its pelt, skeleton and entrails are all intact. There is a small incision on its chest. The number 145b is scratched on the side of its jar, not near the base as Whitehead said, and the letter 'b' now appears at the end of this number. The jar also contains two labels, one that reads 'Macropus major' and another that reads 'Macropus giganteus'. Both labels are marked 'yg', probably for the word 'young', and 'Cook's voyages'. Taken together these things suggest that this is not an *Endeavour* specimen, and certainly not the the young female caught by Banks's greyhound (see number 116 above).

120

121 'An animal found on the coast of New Holland called Kanguroo.'

John Hawkesworth, ed., An account of the voyages undertaken by the order of His Present Majesty for making discoveries in the southern hemisphere, and successively performed by Commodore Byron, Captain Wallis, Captain Carteret, and Captain Cook, in the Dolphin, the Swallow and the Endeavour: drawn up from the journals which were kept by the several commanders, and from the papers of Joseph Banks, Esq, 3 vols (London: Printed for W. Strahan and T. Cadell, 1773). Vol. III, plate 20. Original painting by G. Stubbs. Engraver unknown.

The British Library, London. 455.a.23. Rebound in half blue leather and blue cloth. Title, place and date of publication, gilt lettering on spine. Editor and volume given. 29 × 24.2 cm (closed). Plate 22.5 × 26.8 cm.

This print engraving is the first published image of a kangaroo. It appeared in the official voyage account of the *Endeavour* expedition, edited by Dr John Hawkesworth, and is based on the painting of a specimen skin from the voyage by George Stubbs (see number 120 above). Hawkesworth accompanied the engraving with a description of the animal derived from the journal comments of Cook and Banks, following Banks in likening its form and motion to that of a jerboa, and he appended the engraving to the entry for 14 July, the date that Third-Lieutenant John Gore shot his first male of 38 lbs. Note that the engraving is a reverse image of the original painting due to the engraved plate being turned over to print the image on to paper. The plate is a reliable reproduction of the Stubbs original. The animal is given prominence on a rocky outcrop, as it was by Stubbs, and in the background is a mountainous scene evidently intended to reflect the Australian landscape as encountered at this point during the voyage. Similar native flora appear in both the painting and its engraving. On the tail-side of the kangaroo there are in each, for example, plants of the grass tree (*Xanthorrhoea sp.*), which was used by Aborigine tribesmen to make spears of the kind featured in number 106. A palm is visible in the middle distance in the same direction, and what may be eucalypts are scattered on the hills ranged behind.

This was how many Europeans gained their first impression of the kangaroo. Kangaroos fascinated scientific and polite society in Georgian England. According to James Boswell, Samuel Johnson was unimpressed by the findings from the voyage when he read about them in the official account, with the notable exception of kangaroos, which the bulky scholar imitated by hopping about a room (reputedly at Inverness during a tour of the Hebrides late in 1773). A Staffordshire pottery responded to the discovery by producing mugs decorated with the Stubbs kangaroo, and when a live specimen arrived in London late in 1791 it attracted huge crowds of viewers paying an entrance fee of a shilling each to see it. Not only that, the publication of the official account introduced into the English language an Aboriginal word for this animal, based on the Guugu Yimithirr term *gangaru*, which was rendered by Banks and Cook in their journals as 'kangooroo', 'kanguru' and by other similar spellings (see number 129). The Guugu Yimithirr tribe resided in the area where *Endeavour* was landed for repairs. The adoption of a local name into popular and natural history usage was relatively rare for this voyage and must partly reflect the fact that kangaroos were essentially new to Europeans, who had no existing word for them and lacked a suitable systematic category into which to place such an animal. According to Raphael Cilento in a survey of Guugu Yimithirr words still in use in 1970, it appears likely that the specimen identified to Banks as *gangaru* was an Eastern wallaroo, *Macropus robustus* (see number 118), or an Eastern grey kangaroo, *Macropus giganteus* (see number 117), two of five kangaroo, wallaroo and wallaby species indigenous to the Endeavour River area. Guugu Yimithirr tribespeople at Hopevale in North Queensland who were interviewed on behalf of Cilento had various specific words for macropods found in the vicinity and gangaru was their name for 'the big black kangaroo'.

Australia's often perplexing fauna has raised many taxonomic questions that have been the subject of lively debate among naturalists since the late eighteenth century, not least among these being those concerning the monotremes platypus and echidna, and those to do with the marsupials kangaroo, wombat, koala and the Tasmanian devil.

122 *Callistemon viminalis* (Solander ex Gaertner) G. Don in Loudon, *Hort. brit.*: 197 (1830)

The Natural History Museum, London. Diment et al., Catalogue, Part 1 Botany (1984), A 3/130. 44.5 × 28.5 cm (mount).

This specimen was collected in the Endeavour River area during the period that HM Bark *Endeavour* was being repaired and supplies gathered there. It is the holotype specimen, the specimen to which the species name is attached.

Callistemon is a genus of more than 30 species of shrubs in the myrtle family, Myrtaceae, all of which are endemic to Australia. The species are commonly known as bottlebrushes because of the brush-like form of their cylindrical inflorescences, each flower bearing a mass of usually brightly coloured stamens. The petals are inconspicuous. Callistemons are widely grown, highly ornamental plants, and have been known in Europe since *C. citrinus* (which was also collected at Botany Bay) was introduced to Kew Gardens by Banks in 1789 and there cultivated. Many species are winter-hardy in cool temperate climates. *C. viminalis* is commonly known as the Weeping Bottlebrush and is a shrub or small tree growing up to 8 m in height. Many cultivated varieties of this species have been selected, including *C.* 'Captain Cook', this name marking the 1970 bicentennial of Cook's voyage to Australia.

The ensuing sequence shows the stages of preparation leading to publication, from the specimen gathered by Banks and Solander in 1769 through to a print from the original engraved plate of the plant, which was produced in 1982. It took some two hundred years from the time that Banks first embarked on publication of his planned *Florilegium* of new plants discovered on the *Endeavour* voyage for this enormous work to be fully published in the 1980s using the copper plates that he had engraved for the purpose. As conceived by Banks, the *Florilegium* was to be a folio work in fourteen volumes, separate from the official account of the voyage as edited by John Hawkesworth, and organized and funded privately by Banks himself. For a variety of reasons the work was not published by Banks and his co-worker Daniel Solander (see below), but the illustrations and engraved plates that were compiled by them remained among Banks's collections and eventually passed to the Natural History Museum, London.

123 *Callistemon viminalis* (Solander ex Gaertner) G. Don in Loudon, *Hort. brit.*: 197 (1830)

By Sydney Parkinson.

The Natural History Museum, London. Diment et al., Catalogue, Part 1 Botany (1984), A 3/130. Height 54.5 × width 36.5 cm; plant ill. height 34 cm.

This is an outline pencil drawing by Sydney Parkinson. Many of his sketches were signed and dated, and the place of collecting noted in ink on the reverse by Banks. Parkinson often lacked time during the *Endeavour* voyage to make completed illustrations of the many plants that Banks and Solander were collecting, and so after Tierra del Fuego he increasingly resorted to outline sketches of specimens annotated with notes as to their colours. The colours of plant specimens rapidly faded and each was soon pressed for preservation in Banks's herbarium. Using his sketches and notes, as well as the herbarium specimens, Parkinson could later produce finished illustrations. He died, however, on the way back to England and so his preparatory sketches and notes provided material for completion by others.

After the return of *Endeavour* a large number of illustrations of new plant species were finished in London, and then line-engraved on copper plates for Banks's planned *Florilegium*, a work not actually printed in full until the 1980s. This modern printing included all of the 738 extant *Florilegium* plates now held at the Natural History Museum, London. Not every detail of the original sketch was exactly copied through to the final plate, but a close similarity was maintained and the essential botanical facts preserved.

As a guide to its appearance Parkinson applied some colour to his sketch of the present plant, and he also indicated the plant's colours in pencil on the verso, the side normally chosen by him for such notes. These notes may be compared with the finished watercolour and print, 'the Petala & stamina carmine anthera cream colour. leaves above grass green faintly vein'd with w^t lighter below more pale vein'd the same. Capsula sordid brown. buds green ting'd w^t red.' On the same side in an unknown hand is noted 'Metrosideros viminalis', and there is an ink note by Banks 'Endeavours River'. The recto carries a Parkinson pencil annotation '70'.

124 *Callistemon viminalis* (Solander ex Gaertner) G. Don in Loudon, *Hort. brit.*: 197 (1830)

Watercolour by Frederick Polydore Nodder after Sydney Parkinson.

The Natural History Museum, London. Diment et al., Catalogue, Part 1 Botany (1984), A 3/130. Height 54.5 × width 36.5 cm; plant ill. height 33.5 cm.

This is a finished watercolour by Frederick Polydore Nodder after Sydney Parkinson. An anonymous ink annotation on the recto reads 'Fred^k· Polydore Nodder. Pinx^t· 1777.' It was engraved for Banks's planned *Florilegium*.

During the *Endeavour* voyage, Banks and his party worked in the ship's great cabin, each day busily describing and illustrating the natural history collections until dark. Banks and Solander selected specimens that were new or noteworthy for illustration. They guided Parkinson in his mainly botanical illustrations, which are always elegant and vital. Parkinson was able to complete most of his illustrations during the Atlantic stage of the mission, but thereafter he mainly produced annotated sketches due to the great number of new plants being collected. More than one manuscript listing of the specimens that were gathered was compiled during and after the voyage, with systematic descriptions also being made in Solander's slip catalogue. The classificatory system employed was that of Carl Linnaeus. Hand copies of much of this material were prepared, but none of it would be published. Banks later wrote of his party at work: 'We were well supplied with books on the natural history of the Indies. Storms were seldom strong enough to interrupt our study time, which lasted each day from about 8 a.m. to 2 p.m., and after the smell of food had disappeared, from 4 or 5 p.m. until dark. We worked at the great table in the cabin with our draughtsman opposite. We directed his drawing, and made rapid descriptions of our natural history specimens while they were still fresh. When we had been at sea a long time, and they were exhausted, we

123

124

completed our descriptions, and added synonyms using our library' (Chambers, ed., *Select Letters*, letter 23).

After his withdrawal from the *Resolution* voyage and trips to Iceland and then Holland, Banks started to prepare the botany of the voyage for publication, which entailed completion of the descriptions of specimens selected for inclusion, and finishing illustrations of them for engraving. He anticipated a fourteen-volume *Florilegium* that would carry his and Solander's names as authors. Before his untimely death, Sydney Parkinson had finished 269 plant drawings as well as 674 annotated plant sketches, making a total of 943 plant illustrations. In order to complete the required draft sketches, Banks employed five artists from 1773 to 1784, when work tailed off. Based on the charges arising from similar botanical works, among them William Roxburgh's *Plants of the Coast of Coromandel*, the estimated average cost of each final illustration was about £3. The initial artists were J.F. Miller, his brother James and also John Cleveley, who together finished 210 illustrations, and Thomas Burgis, who finished just three. Frederick Polydore Nodder joined the project when Banks moved from his London residence in New Burlington Street to Soho Square in 1777. Nodder was an assured botanical artist and engraver who contributed to works by Thomas Martyn, Erasmus Darwin and George Shaw among others. The former illustrators having left to participate in other works, Nodder went on alone to complete 272 drawings for Banks. The contribution of these London artists when added to the finished Parkinson drawings along with 110 unattributed illustrations amounts to a grand total of 864 botanical drawings. Of these Banks selected 753 to go forward for engraving and 738 of the engraved plates are today extant.

251

125

126

125 *Callistemon viminalis* (Solander ex Gaertner) G. Don in Loudon, *Hort. brit.*: 197 (1830)

Engraving by Gerald Sibelius.

The Natural History Museum, London. Diment et al., Catalogue, Part 1 Botany (1984), A 3/130. R. Brown Ms. 19/458. Plate height 46 × width 29.5 cm; plant ill. height 33 cm.

This copper plate was prepared by the Dutch engraver Gerald Sibelius, whom Banks first met during a short visit to Holland in early 1773. Sibelius was working for Banks on his planned *Florilegium* by the autumn of 1774.

Like all the plates engraved for the *Florilegium*, the present one was used to make a black ink proof impression, of which about three sets were produced, and some proof sheets sent by Banks to fellow botanists. The proof of the present plate is not included here. Work engraving the copper plates for the planned *Florilegium* commenced in earnest on Banks's return to England from trips to Iceland and Holland. Selection of the plants to be engraved lay with Banks and Solander, the latter of whom also ensured completion of the taxonomic descriptions. They supervised progress together, along with Banks's new librarian from 1777, the Swede Jonas Dryander. Importantly, it was in 1777 that Banks moved from his New Burlington Street residence to 32 Soho Square, his London home and the base for his natural history collections for the remainder of his life. Sigismund Bacstrom, who acted as Banks's secretary, compiled a master list of the species chosen for possible inclusion, indicating where each specimen was collected, their manuscript plant name, who drew and engraved them, whether this work was completed or not, and which species were selected for final illustration and engraving, and he also transcribed the botanical descriptions. His list indicates that more than 800 plants were considered for publication.

From 1773 to 1784 eighteen different engravers were employed on the *Florilegium*, few of whom are well known today since London's leading botanical engravers were mostly occupied on other projects. Johann Sebastian Müller's *Illustratio Systematis Sexualis Linnæi*, issued in twenty parts from 1770 to 1777, and William Curtis's *Flora Londinensis*, appearing in six fascicles from 1775 to 1798, were the chief illustrated botanical works then absorbing such talents as Müller himself, Kilburn, Sansom and Sowerby, but at least these served as useful examples for Banks to consider when deciding how to render his own collection in copperplate engravings. Of those men engaged by Banks, the major contributors were Daniel MacKenzie, Gerald Sibelius and Gabriel Smith, who engraved 251, 195 and 118 plates respectively. The remaining 189 plates were the work of the other engravers.

Assuming an average cost of at least £3 per engraving based on the later *Plants of the Coast of Coromandel*, this and the drawings for the *Florilegium* at the same cost per plant of about £3 each, as well as printing and binding, would have amounted to a total publication outlay by Banks of more than £12,000. Publication of the *Florilegium* raised a number of technical problems too. Representing the plants and their structure accurately was one, particularly where subtle or indistinct features were to be rendered, as was the question of colour in the final prints, something for which Parkinson laid the groundwork in his sketch notes. Despite the offer in November 1773 by Peter Perez Burdett of a possible method for achieving greater tonal variation and delicacy in the plates, using what was in effect the technique then new to Britain of aquatint, in the end Banks preferred to show the outline, structure, texture and shading of his *Florilegium* plants in line engraved prints. And while in future years colour was applied to other illustrated works that he oversaw, for the *Florilegium* he opted for black ink only.

126 *Callistemon viminalis* (Solander ex Gaertner) G. Don in Loudon, *Hort. brit.*: 197 (1830)

Modern print from copperplate engraving by Gerald Sibelius

The Natural History Museum, London. Diment et al., Catalogue, Part 1 Botany (1984), A 3/130. col. engraving 1982 BF: pl. 114. Plate height 46 × width 29.5 cm; plant ill. height 33 cm.

This is a 1982 coloured engraving print from the original plate that was intended for Banks's *Florilegium* but never actually published by him.

The reasons for Banks's non-publication of a work far advanced by 1784 have been much debated. Botanical descriptions for the first half of the voyage and for New Zealand were more or less ready for publication. Many of the drawings and plates were also done. The first stages of the voyage might therefore have appeared since this was a work that could be issued in parts. Clearly, technical and other production issues may have had a bearing on its non-appearance, and it should be added that staffing and the sheer organization of such a large project were formidable obstacles in themselves. However, the three main reasons usually given for Banks's failure to publish are the sudden death in May 1782 of his great friend and collaborator Daniel Solander, although work did not altogether cease with his demise. Then there were the multiplying commitments to other organizations that Banks increasingly shouldered, foremost among them the Royal Society, of which he was president from November 1778. There was, too, his attendance at the British Museum, of which he was a dutiful *ex officio* trustee as a result of his election to the presidency. At Kew Gardens he also acted as unofficial director under the king from 1773. In particular, it should be noted that the year 1784 was one of upheaval at the Royal Society. Over the Christmas and New Year period a number of fellows combined to try to unseat Banks as president. He survived what was a bruising encounter, but if a reminder were needed of the attention that such positions really require then this was an especially sharp one. Moreover, Banks increasingly oversaw publications associated with these and other bodies, not least that for Cook's last voyage, a major work which appeared in 1784. This range of activity, along with the organization of new expeditions and their incoming collections, seems to have eclipsed his own publishing venture. Finally, economic problems caused by the American Revolution and a slump in the wool trade badly affected the income of landowners like Banks, as well as damaging the likely market for expensive botanical works. Thus, by 1784 the *Florilegium* had all but come to a halt.

In his will Banks bequeathed his Soho Square herbarium and library to the lifetime use of Robert Brown, the last of his librarians, both to be passed to the British Museum on Brown's death or beforehand if he agreed to do so. Brown agreed to a transfer to the museum in 1827, where he continued to oversee the collections as their keeper, including all the materials for the *Florilegium*. These and the herbarium were transferred to the newly created Natural History Museum in 1881, but Banks's library, their working partner, remained at the British Museum.

Hence the *Florilegium* materials now reside in South Kensington, whereas the library is to be found at the British Library, St Pancras, having in recent years been moved from the British Museum. Since Banks's death in 1820 there have been a number of proposals and some attempts to publish at least part of his *Florilegium*, but none succeeded in producing the complete work until the 1980s. Alecto Historical Editions then undertook this monumental task, which was completed between 1980 and 1990 using a printing technique employing colour *à la poupée* in conjunction with stipple engraving. One hundred numbered sets were made comprising the entire 738 extant plates, published in 35 parts. The *Florilegium* contains plates depicting the botany of Madeira (11 plates), Brazil (23), Tierra del Fuego (65), the Society Islands (89), New Zealand (183), Australia (337) and Java (30). The full modern work was also accompanied by a superbly detailed three-volume catalogue. Note that the final print shown here is a reverse image of the finished drawing, since the copper plate was turned over in order to print on to the paper.

127 *Trichoglossus haematodus moluccanus* (Gmelin)

The Natural History Museum, London. 1909.3.23.3. Length 29, width (side to side) 7 and height (back to stomach) 5.5 cm.

This specimen was probably not the subject of the engraved plate printed by Peter Brown in 1776 (see number 128 overleaf). The bird depicted by Brown was given by Banks to the ornithologist and collector Marmaduke Tunstall, and kept in his menagerie in London, and afterwards at his estate at Wycliffe in North Yorkshire. A year after Tunstall's death in 1790 his friend, the antiquary George Allan of Darlington, purchased part of the Tunstall collection, mainly the birds and some ethnographic objects, which he added to his museum. Allan died in 1800, but his son continued the museum until 1822, when the Literary and Philosophical Society of Newcastle purchased everything. The collection then formed the basis of the Newcastle Museum, which became the Hancock Museum in 1891, and is currently part of the Great North Museum: Hancock. The present whereabouts of the Tunstall lorikeet specimen, if it still exists, is unknown. It was last recorded in the society's collection in 1827 by G.T. Fox (*Synopsis*, p. 129).

Scholars believe that the specimen shown here was collected on a Cook voyage, from where it went into the private museum of the eminent collector Sir Ashton Lever, which from 1774 was housed in a former royal residence in London's Leicester Square. There Lever exhibited his collections under the name Holophusicon to signify that they embraced all of nature, and he included in them large numbers of natural history specimens and artefacts from Cook's second and third voyages. After the sale of the Leverian museum in 1806 this specimen passed into the collection of the Royal College of Surgeons, entering the British Museum in 1909 among a batch of 29 specimens from the college's collection. Since the natural distribution of this particular subspecies is eastern Australia and Tasmania, it is possible that the specimen originates from the *Endeavour* voyage. It may therefore be another specimen brought back by Banks or one of his companions since it cannot be the bird given by Banks to Tunstall. This skin is of considerable taxonomic importance as a type specimen and a rare survivor of zoological material from a Cook voyage.

128 'THE BLUE-BELLIED PARROT.'

Peter Brown, New illustrations of Zoology, containing fifty coloured plates of new, curious, and non-descript birds, with a few quadrupeds, reptiles and insects. Together with a short and scientific description of the same (London: Printed for B. White, 1776). Plate 7. By Peter Brown.

The British Library, London. 435.g.2. Rebound in half green leather and green cloth. Title, place and date of publication, gilt lettering on spine. Author given. 25 × 29.8 cm (closed). Plate 24.2 × 18. 7 cm.

Here in Banks's library copy of *New Illustrations of Zoology* by the natural history artist Peter Brown is the first published illustration of the Australian Rainbow Lorikeet, *Trichoglossus haematodus moluccanus* (Gmelin). Beautifully produced, it appears as plate 7 in that work along with an accompanying description of the 'Blue-Bellied Parrot'. In his comments on the parrot Brown states that this bird came from New South Wales and was 'very numerous in Botany Bay'. He adds that it 'was first brought over by JOSEPH BANKS, Esq.'. The Rainbow Lorikeet *Trichoglossus haematodus* is a parrot with spectacularly coloured plumage, various races of which are found in Australia and a wide area of the Pacific. This subspecies, *T.h. moluccanus*, has a distribution in eastern Australia and Tasmania, and is distinguished by its blue belly and in lacking the barring found on the breast of the nominate race, *T.h. haematodus*. It is presently known by a variety of common names, including Swainson's Lorikeet, the Blue Bellied Lorikeet and the Blue Mountain Parrot.

Sydney Parkinson devoted much of his time while in Australian waters to depicting botanical subjects, and did not complete any bird illustrations. Indeed, there is only one surviving Parkinson pencil sketch of a bird from this stage of the mission, that of a female red-tailed cockatoo, *Calyptorhynchus magnificus* (Shaw, 1790). Consequently, work on Australian bird species from the voyage had to wait until the expedition returned to Britain. It would seem that Tupaia kept a live lorikeet as a pet before his untimely death at Batavia. Banks later made a gift of the bird to the naturalist Marmaduke Tunstall, although precisely when, and even whether it was alive or dead by that stage, is not clear (see number 127 above). Brown painted this Lorikeet specimen, and a copper plate dated 3 November 1774 was produced, which he published in his *New Illustrations of Zoology*. It was quite typical of Banks to allow other naturalists to publish from his collections, not least at a time when he was contemplating another Pacific voyage, and forming his own publication plans concentrated on new plant species rather than animals from the *Endeavour* mission.

There is an earlier, unpublished painting of the rainbow lorikeet by the Welsh natural history and antiquarian artist Moses Griffith. Griffith worked for the naturalist and travel writer Thomas Pennant, who was allowed especially generous access in order to research and publish material from Banks's collections. Griffith's finished painting, in gouache on vellum, is signed and dated 1772. It appears to be the result of a visit with Pennant to Banks in London in September 1771, shortly after the return of *Endeavour*. Griffith may have sketched the same specimen used by Brown, or another like it, and completed his painting in the following year. The National Library of Australia, Canberra, purchased Griffith's painting from a private owner in 2012.

129 'Language of New Holland' by Joseph Banks

School of Oriental and African Studies, London. Library MS 12153. Part 11, 'Language of New Holland' in three folded folio sheets, when open each of 41.5 × 32.5 cm.

Volume rebound in brown cloth, a single volume containing 14 parts, these being vocabularies written on folios of different sizes by a variety of hands. Title in gilt lettering on spine 'Banks Vocabularies MS. 12153.' 34.7 × 24.7 cm (closed).

During the *Endeavour* voyage Joseph Banks and his team collected words as well as plants and animals. These formed part of the study they made of the cultures that were encountered, their social organization, beliefs and customs, dress and crafts, economic life and natural surroundings. Surviving mission word lists reflect this range and organization of approach. Often they include groups of words for such aspects as parts of the human body, names for man and woman, time divisions of day, month and year, the sky and celestial bodies seen in it, the names of gods and of certain individuals, words for animals, those for trees and parts of them, those for fruits and spices, also various man-made items, colours, numbers and selections of verbs and larger phrases. Arrangement is generally alphabetical using the English equivalent term for an indigenous word or else according to an indigenous word and its English definition grouped broadly by meanings as above. Like his other collections, the vocabularies that Banks compiled, along with those that he obtained from later voyagers, were placed at the disposal of scholars in London.

The present sheet, bound within a manuscript volume MS 12153, is one side of a list of 180 Aboriginal words written by Banks under the title 'Language of New Holland', a list enlarging considerably on the 38 words covered in his journal comments made immediately after *Endeavour* had departed the east coast. Scholarly consensus today is that the 'Language of New Holland' comprises words gathered at Endeavour River from the local Guugu Yimithirr people, with different columns recording variations in dialect within the group encountered there. Three obvious sources for this list are Banks's and James Cook's much shorter journal lists of Aboriginal words as well as Sydney Parkinson's vocabulary, which was later published posthumously by his brother (see *Banks's Journal*, vol. 2, pp. 136–7, *Cook's Journals*, vol. 1, pp. 398–9 and *Journal of a Voyage*, pp. 148–52). In this sense the list is interesting since it would appear to be a neatly written compilation of words and meanings gathered at Endeavour River rather than a working field list of the kind also found in MS 12153. This volume contains 14 parts, each part being a manuscript word list in Banks's or in a different hand, most of which were gathered during the *Endeavour* voyage.

The early parts comprise extensive lists of Tahitian words. The ensuing parts cover languages encountered later in the voyage but in somewhat less detail. This volume was formerly in the possession of the linguist and numismatist William Marsden, who in January 1835 presented his books and manuscripts to King's College, London. It was among a selection of Marsden library material that was transferred to the School of Oriental and African Studies shortly after its foundation in 1916.

William Marsden served as an East India Company official at Fort Marlborough, Sumatra, from 1771 to 1779 and there rose to the position of secretary to the company's government, but his real interests lay in oriental scholarship. Having accumulated enough wealth to live independently, and being inspired by accounts of the voyages of James Cook, he decided to return to Britain to pursue his studies in the company of the group centred on Banks and the Royal Society. His first publication in the journal of the Society of Antiquaries, in 1782, was a letter to Banks dated 5 March 1780. In it Marsden noted common words of similar pronunciation used across the southern oceans from Madagascar to the Marquesas, with Sumatra being his main area of focus. It was an early attempt to understand the links that exist between the widely scattered languages of this vast intertropical region. In 1783 Marsden was elected a fellow of the Royal Society, having that year published his *History of Sumatra*, a work which established his reputation as a writer on natural history, geography and languages. In 1795 he accepted a post as second secretary at the Admiralty, and in 1804 became first secretary, a post he held until 1807. Marsden's Admiralty connections no less than his intellectual pursuits made him a natural companion to Banks. He served as treasurer to the Royal Society from 1802 to 1810, and deputized in the chair for Banks when he could not preside at society meetings. Banks in his turn proved an especially useful contact for Marsden, whose memoirs state: 'Sir Joseph Banks considers me the repository of everything respecting languages that comes to his hand, and when a ship arrives from a discovery voyage, the vocabularies are delivered to me, as regularly as the Journals to the Admiralty' (*A brief memoir of the life and writings of the late William Marsden* (London, 1838),

41. the Nails	Molke		Mulke
42. a Stone	Walba	Walba	Walba
43. a Plantain stalk	Wolbit maiye (175)	Molbel-mayer	Molen-badje/Madje
44. Sand	Joowal	Yowall, Chóska	Joval-ta
45. Water	Poorai		
46. Wood	Yoocoo	Yacui	Joho (Gjoco)
47. a Basket	Gjendoo	Yangoo	Gjandie (Jandjie)
48. the scars on the body	Moro	Shkey, Wakje Faky	
49. a rope or line	Goorga	Gurka	Gurka
50. Fire	Maianang (29)	Meangal or Myangal	Meanang
51. Bones	Baityebai		
52. a Bag	Charngala	Charngala	
53. Earth	Joapoa		
54. Hole in the nose	Jennapuke		
55. a sore	Pandal	Combaky or Pander	
56. the Sun	Galan	Gallan	Gallan
57. the Clouds	Wulgar		
58. the Sky	Kere	Kearre	
59. the bone in the nose	Ja'pool		
60. a Father	Dunjo		
61. a Son	Jumurre		
62. a Man	Bamma	Bāma	
63. a Woman	Mootjel	Awhaore?	
64.	Kangooroo	Wultol (173)	Kanguru
65.	Jaguol	Quol	Ge-Kuol
66. a Dog	Cotta	Kota	
67. Lowry quet	Perpere	Peer-peer	Pier Pier
68. Cocatoo	Wanda (71)	Wanda	Wanda-Wandra
69. his crest	Waouwa	Waouwa	
70. a Hawk	Goromoco	Koromoco	
71. a Feather	Poetjo	Joitie Wanda 68	Warndia (69)
72. a Fly	Jabigga	Chepaca	Ge-Paga
73. a Butterfly	Walboolbool		
74. Male Turtle	Poenja	Poinja	Ge-Poinja
75. Female	Mameingo	Mameingo	Ge-maningo
76. Great cockle	Moenjo	Moinje	Moinge
77. Telescope shell	Meticul	Meture	Ge-Makiul
78.	Ebapee	Bapea	Ge-Papie

pp. 77–80). Marsden produced various academic papers, further editions of his *History of Sumatra*, and in addition to these *A Catalogue of Dictionaries, Vocabularies, Grammars and Alphabets*, privately printed in 1796, and *A Dictionary of the Malayan Language* (London, 1812). He later published a translation of Marco Polo's travels, two volumes on his coin collection, a catalogue of his library and in 1834 a volume of miscellaneous *Works* containing an important essay on Polynesian languages (see below).

Note that the present sheet includes at 64 the 'Kangooroo' or 'Kanguru', which is cross-referenced through 'Wultol' to a list entry for the word 'skin' at 173, skin there being defined as 'Wottol'. There was, of course, no existing English word for the kangaroo and so one is not given in Banks's list. The present page also contains the term for lorikeet at 67, given as 'Perpere', 'Peer-peer' or 'Pier Pier', a good approximation to the modern Guugu Yimithirr word for this animal. For the kangaroo and the lorikeet see numbers 116–21, and 127–8. The volume in which the present list is contained might have been lent by Banks to Marsden as late as 1811 (see Chambers, ed., *Indian and Pacific Correspondence*, vol. 8, letter 11). It was evidently not returned by Marsden since, as explained, it today resides among the Marsden library collections held at the School of Oriental and African Studies. There, too, are a small group of similar volumes probably from Banks containing bound manuscript vocabularies in various hands, including Banks's, which cover words collected not only on the *Endeavour* voyage but also on Cook's ensuing voyages and those of later explorers and colonists, see MSS 9882, 12023, 12156 (which largely reproduces in fair copy the contents of 12153) and 12892.

Interestingly, it would appear that some of the vocabularies in MS 12153 were used in contemporary and later publications, for example the word lists for Savu, Anjenga and Seram, part 8, which match the lists published posthumously in Sydney Parkinson's *Journal of a Voyage*, pp. 163–70, 195–7 and 200. These and other lists in MS 12153 are in Parkinson's hand, and it is possible that Banks obtained this material from Dr John Fothergill after the death of Stanfield Parkinson, who published the illicit edition of his late brother's voyage papers (see numbers 5 and 84). Moreover, Marsden himself clearly made use of Banks's vocabularies in his *Works* essay 'On the Polynesian, or East-Insular Languages', there building on his earlier exploration of Polynesian language by again comparing various words from its different branches. He drew in particular on those parts in MS 12153, covering New Zealand (part 13), the Endeavour River (part 11) and Savu (parts 7 and 8) (see *Works*, pp. 102, 112 and 110 respectively). Alongside these he used Daniel Solander's manuscript vocabularies for Tahiti at MSS 12023 and 12892, p. 101. In places he also made use of the published accounts by Hawkesworth and Stanfield Parkinson, copies of which were in his library. By this time Marsden had a large range of publications and other sources on which to rely as well as his own direct knowledge of language in the East Indies.

In the Society Islands Banks started to speculate on possible relationships between widely dispersed languages using his library aboard *Endeavour* and Tahitian numbers from one to ten that he himself had gathered. His journal contains words for these numbers from Madagascar, Tahiti, the Cocos Islands and Papua New Guinea that indeed appear similar, although in the case of the Madagascan numbers this seemed to Banks to be 'almost if not quite incredible'. He looked forward to reaching Java to pursue the idea already forming in his mind regarding 'the East Indian Isles, which from their situation seem not unlikely to be the place from whence our Islanders originaly have come' (*Banks's Journal*, vol. 1, pp. 370–2). Number lists covering the locations mentioned above and other places are to be found in part 5 of MS 12153. Further similarities are apparent between the word lists in the MS 12153 and those later recorded in Banks's journal at locations visited on the east coast of the North Island of New Zealand (see part 13 and *Banks's Journal*, vol. 2, pp. 35–7). In his journal Banks noted the resemblance between many words spoken in New Zealand and their Tahitian equivalents, deeming the two languages to be 'almost precisely the same at least in fundamentals'. He noted as well similarities in 'manners and customs' between what he recognized were related cultures. Banks seems to have relied on Sydney Parkinson for Māori words from Queen Charlotte Sound on the South Island, although he expressed doubt as to the accuracy of Parkinson's attempts to capture their sounds.

Banks continued to collect words in the Dutch East Indies, briefly noting in respect of the language at Savu that 'The genius of it seems much to resemble that of the South Sea Isles' (*Banks's Journal*, vol. 2, p. 174). At Panaitan he noticed basic similarities between a selection of words as spoken at that island and their equivalents in the Javan and Malay tongues. Further comparison of each of these with the language of the South Sea islands also revealed similarities, much as he suspected it would. He remained deeply perplexed, however, by the resemblance apparent in certain words and numerals obtained at Batavia from a Madagascan sailor of African race when set alongside their counterparts from the foregoing languages. An explanation might lie, he thought, in the then current theory that Egypt was an ancient cradle of language, from whence words might have spread by at least two routes running into Africa and also to Asia, thus accounting for these baffling findings. He could little have guessed that Madagascar had first been colonized from across the Indian Ocean well over two millennia before. At this point in Banks's journal words from Java, Malaysia, Madagascar and the South Sea are arranged in matching columns to illustrate his ideas concerning what are today held to be members of the extensive Austronesian language family. Word lists for Java and Panaitan in parts 10 and 14 of MS 12153 are almost certainly ones that he used when compiling them (see *Banks's Journal*, vol. 2, pp. 238–41). The voyagers did grasp, however, that the language of the Aborigines was unrelated to those of Polynesia, not least because Tupaia was unable to converse with any of the people on Australia's east coast.

Banks recognized that misunderstandings were highly likely when collecting words and their meanings, and compared his findings with those of his companions in an effort to eliminate obvious errors, as in the case of those words from New Holland that are included in his journal. It is clear, then, that the voyagers drew on each others' word lists, and further that Cook used Banks's journal vocabularies, sometimes very closely, in his own journal account of the voyage (see, for example, the lists for New Zealand in *Banks's Journal*, vol. 2, pp. 35–6 and *Cook's Journals*, vol. 1, pp. 286–7). By the later stages of the mission, however, Cook had largely lost interest in collecting words, whereas Banks's pursuit of them remained strong. Dr John Hawkesworth, as is well known, used both Banks's and Cook's journals to compile the official account of the voyage, and it would appear that he relied heavily on Banks's journal word lists, supplementing them with items from those of Cook as necessary.

130 'A Chart of NEW SOUTH WALES, *or the East Coast of* New-Holland. *Discover'd and Explored* BY *Lieutenant J:* Cook, COMMANDER *of his* MAJESTY'S BARK ENDEAVOUR, *in the Year* MDCCLXX'

John Hawkesworth, ed., An account of the voyages undertaken by the order of His Present Majesty for making discoveries in the southern hemisphere, and successively performed by Commodore Byron, Captain Wallis, Captain Carteret, and Captain Cook, in the Dolphin, the Swallow and the Endeavour: drawn up from the journals which were kept by the several commanders, and from the papers of Joseph Banks, Esq, 3 vols (London: Printed for W. Strahan and T. Cadell, 1773). Vol. III, facing page 481. By J. Cook and I. Smith. Engraved by W. Whitchurch.

The British Library, London. 455.a.23. Rebound in half blue leather and blue cloth. Title, place and date of publication, gilt lettering on spine. Editor and volume given. 29 × 24.2 cm (closed). Plate 36.9 × 78.8 cm. Scale approx. 1:4,500,000.

Following an epic voyage northwards along the east coast of Australia, the expedition entered a passage that James Cook named Endeavour Strait. He halted near Possession Island and, on 22 August, this was where he claimed for the Crown a coast that he now called New South Wales. The present impressive chart of the east coast was engraved and published in volume three of John Hawkesworth's 1773 edition of the *Endeavour* voyage account. A manuscript chart of the coast by James Cook and Isaac Smith is held at BL Add MS 7085, f. 34. There are a number of other charts and coastal views of New South Wales from this mission (for which see David, *The Charts & Coastal Views of Captain Cook's Voyages*, vol. 1, pp. 260–312).

Sighted by the consumptive Second-Lieutenant Zachary Hicks from the foremast rigging early on the morning of 19 April 1770, the east coast of Australia had never before been explored and charted by Europeans. At the southernmost point of the chart, Cape Hicks commemorates the voyager who first saw lands then broadly known as New Holland, but not yet understood to be part of a single landmass. The Dutch had already visited the other three coasts of the continent and, for a while yet, their choice of name persisted, but its days were numbered. At the northern end of the chart lies Possession Island, where Cook landed and claimed the coast he had just surveyed for Britain. This would henceforward be known as New South Wales. He acted in accordance with his instructions, and on the assumption that it had not before been visited by Europeans, although smoke from the shore and other nearby islands clearly indicated that these lands were inhabited.

Contacts with Aboriginal people had been limited since they cautiously shunned the voyagers. Partly as a result of this, local names are lacking in Cook's charts at this stage of the voyage. Most of those used derive from naval personages or other individuals, voyage events, calendar or festival days, or places in Britain, or they are descriptive in some way of the locations themselves. Cook rarely if ever named places after himself or his family – Cook Strait in New Zealand is said to have been named at Banks's suggestion.

Note Botany Bay and Port Jackson, the latter of which Cook named but did not enter. Here, after rejecting Botany Bay, Arthur Phillip would later situate the first British settlement on this coast, at what he chose to call Sydney Cove. Note, too, Endeavour River, near which it is recorded that 'On this Ledge the Ship laid 23 Hours'. The 'Ledge' concerned was Endeavour Reef. The near wreck of *Endeavour* necessitated a lengthy stop for repairs during which Banks and Solander explored and collected ashore, and contacts with local inhabitants took place. Putting back out to sea through the Great Barrier Reef proved both difficult and dangerous but, once achieved, the most hazardous moment in the entire voyage occurred when the ship was suddenly becalmed and almost carried by the incoming tide back on to the reef, at that time awash with a cauldron of cascading water. Providential Channel was so named because it was where *Endeavour* found a passage back inside the Barrier Reef, thereby avoiding certain destruction – a second lucky escape for Cook on a coast to which he never returned. Banks gives a vivid account of the episode in his journal, concluding in obvious relief at the ship's deliverance: 'How little do men know what is for their real advantage: two days [ago] our utmost wishes were crownd by getting without the reef and today we were made again happy by getting within it' (*Banks's Journal*, vol. 2, p. 108). In the ensuing Cook missions, two vessels were employed as a precaution against one being damaged or wrecked, although it was not always possible for these ships to remain together.

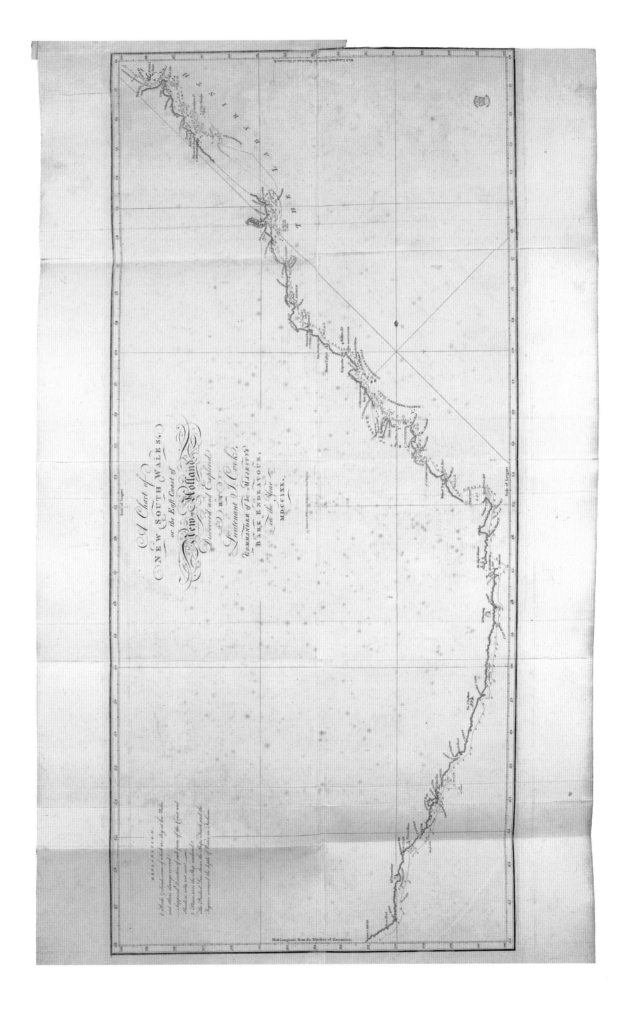

HOMEWARD BOUND:
NEW GUINEA, SAVU, JAVA, PRINCES ISLAND, CAPE OF GOOD HOPE AND ST HELENA

From Endeavour Strait, itself a notable discovery among the islands and reefs lying between Australia and New Guinea, Cook passed out of Torres Strait and entered the Arafura Sea, briefly visiting the New Guinea coast, then making his way first to the island of Savu and from there to Batavia to repair and restock his vessel. Thereafter his course lay across the Indian Ocean to the Cape of Good Hope, into the Atlantic and homeward via St Helena. Many aboard the ship now yearned for home, and Banks commented perceptively in his journal on the psychological effects of prolonged stress and absence: 'The greatest part of them are now pretty far gone with the longing for home which the Physicians have gone so far as to esteem a disease under the name of Nostalgia; indeed I can find hardly any body in the ship clear of its effects but the Captn Dr Solander and myself, indeed we three have pretty constant employment for our minds which I beleive to be the best if not the only remedy for it' (*Banks's Journal*, vol. 2, p. 145). He little knew that the mission's latter stages, so keenly anticipated by everyone, would bring worse afflictions than homesickness and exhaustion. A large number of the crew and much of his own party died of diseases contracted at Batavia, the first European port encountered since the Atlantic stage of the voyage. The survivors reached Deal on 12 July 1771.

131 Sketches of various objects

By Sydney Parkinson.

The British Library, London. BL Add. MS 9345, f. 20 verso. 18.4 × 23.5 cm.

This undated pencil sketch by Sydney Parkinson shows an Aboriginal man paddling in a bark canoe, two Aboriginal huts, a palm tree, a shrub and a dwelling annotated by Parkinson as 'Javanese house'. On the recto are other Parkinson sketches of a *woomera*, the head of an Aboriginal spear, part of a necklace, a man holding a spear beneath a coconut palm and a partly erased fish. Cook decided to repair and resupply at the Dutch port of Batavia (now Jakarta), a notoriously unhealthy place due to its canals and swamps, which provided an ideal breeding ground for diseases like malaria. The health of his men rapidly deteriorated during a stop lasting from 9 October to 26 December 1770. This was in stark contrast to their health beforehand, which had been excellent throughout the Pacific stage of the voyage as a result of Cook's measures to maintain cleanliness and prevent scurvy.

Cook was well aware of the dangers of scurvy, which could develop to such an extent on long sea voyages that a majority of the crew could be lost to it. Scurvy was a killer disease leading to ulceration, the breakdown of skin tissue, rictus of the limbs and, ultimately, death if untreated. Since the early part of the twentieth century it has been known that the symptoms of scurvy result from a dietary deficiency of vitamin B and primarily vitamin C (ascorbic acid). Before then it was, however, apparent that once sufferers returned to shore, eating certain plants such as 'scurvy grass', wild celery, wood sorrel and Kerguelen cabbage could bring about recovery. Perhaps most importantly, James Lind published in 1753 his *Treatise of the Scurvy* demonstrating that citrus foods such as lemons and limes had a rapid beneficial effect upon the disease. Concentrated fruit juice was used as a preventative on Cook's and other voyages in the 1760s. However the 'rob' (lemon concentrate) carried on *Endeavour* had been boiled to concentrate it, and in the process would have lost its vitamin C. Cook's success in managing the health of his crew probably resulted from a combination of measures. He also employed malt together with portable soup (a preparation of dried vegetables), vinegar, mustard, wheat and sauerkraut. Additionally he paid strict attention to the general condition of his ship and crew, particularly with regard to the airing and drying of the lower decks. There were only five reported cases of scurvy on the *Endeavour* and reputedly no deaths from it, something Cook attributed to the regular doses of malt taken, sometimes unwillingly, by his crew. By overcoming obstacles like scurvy it became possible in the eighteenth century for Europeans to undertake long sea voyages of exploration, although not every commander matched Cook's exemplary performance in defending against its deadly effects (see number 96).

Venereal diseases may have been a contributing factor to the high mortality rate during the later stage of the voyage. Upon landing in Tahiti Cook found that venereal diseases, mainly gonorrhoea but also syphilis, were established among the native population, presumably from earlier European visitors, most likely the French under Bougainville. Everyone on board *Endeavour* was examined by the ship's surgeon before arrival at Tahiti and found to be infection-free, though by July 1769 more than 40 crew members were found to have contracted venereal disease. One complicating factor in establishing the probable introduction and spread of syphilis in Polynesia is that its symptoms are similar to those caused by yaws, a tropical disease related to syphilis that is endemic to the region but not normally spread by sexual contact. What is quite clear is that such diseases, once present, caused devastation among the indigenous population.

At Batavia the surgeon William Monkhouse died of fever, as did Tupaia and Taiato, the two Society Islanders. Banks and Solander both contracted malaria, the latter becoming seriously unwell, and as a result Banks took the decision to move out of the town and into the countryside. With help he nursed Solander through his worst bouts of illness, and continued to support his ailing companion during the ensuing passage to and stay at the Cape of Good Hope. When Cook eventually departed Batavia, he commanded a sick boat, with a number of men suffering from fevers. Seven people died at Batavia. A further 23 died on the way to the Cape of Good Hope, most of them apparently from dysentery, possibly caused by contaminated food and water taken on at Princes Island (although Cook met with other captains at the Cape who suffered similar losses not having visited Princes Island after leaving Batavia). Banks's remaining artists, Sydney Parkinson and Herman Diedrich Spöring, were among the fatalities. Charles Green, the mission astronomer, died during the Indian Ocean crossing, his addiction to alcohol perhaps accelerating scorbutic and other symptoms. Thus the mortality rate during the voyage as a whole was much higher than any other under Cook. Of the original party of nine that Banks took on the voyage, just four survived – Banks, Solander albeit narrowly, Peter Briscoe and James Roberts.

132 'A Chief's house in the island of Savu near Timor'

By Sydney Parkinson.

The British Library, London. BL Add. MS 23920, f. 31.

29.8 × 48.3 cm.

This pen and wash drawing is annotated in pencil on the folio 'Drawn by S. Parkinson' and in ink as above. It is annotated in pencil on the verso 'Savu'. The expedition was at Savu for three days from 18 to 21 September 1770. In his journal account of Savu, dated 20 September 1770, Banks recorded that: 'Their houses are all built upon one and the same plan differing only in size according to the rank and riches of the proprietors, some being 3 or 400 feet in leng[t]h and others not 20. They consist of a well boarded floor raisd upon posts 3 or 4 feet from the ground; over this is raisd a roof shelving like ours in Europe and supported by pillars of its own independent of the floor; the Eaves of this reach within 2 feet of the floor but overhang it as much; this open serves to let in air and light and makes them very cool and agreable. The space within is generaly divided into two by a partition which takes off one third. From this partition forward reaches a loft shut up close on all sides and raisd about 6 feet from the ground, which occupies the centre third of the house; besides this are sometimes one or two small rooms taken off of the sides of the house. The uses of these different apartments we did not learn only were told that the loft was appropriated to the women' (*Banks's Journal*, vol. 2, p. 167).

133   A Javanese proa

By Sydney Parkinson.

The British Library, London. BL Add. MS 9345, f. 65 verso.

18.4 × 23.5 cm.

After a stop at the island of Savu, James Cook made for the Dutch port and trading centre of Batavia (Jakarta), which was reached on 9 October 1770. There, amid the filthy canals of a town surrounded by swamps, disease ravaged the crew and Banks's party, whose health in the Pacific had previously been so well preserved under Cook's leadership. *Endeavour* departed on 26 December but made slow progress in poor weather. She was off Princes Island (Panaitan), north of Java Head, from 6 to 16 January to take on additional supplies and thereafter a violent outbreak of dysentery was added to the list of afflictions aboard the vessel. On 24 January Herman Diedrich Spöring died, to be followed two days later by Sydney Parkinson. Then, on 2 February, Cook noted in his journal the arrival of a fresh south-east trade wind and the sadly depleted and still ailing mission at last made off.

While in Batavia Bay, Parkinson drew the last of his sketches, among which he recorded the local boats that he saw. This pencil sketch by him, annotated on the recto 'Java Proe', shows a traditional fishing and transport boat used on the north coast of Java and in the Sunda Straits. The bow and stern shape of such vessels varied from place to place along the coast, but that featured here was probably seen at or near Batavia. Called *mayang*, this sort of vessel could carry a central deckhouse with a light roof of palm matting, as shown. When fishing, the deckhouse was removed to allow space to handle the nets. The presence in Parkinson's sketch of a deckhouse but not of nets suggests that this vessel was being used to transport provisions and trade goods. The round plates of radiating strands near the masthead are magical totems obtained from the *dukun* or medicine man to safeguard the vessel and bring good luck (see Carr, *Sydney Parkinson*, p. 253).

After the Endeavour: *What Next for Joseph Banks?*

ANNA AGNARSDÓTTIR

University of Iceland

THE ENDEAVOUR VOYAGE was a triumph in the geographical and natural scientific sense and the primary object of the voyage – observing the transit of Venus – had been completed as instructed. For Joseph Banks personally the voyage had taken a grim toll of his party, leaving only four survivors. Considerable strides, however, had been made in exploring 'the Great South Sea', as Cook called it, and needless to say the *Endeavour* voyage was of great importance to Banks's position in eighteenth-century society. In John Gascoigne's estimation it elevated him to 'a figure of international scientific significance'.[1]

Banks and his assistants had collected over 30,000 botanical specimens, including 110 new genera and 1,400 new species. Among the zoological collections were over 1,000 animal specimens, famously including the kangaroo (see numbers 16, 116–21 and 129). Besides this, Banks had shown great interest in ethnology, actually learning Tahitian, making wordlists and recording native customs in his journal. Banks's 200,000-word journal of the voyage was more detailed than Cook's and proved invaluable to the commander when he sat down to write his official report to the Admiralty.

The *Endeavour* arrived in the Downs on 12 July 1771, bringing the first collections from the South Pacific to be seen in Britain. Public interest was great. On their return it was 'the gentlemen', that is to say Banks and Dr Solander, who were acclaimed for the achievements of the expedition, Cook being cast in their shadow for the time being. He barely received a mention in the newspapers. *The Gazetteer and New Daily Advertiser* of 26 August 1771 announced that 'Mr Banks is to have two ships from government to pursue his discoveries in the South Seas', sailing the following March. Cook's biographers describe Banks as 'the fêted icon … ebullient, confident and basking in the limelight', being happy enough to play the 'dashing hero' in the salons and great houses of England, and they further accuse him of being an 'unashamed' publicity-seeker.[2] Lady Mary Coke, daughter of the Duke of Argyll, wrote in early August that 'the people who are most talk'd of at present are Mr Banks & Doctor Solander: I saw them at Court & afterwards at Lady Hertford's, but did not hear them give any account of their Voyage round the world, which I am told is very amusing.'[3]

The thirty-one-year-old Banks was invited to join the Society of Dilettanti, a dining club and an influential society of the British Enlightenment. He and Solander dined with Boswell and Johnson, the latter becoming a particular friend (Banks was one of his pall-bearers in 1784), as well as with Sir John Pringle, then the president of the Royal Society, and Dr Benjamin Franklin, who was especially interested in hearing about the Pacific peoples. Both Banks and Solander received honorary doctorates from the University of Oxford, the only academic degree Banks ever attained.

On 10 August George III and Queen Charlotte formally received Banks and Solander at Kew, where they were introduced to the royal couple by Pringle. Banks apparently gave the king 'a coronet of gold, set around with feathers', originally a gift to Banks from a Chilean chief.[4] The king inspected the natural history collections and subsequently he and Banks became close friends. They were of a similar age and shared such interests as the rearing of merino sheep and the development of the Royal Botanical Gardens at Kew. In 1781 the king bestowed a baronetcy on Banks. Cook was not received by the royal couple until four days after Banks, when John Montagu, the 4th Earl of Sandwich and the current first lord of the Admiralty, presented him to the king. On that occasion Cook received his well-deserved promotion from lieutenant to commander.

Queen Charlotte was interested in botany and with her two daughters, the Princesses Augusta and Elizabeth, was in later years privately tutored by the president of the Linnean Society James Edward Smith. One of the plants found at the Cape of Good Hope by the Kew collector Francis Masson was named after her, *Strelitzia reginae*, the queen being a member of the Mecklenberg-Strelitz family. Banks and Solander, those devoted students of Linnaeus, the world's foremost botanist at the time, neglected, however, to send any plants to the great man himself, or indeed to honour him in any way. Despite his disappointment the Swedish botanist generously suggested that New South Wales should be named 'Banksia' and that a statue be erected to the 'immortal Banks'. In a letter Linnaeus wished him well on his next voyage, ending with the stirring words 'Vale vir sine pare' (Farewell O unequalled man).

As befitted his position in society Banks had two portraits painted at the time, one by Benjamin West, 1771–2, where he is dressed in a Māori cloak and surrounded by Pacific artefacts (see number 143) and the other by Sir Joshua Reynolds (1772–3), where the main object depicted is, symbolically, a globe. In the portrait Reynolds includes a well-known saying of Horace, 'Cras ingens iterabimus aequor' (Tomorrow we set out once more upon the boundless main). And that was indeed what Banks proposed to do.

Given the success of the *Endeavour* voyage it comes as no surprise that another expedition to the South Pacific was soon being planned for the spring of 1772 (see number 134). This can be seen both as a natural sequel to the *Endeavour* voyage and as the next step in British imperial ambitions. Following the Seven Years' War (1756–63) there was intense rivalry between the French and British regarding exploration and the acquisition of colonies. Despite Britain's military victory, the French were undaunted and sought in the Pacific to revive their fortunes. Samuel Wallis in HMS *Dolphin* had been the first European to visit and describe Tahiti in June 1767, but he was closely followed by the French admiral and explorer Louis Antoine, comte de Bougainville. Bougainville – the first Frenchman to circumnavigate the globe, in 1766–9 – published in October 1771, the year of *Endeavour*'s return, the sensational book *Le voyage autour du monde*, his description of Tahiti attracting particular attention as an Edenic garden peopled by noble savages. (Incidentally, Bougainville had been elected a member of the Royal Society at the age of twenty-five in 1755, while Banks had received the same honour at the tender age of twenty-three.) The British would have to do something spectacular. The *Endeavour* voyage had failed to find a southern continent and, as 'no-one had yet made what could be properly called an antarctic voyage',[5] another British expedition to the South Seas was launched.

The prime aim of the second Cook voyage on the *Resolution* was thus to search for the mythical *Terra Australis Incognita*, a territory as yet undiscovered, as blank spaces on maps showed clearly. Did it exist? And if so, did it circle the world? Did it border the Indian, Pacific and Atlantic oceans? Banks, convinced that a southern continent existed, was overjoyed when his old fishing companion Lord Sandwich, first lord of the Admiralty, invited him in September 1771 to be the scientific leader of the expedition: 'O, how Glorious would it be to set my heel upon ye Pole and turn myself round 360 degrees in a second!', he wrote.[6] Cook was more prosaic but felt that it was important to settle the question of the continent's existence. In his opinion: 'I think it would be a great pitty that this thing which at times has been the object of many ages and Nations should not now be wholy clear'd up, which might very easily be done in one Voyage without either much trouble or danger'.[7]

Throughout the winter of 1771–2 Banks and Solander sorted the *Endeavour* collections to make them 'usefull to the world even in Case we should perish in this', that is to say the coming voyage of the *Resolution*.[8] As before, Banks was assembling a party of scientists, draughtsmen and assistants at his own expense. Numerous volunteers wrote to Banks eager to join the expedition. Due to the high rate of mortality among Banks's party of eight on the *Endeavour*, however, he took no chances and organized a larger party than before, choosing his companions with care. Of course Solander would accompany him, as well as his field assistants and collectors Briscoe and Roberts, survivors of the *Endeavour* voyage. Four artists or draughtsmen were now hired – the celebrated Johann Zoffany along with John Cleveley Jr, James Miller and his brother John Frederick Miller, the latter having been hired to make illustrations of artefacts brought back in *Endeavour*. Two secretaries were now considered necessary, namely Sigismund Bacstrom and Frederick Herman Walden. Other members of the party were Dr James Lind, an astronomer and physician who received an extraordinary Parliamentary grant of no less than £4,000 for his participation; Lieutenant John Gore, who had already circumnavigated the globe three times; John Riddel, a young seaman and traveller, and a number of assistants and servants, sometimes clad in 'Scarlet and Silver', two of whom doubled as horn-players. The Banks entourage numbered sixteen and their combined salaries for a three-year expedition was no less than £3,000.

The authorities were meanwhile engaged in planning the practicalities of the voyage. First, sturdy serviceable ships were required. On 25 September the Admiralty ordered the Navy Board to purchase two ships 'for service in remote parts'.[9] This was to be a larger undertaking than the *Endeavour* voyage and the fact that the bark had almost been destroyed within the Great Barrier Reef made a second vessel a near necessity. Cook was invited to choose the two ships. He knew what he wanted, quite simply the same kind of ship as the *Endeavour*, preferably built by the same shipyard – a Whitby collier or bark, a vessel that had the advantages of being able to sail close to shore to facilitate mapping without fear of running aground, was easy to beach for repairs and capable of carrying great quantities of supplies. At 462 tons the newly built *Resolution* was 100 tons larger than the *Endeavour*. Cook described her as 'the fitest for the Service she was going upon of any I had ever seen'.[10] The other ship was the 340-ton *Adventure*. No expense would be spared. Cook was pleased that the Admiralty had ordered that the ships should be 'fitted in the best manner possible … every standing Rule and order in the Navy … dispenced with, every alteration, every necessary and usefull article … granted as soon as ask'd for'.[11] This included, for instance, 'Ice Anchors and Hatchets', useful in Antarctic waters. Cook was content to welcome the company of his two erstwhile shipmates, writing in his journal that Banks and Solander 'intended to embark with me … in order to prosecute their discoveries in Natural History & Botany and other usefull knowlidge'.[12]

But when Banks came to view the *Resolution*, on which he would travel, he was severely disappointed. The vessel, he thought, was simply not large enough to accommodate his entourage and he promptly demanded that modifications be made or else he would be forced to abandon the expedition. He got his way. Lord Sandwich decreed that Banks's wishes would be complied with, so extra deck was added and superstructures built in order adequately to house Banks, his men and equipment. This meant a delay of at least three months.

As many Cook biographers have observed, this was 'Banks's voyage'. Cook noted that almost every day 'Strangers', both men and women of all ranks came on board simply 'to see the Ship in which Mr Banks was to sail round the world'.[13] On 2 May 1772, for instance, Banks entertained guests on board the *Resolution* to view the alterations, inviting among others the Earl of Sandwich and the French ambassador. Shortly afterwards the ship set sail for a test run, but at the Nore the pilot gave up – the ship was disastrously top-heavy and was in danger of capsizing. Banks's friend Charles Clerke, a veteran of the *Endeavour* voyage, felt obliged to inform him candidly that the ship was 'so very bad', but to add that he would 'by God … go to Sea in a Grog Tub if desir'd, or in the Resolution as soon as you please; but must say, I do think her by far the most unsafe Ship, I ever saw or heard of'.[14] When Cook declared that 'she was found so crank that it was thought unsafe to proceed any further with her',[15] the naval authorities decided that the *Resolution* be restored to her original state. On 24 May Banks and Solander came to inspect the ship at Sheerness, where the alterations were being dismantled. A young midshipman John Elliot wrote in his memoirs a much-quoted passage that when Banks saw the ship in its modified state he 'swore and stamp'd upon the Warfe, like a Mad Man; and instantly order'd his servants, and all his things, out of the Ship'.[16]

Banks demanded the Admiralty provide a new ship, for instance a frigate or an East Indiaman, but in Cook's opinion his arguments were 'highly absurd', being in no doubt that the *Resolution* was the 'properest' for the purpose of the voyage.[17] Lord Sandwich's patience was at an end and he refused Banks his request, replying 'it is a heavy charge against the Board to suppose that they mean to send a number of men to sea in an unhealthy ship … and that her crew will be in danger of losing their lives if they go to sea in her'.[18] The Navy Board was entirely in agreement and stated bluntly:[19]

> Mr. Banks seems throughout to consider the Ships as fitted out wholly for his use; the whole undertaking

to depend on him and his People; and himself as Director and Conductor of the whole; for which he is not qualified and if granted to him would have been the greatest disgrace that could be put on His Majesty's Naval Officers.

So Banks was forced to withdraw, earning himself negative epithets both from his contemporaries that he did not 'chuse to go the voyage, unless he could ride the waves triumphantly, in all the pomp and splendour of an Eastern Monarch',[20] as well as from modern scholars and biographers, who have described the 'extraordinary behaviour' of the 'big-headed botanist' (Richard Hough), or noted his 'absurdly swollen head' (J.C. Beaglehole), or declared his behaviour as 'juvenile' and 'self-important' (Patrick O'Brian; see number 138, and numbers 135–7).

Thus Banks withdrew in a fit of pique from the expedition. But Banks cared about botany and made sure that despite the fact that he himself would not sail on the *Resolution* a botanist would. The man chosen was Francis Masson, called 'the King's Gardener', the first plant collector to be sent from what are now the Royal Botanic Gardens at Kew in 1772, sending back over five hundred plant specimens from the Cape of Good Hope. With Banks's withdrawal the *Resolution* voyage lacked a scientific leader and two men were hastily appointed to replace the Banksian party, the German scientist Johann Reinhold Forster and his seventeen-year-old son Johann Georg Adam. There is a general consensus that this was an unhappy choice. As James Lind was staunchly loyal to Banks and had left the *Resolution*, his £4,000 parliamentary grant was transferred to the Forsters. Banks must have been severely disappointed, and so, according to the *Gentleman's Magazine* of June 1772, were 'the Literati throughout Europe', whose high-pitched expectations regarding the *Resolution* expedition had now been dashed.

The expedition now became 'Cook's voyage'. It would be a three-year voyage of five vast ocean sweeps and cruises, as Beaglehole described it, wintering in Tahiti and Queen Charlotte Sound in New Zealand. Cook's secret instructions from the Lords of the Admiralty were dated 25 June 1772. His task was to be one of exploring, mapping, gathering wildlife, minerals and plants and sailing as near as possible to the South Pole. He was further ordered to observe:

the Genius, Temper, Disposition and Number of the Natives or Inhabitants, if there be any, & endeavour by all proper means to cultivate a Friendship and Alliance with them, making them Presents of such Trinquets as they may value, inviting them to Trafick, & shewing them every kind of Civility & Regard; but taking care nevertheless not to suffer yourself to be surprized by them, but to be always on your guard against any Accident. You are with the consent of the Natives to take possession of convenient Situations in the Country in the Name of the King of Great Britain, and to distribute among the Inhabitants some of the Medals with which you have been furnished to remain as Traces of your having been there. But if you find the Country uninhabited you are to take possession of it for His Majesty by setting up proper Marks & Inscriptions as first Discoverers & Possessors.[21]

The voyage would also provide an opportunity to test the chronometer as a means of calculating longitude, and the Board of Longitude paid for two men to go on the expedition, one on each ship, the astronomers William Wales and William Bayley. *The Endeavour* voyage had made great strides in battling scurvy, but the Admiralty was interested in further experiments and 20,000 lbs of sauerkraut were stowed on the ship.

After a short bout of coolness Cook and Banks continued as friends – the *Resolution* accommodation fiasco was, after all, between Banks and the Admiralty. In 1776 Cook was unanimously elected a fellow of the Royal Society, while in 1784 Banks, now president of the Royal Society, commissioned a commemorative medal of the then dead Cook (see number 140).

The voyage of the *Resolution* established the fact that there was no southern continent, as Cook wrote in his journal: 'I flater my self that the intention of the Voyage has in every respect been fully Answered, the Southern Hemisphere sufficiently explored and a final end put to the searching after a Southern Continent', although he felt certain that there was 'a large tract of land near the Pole'.[22] Banks missed out on this momentous discovery, though it may have been of some comfort to him that when Cook reached Tahiti again he was greatly missed by the inhabitants, who were constantly asking after him.

But Banks was not one to give up. Though 'disagreeably disappointed', he had already assembled an impressive scientific party at his own expense and it was of prime necessity to engage them in a new project. He thus quickly 'resolved upon another excursion'.[23] By early June Banks had settled on his new destination. Instead of searching for a massive southern continent, he decided to head north, his choice falling on Iceland, an island in the mid-Atlantic and a dependency of the King of Denmark since 1380. Between the fifteenth and seventeenth centuries the coast had been frequented by English seamen, especially from East Anglia, after which British contact had almost ceased. A question begging to be answered is why did Banks choose Iceland?

Scholars have been only too keen to explain Banks's choice, suggesting a follow-up voyage to the Labrador-Newfoundland expedition of 1766, for instance, or a burgeoning English interest in Iceland's medieval literature emerging in the second half of the eighteenth century. Samuel Johnson had apparently been on the brink of visiting Iceland for this reason. It is, however, rather difficult to reconcile Banks with an interest in the Sagas. Beaglehole believed Banks had always been interested in visiting the north, which now became a real option. However, in his Iceland journal Banks more than adequately explained the reasons for his decision. He wrote that as the sailing season was much advanced he:[24]

> saw no place at all within the Compass of my time so likely to furnish me with an opportunity as Iceland, a countrey which ... has been visited but seldom & never at all by any good naturalist to my Knowledge. The whole face of the countrey new to the Botanist & Zoologist as well as the many Volcanoes with which it is said to abound made it very desirable to Explore.

Accordingly, Iceland had the advantages of being relatively near and unexplored, its chief attraction being that it was the site of many active volcanoes.

Banks was, however, wrong in his assumption that no 'good naturalist' had ever visited Iceland. The King of Denmark and the Danish equivalent of the Royal Society had in the spirit of the Enlightenment been eager to explore their dependency, and several scientists had been sent to the island, including the botanist Johan Gerhard König, a pupil of Linnaeus and acquaintance of Solander, who collected plants there in 1764–5. Moreover, the French had already been to Iceland. At the beginning of 1767 a Breton nobleman, Yves Joseph de Kerguelen-Trémarec, was summoned to Versailles to sail off to the North Atlantic to explore not only the islands there, including Iceland, Greenland and the Faroes, but also the Orkneys, Shetlands and Norway. He published a voluminous account of his voyage in 1771, remarking that a botanist would find much to observe in Iceland. It is known that Banks possessed a copy of this recently published book and took it with him to Iceland in his select reference library, like the one he had on the *Endeavour*. Furthermore, a couple of months earlier, in April 1772, Banks had been given some specimens of Icelandic lignite. This would have strengthened his belief that Iceland had much to interest a naturalist.

At the time there was a growing interest in vulcanology. The documentary evidence certainly points to the fact that seeing 'burning mountains' was the major aim of the voyage, as well as collecting mineral specimens and other curiosities of nature. In Banks's passport, quickly issued on 2 July by Count von Diede, the Danish envoy in London, the main purpose of Banks's visit was recorded as 'observing Mount Hekla', the most famous of the Icelandic volcanoes.[25] The ascent of Hekla is the highlight of the Iceland part of Banks's 1772 journal, the measurements of the spouting hot springs, described by Banks as 'volcanoes of water' (the word *geyser* was coined later, Geysir being the proper name of the most magnificent of the Icelandic hot springs), coming a close second. On their return *The Scots Magazine* of November 1772 reported that they had 'applied themselves in a particular manner to the study of volcanoes'.

Thus instead of participating in resolving the issue of the existence of the southern continent, Banks's attention was turned to examining volcanoes and hot springs. Banks prepared his voyage as best he could within the limited period of time he had. Understandably he found no one in London who had been to Iceland, but Claus Heide, a Dane resident in the capital, gave Banks information 'Chiefly out of books', finding out 'the names of People of the greatest note', as well as the best advice on 'the most Proper places for you [Banks] to go to Vieuw Mount Heckla, or what place is burning at present'.[26] The King of Denmark was notified of Banks's wish to visit Iceland and was only too happy to sanction the 'celebrated' English 'Lords' journey to his island. The Governor of Iceland was enjoined to extend every courtesy and do everything in his power to assist them. On their departure *The Annual Register* noted that they were equipped with everything necessary to examine the natural history of the places they proposed to visit, and off they went 'at their own private expence' to Iceland, 'to prosecute new discoveries in the science of botany', as *The London Magazine* of 1772 worded it.

The Banks Expedition was about twenty strong. Only Zoffany was not among the original group Banks had engaged to sail with him on the *Resolution*. In Harold Carter's opinion Zoffany had 'perhaps been the chief sufferer' when Banks withdrew from the *Resolution* expedition, but Banks paid him one year's salary, a sum of £300,[27] and he went off to Florence to copy pictures for the king. Another Swede, however, had joined Banks's party. This was the twenty-six-year-old Uno von Troil, a friend of Solander's, later to become archbishop of Uppsala. Troil had come to England directly from Paris, where he had met such luminaries of the Enlightenment as Rousseau, d'Alembert and Diderot, and attended King Louis XVI's levée at Versailles. He was invited by Banks to join the expedition at the last moment, and luckily so, as he afterwards produced a book on the expedition, *Letters on Iceland*, first published in Uppsala in 1777 and translated into English in 1780. As Banks was now the undisputed leader of the Iceland expedition, the master of all decisions, he was free to add more men to his retinue, including a gardener, the astronomer James Hay and a French chef, Antoine Douvez, who was to cook delicious meals for the Icelandic elite.

Banks's chartered ship, the *Sir Lawrence*, a brig of 190 tons, captained by James Hunter with a crew of twelve, was hired by Banks at a cost of £100 a month. The *Sir Lawrence* eventually left Gravesend on 12 July, ironically the same day as Cook started on his second voyage with his two ships *Resolution* and *Adventure*. Banks was in no especial hurry to reach Iceland. He had all the time in the world and the *Sir Lawrence* progressed in a leisurely manner along the south coast of England and up through the Western Isles, where Banks visited the island of Staffa with its strange basaltic columns, his description of Staffa when published by Thomas Pennant winning instant recognition. *The Scots Magazine*, naturally enough, considered the island 'one of the greatest natural curiosities in the world', the wonders of Stonehenge paling into comparison as 'trifles when compared to this island'.[28] This may have been some small compensation for Banks.

After suffering extreme bouts of seasickness, the Banks expedition finally arrived in Hafnarfjörður in south-west Iceland on 28 August 1772, too late to catch the flowers blooming. After an eventful stay of six weeks, they left Iceland on 8 October 1772, loaded down with specimens of lava and old Icelandic manuscripts. But that is another story.[29]

AFTERMATH

Following his voyage on *Endeavour*, and his withdrawal from James Cook's ensuing Pacific mission in HMS *Resolution*, Joseph Banks settled down to the management of his country estates and to life among London's learned societies. His *Endeavour* collections were first housed in New Burlington Street and then at 32 Soho Square, which became a centre for naturalists and explorers wishing to learn about the Pacific. There Banks built up an unrivalled library of natural history and travel literature to complement his herbarium, rich in Pacific specimens from the *Endeavour* voyage and from later settlers and collectors in the region. His collections were eventually bequeathed to the nation and transferred by Robert Brown, his last curator, to the British Museum in 1827. From 1773 onwards Banks was unofficial director of Kew Gardens for George III, raising the gardens to international eminence as a centre for plant exchange and economic botany. In 1778 he was elected president of the Royal Society, a position he held until his death in 1820, the longest tenure of any president. Banks became, too, a Privy Councillor in 1797. Acting in an unofficial capacity he advised the government on matters relating to trade, exploration and collecting. The missions of William Bligh to transfer breadfruit from the Pacific for use as a food source on plantations in the West Indies were organized by Banks. He gave advice on the settlement of New South Wales. Notably, Banks also organized the scientific party that sailed on HMS *Investigator* in 1800 under his Lincolnshire countryman Matthew Flinders to chart the coasts of Australia, proving for the first time that it is one great island continent. By his efforts Banks helped to establish the tradition of scientific staff travelling aboard Royal Navy vessels on discovery and survey missions leading, ultimately, to the historic voyage of Charles Darwin in the *Beagle*. Banks also assisted in the foundation of important new societies that still flourish today, such as the Linnean Society of London in 1788, and the Royal Horticultural Society in 1804.

134 'THE GREAT PACIFIC OCEAN'

Engraved by John Bayly for Joseph Banks, February–March 1772.

Dr Neil Chambers, Surrey.
28 × 34.5 cm.

Commissioned by Joseph Banks early in 1772, this map was not published, but the original copper plate from which this modern strike was taken is held at the Natural History Museum, London. In Banks's lifetime it was stored among the plates prepared for his unpublished *Florilegium* (see numbers 122–6), and it eventually came with them from the British Museum to the Natural History Museum when the latter was completed in the early 1880s. The plate lay hidden among Banks's *Florilegium* plates until H.B. Carter identified it in January 1975. The lower circular portion shows part of the southern hemisphere on a polar stereographic projection from the South Pole to 30° S. The upper rectangular portion is a cylindrical or Plate Carrée projection showing the equatorial Pacific from 30° S to 30° N, the two portions being joined tangentially at the meridian of 180°. The islands of New Zealand and the east coast of Australia, named 'New South Wales', appear on this, the earliest printed map to show them. Banks had the map made apparently with the intention not only of recording the major discoveries of the *Endeavour* voyage, but also with a view to exploration of the vast blank in geographical knowledge so clearly apparent in the Antarctic region. At the time, a second circumnavigation under Cook was being contemplated to probe these high latitudes for an unknown southern continent.

The map was based on one published in 1770 by Alexander Dalrymple as the frontispiece to volume 1 of his two-volume *An Historical Collection of the Several Voyages and Discoveries in the South Pacific Ocean* (London, 1770 and 1771). The inscribed copy of this work as given to Banks after his return from the *Endeavour* voyage is held at the British Library, 454.h.5–6, and was itself an edition of an earlier Dalrymple account (1767) of discoveries made in the South Pacific prior to 1764 that Banks actually took on the mission. With its two stereographic projections of the north and south polar regions set either side of but not connecting to a central cylindrical projection of the equatorial zone, Dalrymple's frontispiece map was evidently discussed with the voyagers in light of their recent discoveries. These it lacked, since *Endeavour* was still at sea in 1770, but as adapted by Banks it nevertheless provided a good visual base on which to indicate the ultimate course and discoveries of any ensuing venture into the 'Great Pacific Ocean'. Each portion of Banks's version covers the same latitudes as its equivalent in the Dalrymple original. Moreover, the depth of the cylindrical projections are the same physical measurement, as are the diameters of the stereographic projections, and Dalrymple's titles were also employed by Banks for the Pacific area on which he concentrated. Dalrymple was a keen advocate of the search for an undiscovered southern continent, something his latest publications were intended to promote, but the results of Cook's first Pacific voyage confined any hope of finding such a landmass to increasingly remote polar latitudes.

John Bayly engraved Banks's map in February or March 1772. Located in Red Lion Square, Bayly was engraver to the nearby British Museum, and he also engraved Cook's chart of New Zealand for the official account of the *Endeavour* voyage. Banks ordered 100 prints from his plate, of which one is known to survive in the British Library at BL MAPS 181 m 1, vol. 3, pl. 61, with others doubtless surviving elsewhere. Copies were almost certainly distributed among the voyagers on Cook's second Pacific expedition, from which Banks withdrew after a disagreement with the Admiralty, and it seems clear that his and Dalrymple's maps influenced the cartography of the second voyage, not least in the case of two of Cook's own manuscript charts of the southern hemisphere showing the track and discoveries of *Resolution*, NA MPI 94 (Adm. 55/108 1b) and BL Add. MS 31360, f. 7. These charts are of a similar construction to Banks's map, showing the Pacific in a stereographic projection from the pole to 30° S, and then an upper cylindrical projection of the remaining hemisphere from 30° S to the equator, the two being joined at a meridian of 180°. Cook later opted for a single stereographic projection of the southern hemisphere on which to record the tracks of various Pacific explorers up to and including his own voyages, publishing this as the frontispiece to the official account of the second voyage. It was the first polar chart to appear in a Cook account and its publication effectively marked an end in his generation to the search for a hidden continent in the deep southern oceans. Any that did exist, Cook now thought, must lie beyond reach near the pole (see *Cook's Journals*, vol. 2, p. 643). Nathaniel Dance featured this map in his

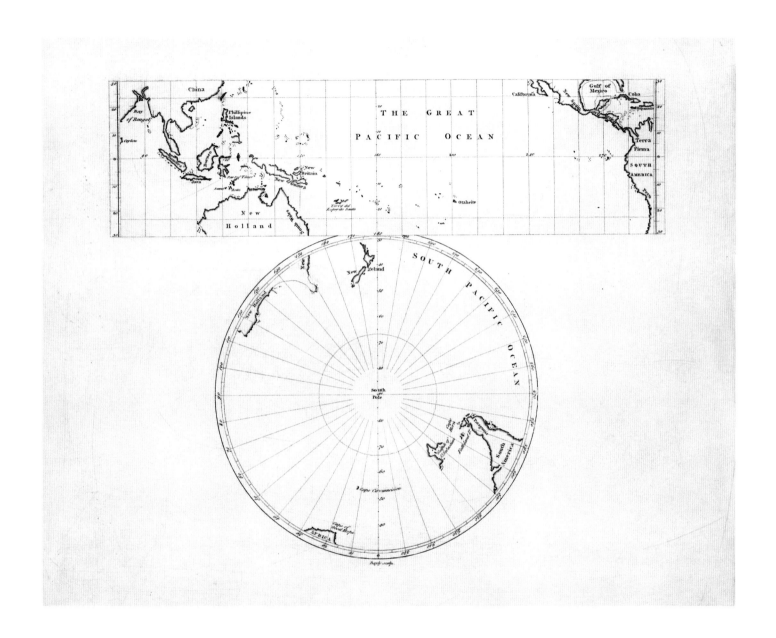

famous portrait of the great navigator as commissioned by Banks.

As explained, it is likely that discussions took place early in 1772 between Banks and Cook regarding the forthcoming *Resolution* voyage. Banks had his map prepared at this time, while Cook also produced a memorandum accompanied by his own polar stereographic chart to outline the route that a second Pacific mission might take. In the memorandum he advocated probing the southern oceans to about 60° S. to establish whether a great landmass lay anywhere in these as yet unexplored higher latitudes. Cook submitted these documents to Lord Sandwich on 6 February 1772 and his ideas were duly used in drawing up instructions for the *Resolution* voyage (see ML, MS Safe 1/82).

135 *The* FLY CATCHING MACARONI, *aet* 29

By Matthias and Mary Darly (Whipcord, pseud.), 39 The Strand.

The British Museum, London. PD 1868,0808.4476. 17.8 × 12.3 cm.

Published on 12 July 1772, this hand-coloured satirical etching targeted Banks the exploring naturalist, brandishing his catching nets astride the world. Under the caption are the lines:

*I row from Pole to Pole, you ask me why
I tell you Truth, to catch a – Fly.*

Banks did not, of course, explore from 'Pole to Pole' as he did not accompany James Cook on his second Pacific voyage to seek a previously undiscovered landmass thought by some to be located in high southern latitudes. Banks did, however, visit Iceland in 1772, thereby encompassing in his travels latitudes that spanned the South Island of New Zealand in the southern hemisphere to Iceland in the northern hemisphere. The Darlys gesture at these exploits in their satirical engravings of both Banks and Solander, who travelled together on these voyages (see number 136 below). In the present print Banks stands astride two poles of the globe, labelled 'Antartick Circle' and 'Artick Circle'. He is shown as a well-dressed man with elaborately curled hair, ass's ears and a large feather in his hat. As the title of the print indicates, he is depicted as an amateur collector and dandy, a common subject of satirical ridicule for the Darlys.

Engraved prints of both portraits and satirical cartoons disseminated to the eighteenth-century public a lively range of images of people whose achievements and interests were celebrated or lampooned to make moral points of one kind or another. Banks was no exception and the satirical imagery surrounding him provides a fruitful point of comparison with official portraits that show him, initially, as a youthful traveller and collector (by Benjamin West) and then, later on, as a formidable patron of the sciences (by Thomas Phillips). Here Darly's playful joke at Banks's expense derives from the enormous lengths to which naturalists went in order to obtain something as apparently trivial as an insect. In this period natural philosophers of most kinds were derided by satirists for their seemingly worthless enquiries into subjects like the weight and composition of air or the study of microscopically small objects. Such activities provided rich material for those seeking to mock Banks and others like him. This was one of the reasons why throughout his life Banks was always so keen to make science appear useful.

It was Banks's sexual exploits during the *Endeavour* mission and his plant and animal collecting that provided the chief ammunition for satirists in the early days. After his election to the presidency of the Royal Society in 1778, his perceived status as a dilettante unfit to fill the chair once held by Sir Isaac Newton attracted especially biting attacks. Critics both in and outside the society accused him of turning it into a gentlemanly club and of being ill-equipped to foster the mathematical and physical sciences. There is, however, no evidence during Banks's presidency that contributions on natural history were given special preference in the society, with those falling outside the sciences being firmly excluded from its meetings, and the society's social mix remaining pretty much unchanged. Perhaps the most famous satirical image of Banks is that by James Gillray on the occasion of Banks being awarded the prestigious Order of the Bath in 1795. Gillray drew on Banks's well-known interest in natural history, showing him transformed from a creeping grub into a colourfully decorated butterfly by the warming influence of the Crown. Banks's wings are adorned with numerous specimens, among which are shells resembling *bonnets rouge*, an allusion to the revolutionary mood of the times. Gillray's caption is in the style of a classificatory description from the *Philosophical Transactions*, but with ironic undertones suggestive of the virtuoso collector, for the Bath Butterfly is only valued for its rarity and beauty rather than as a genuine scientific discovery. Gillray introduces the courtly star and red sash of the Bath that would henceforward denote Banks in satirical prints and commissioned portraits alike (see numbers 1, 142 and 143).

136 *The* SIMPLING MACARONI, *aet* 39

The SIMPLING MACARONI.
Like Soland-Goose from frozen Zone I wander,
On shallow Bank's grow's fat. Sol......

By Matthias and Mary Darly, 39 Strand.

The British Museum, London. PD 1868,0808.4477. 17.8 × 12.7 cm.

One of a pair (see number 135 immediately above), this satirical illustration of Daniel Solander as a plant collector was published on 13 July 1772. Under the caption are the lines:

Like Soland-Goose from frozen Zone I wander,
On shallow Bank's grow's fat, Sol____

Darly was a London designer and printseller. He undertook jobbing engraving of visiting cards, shop bills and bookplates and, more importantly, he produced plates and books of designs for use in furniture, wallpapers, fabrics, pottery and ornamental architecture. He and his wife Mary also published satirical portraits of politicians, with Mary taking charge of this side of their business in 1762. From the 1770s onwards the Darlys concentrated on satires of celebrities, manners and the latest fashions. The dress and behaviour of macaronis (fops and grand tourists who aped foreign habits) were common targets in their satirical prints. Mary was industrious in her output, inviting the general public to submit sketches for engraving, if necessary in secret. The satires of Banks and Solander may therefore derive from an as yet unidentified third party. The Darly shop front at 39 The Strand was itself depicted in an engraved print issued on 14 July 1772, *The Macaroni Print Shop*, in which the newly published satire of Banks appears in a window pane on the left side. Thus the satires of Banks, Solander and the Darly shop were issued on three successive days in July 1772.

Note that the Solan Goose is an old name for the Northern Gannet, *Morus bassanus*, a sea bird with a distribution that includes both Sweden and the British Isles, and so here it may reference Solander's 'migration' from Sweden to England. Gannets have a supposed capacity for eating large quantities of fish, and the present satire of Daniel Solander alludes to his relationship with Banks, in which the rather portly Swede is portrayed as feeding off Banks's patronage to enlarge his own collections and reputation. The pun on Banks's name and a riverbank became a satirical commonplace.

137 An Epistle from Oberea, Queen of Otaheite, to Joseph Banks, Esq. Translated by T.Q.Z. Esq. Professor of the Otaheite Language in Dublin, and of all the Languages of the undiscovered Islands in the South Sea; And enriched with Historical and Explanatory Notes (London: J. Almon, 1774 [1773])

Anon. Attrib. John Scott-Waring.

The British Library, London. 1487.r.5. Rebound in blue cloth. Title, place and date of publication, gilt lettering on spine. 25.7 × 21 cm (closed).

(12)

Experienc'd matrons the young pair survey'd,
And urg'd to feats of love the self-taught maid;
With skill superior she perform'd her part,
And potent nature scorn'd the tricks of art.
Curst be the envious gales that wafted o'er
Those floating wigwams to our peaceful shore:
With specious gifts a crew insidious came,
And left us * *bitter pledges* of their flame.
'Till then was nature free and love sincere,
Nor generous passion quench'd by slavish fear.
No pining maiden knew the venom'd kiss,
But all was genuine extacy and bliss.

Oft have I wish'd, for such you love, that I
Were metamorphos'd to some curious fly;
Beyond the main I'd speed my eager way,
And buz around you all the live-long day.
Nor would I not be some ombrageous tree,
That shades thy grot, and vegetate for thee;
At thy approach I'd all my flowers expand,
And weave my wanton foliage round thy hand.

Think

NOTES.

* I suppose this alludes to the introduction of the venereal disease among them by Monf. Bougainville, which they emphatically call *the Rottenness*. See *Hawkesworth's Voyages*.

† Α βομ βινοσα μελισσα και ις τεον αυλιον ικοιμαν' ΤΗΕΟΣ.

In the period after his return from the *Endeavour* voyage Banks's activities attracted the attention of satirists for the first time. Picking up on his youth and fortune, they initially made much of stories linking him to a sort of sexual tourism, as if his explorations parallelled the exploits of those dissolute grand tourists long known in Europe. Collecting of plants was itself satirically linked to exploration of female sex through imagery to do with flowers. Flowers figuratively referenced love as well as sexual organs or acts (deflowering), and since in the Linnaean system their reproductive structure was used for the purposes of classification, the pursuits of the roving botanist could be portrayed as indecent and invasive. Exploration of the manners and customs of Pacific societies was likewise represented in erotic terms, particularly when it came to 'investigating' the women of Tahiti. Not only did such activity entail a corrupting influence among the indigenous peoples encountered, but for many at home the lurid reports of such 'contacts' were just as dangerous to British morals. The satirists had a field day.

In later years Banks would remain a target for satirists, who focused on his seemingly trivial interest in natural history, his position at the head of the Royal Society, his status as a gentleman amateur, his patronage by the king and, as he aged, his sheer physical bulk and the commensurate breadth of his influence. Rowlandson, for example, showed a bloated Banks greedily feasting on an alligator in 'The Fish Supper' (1788), an allusion to his ravenous acquisition of all manner of specimens, and perhaps also to the fact that he did actually eat plant and animal specimens to test their qualities as food (see number 119). Increasingly, Banks was derided by satirists for being an establishment figure connected to government and to the Crown, linking their interests and those of the landed class to science and learning. The very attributes celebrated in portraits of Banks by Thomas Phillips – his size and dominating presence, his courtly bearing and seniority – became for the satirists objects of comic amusement or scorn (see numbers 135–6 and 1, 142 and 143).

In the present mock *Epistle from Oberea, Queen of Otaheite* (Sept. 1773), Banks's supposed liaison with 'Oberea' (Oboreah or more properly Purea) provided a colourful episode on which to base an attack (see number 68). Purea writes unhappily to her departed lover on page 12:

Curst be the envious gales that wafted o'er
Those floating wigwams to our peaceful shore:
With specious gifts a crew insidious came,
And left us *bitter pledges* of their flame.
'Till then was nature free and love sincere,
Nor generous passion quench'd by slavish fear.
No pining maiden knew the venom'd kiss,
But all was genuine extacy and bliss.
Oft have I wish'd, for such you love, that I
Were metamorphos'd to some curious fly;
Beyond the main I'd speed my eager way,
And buz around you all the live-long day.
Nor would I not be some ombrageous tree,
That shades thy grot, and vegetate for thee;
At thy approach I'd all my flowers expand,
And weave my wanton foliage round thy hand.

To this Banks is afterwards made to reply in *An Epistle from Mr. Banks, Voyager, Monster-hunter, and Amoroso, to Oberea, Queen of Otaheite* (Dec. 1773), page 14:

Yes, justly, dearest object of my flame,
You execrate th' infernal *Gallic* name;
Help ME to curse those too, whose deeds approve
Themselves unworthy of thy country's love;
Whose sordid souls could basely plant disease,
Where all was joy, and liberty, and ease.

And wouldst thou wish, my fair, thyself a fly,
To buz around, and in my bosom die?
My gracious Queen, I thank thee----in one line
Oh let me say that I HAVE DIED on thine!
And wouldst thou vegetate a vernal stump?
No----clasp thy *tendrils* round my brawny rump.

The exchange is closed by *A Second Letter from Oberea* (1774).

It is unlikely that Banks ever had an affair with the real Purea (he described her in his voyage journal rather unflatteringly as having lost her youthful looks). Yet in its satirical way their supposed relations stood for any others that did take place involving him and, to an extent, for contacts between the voyagers and islanders more generally. In the poems, Banks has deserted his erstwhile lover. Purea complains to him in her first epistle that the innocent free love of the Society Islands has been ruined by the introduction of venereal disease caught from the visitors (lust as a flame, punning on inflamed by disease), though it is clear that her character is far from pure. The expanding of her metaphorical flowers at Banks's approach and the weaving and clasping of her 'wanton foliage' suggests Purea's sexual abandon. She wishes to be changed into a 'curious fly' so as to speed to Banks in England, at one and the same time an appropriate but also ironically demeaning transformation for the royal lover of a collecting naturalist. The Banks character responds in his epistle with a further pun that employs 'plant' in the sense of infecting or implanting a disease, a clear reference to the main satirical device in these poems, which is derived from Banks's interest in natural history and especially botany. The play on words and meanings is made explicit in the Banks character's lewd references to physical sex, as when he dies on Purea's bosom (to die for to orgasm), and in the vegetating of his vernal stump (for engorged male genitalia). Such imagery is by no means subtle or original, but it picks up on specific aspects of the voyage, for example Banks spending the night on Purea's canoe and his interest in collecting, and it references, too, European concepts of the 'noble savage' then being brought sharply into focus by reports of island life in the Pacific and the impact of contact with European civilization upon it (see number 131).

138   To John Montagu, 4th Earl of Sandwich

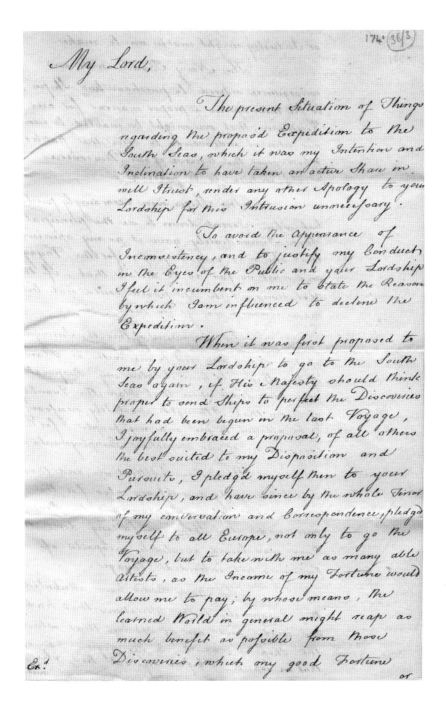

[*From 14 New Burlington Street, 30 May 1772*].

Captain Cook Memorial Museum, Whitby.
WHICC 174.36.3.
31 × 19.5 cm.

Joseph Banks was an immediate celebrity following the arrival home of *Endeavour* at Deal on 12 July 1771. He moved his voyage collections – then including not only botanical specimens and illustrations but also zoological and ethnographic material – to New Burlington Street, there establishing, in effect, the first museum of the South Seas, private or public, to be seen in London. From these collections would grow his later influence among natural historians, explorers and even in due course with government itself. The *Endeavour* voyage left a number of geographical questions unanswered, most notably whether an undiscovered continent still lurked somewhere in the deep southern oceans, and so another voyage under James Cook was planned to resolve this issue and to open up more of the Pacific to scientific enquiry and European contact. Two vessels, both colliers like *Endeavour*, were chosen as a safeguard against one of them getting into difficulties as happened on the first voyage (although they could and did become separated). HMS *Resolution* would sail under Cook, and HMS *Adventure* would accompany her under Tobias Furneaux, a veteran as second lieutenant of the *Dolphin* expedition under Samuel Wallis.

It was accepted that Banks would join this mission to build on the achievements of the first voyage in natural history and ethnography. Lord Sandwich at the Admiralty was keen to encourage both Cook and Banks, but following the success of the *Endeavour* voyage Banks had become more ambitious and with it more demanding. He wanted to expand the mission to include more work on his favourite pursuit of natural history. As a result a number of alterations were made to *Resolution* in order to accommodate his entourage of collectors, artists and even musicians, totalling thirteen people besides himself. *Resolution* was, however, found to be top-heavy when she sailed down the Thames. The additions were therefore removed at Sheerness on the orders of the Admiralty. An acrimonious exchange of letters took place between Banks and various Admiralty officials, foremost among them Lord Sandwich.

In the present letter Banks advanced the view that the changes to *Resolution* were not correctly completed, and further that a different and larger vessel might have been selected for the voyage. He reminded Sandwich of the amount of money that he had spent equipping the mission for natural history exploration, and of the public expectation that he would sail again to enlarge this department. Sandwich received advice from the Navy Board and prepared a long and compelling response to Banks's claims. In it he pointed out the qualities of the vessels chosen for the voyage as against those that Banks had mentioned. He also observed that unjustified criticisms of the safety and healthiness of the mission ships would be highly improper given others were to sail in them with or without Banks, and he further cautioned that publishing such criticisms would necessitate a robust counter from the Admiralty. As a consequence, the exchanges remained a private matter, from which it was obvious that Banks would not have his way. He therefore withdrew from the voyage. It seems that he expected to return to the Pacific in an East Indiaman at a later date, but this never happened. For now, he financed and led an expedition to Iceland, taking with him much of the equipment and personnel that he had assembled for the *Resolution* mission. His ship, the *Sir Lawrence*, sailed from Gravesend on 12 July 1772, the same day that Cook took his ships out of Plymouth Sound.

The *Resolution* fiasco is usually regarded, and not without good reason, as an example of the youthful Banks overreaching himself. Banks should perhaps have known better, for he had previously shown a deft touch with the Admiralty during his preparations for the *Endeavour* voyage, when the then first lord Sir Edward Hawke had refused to allow anyone but Banks and Solander to join the mission. At the time Banks cannily circumvented the first lord by gaining the support of the long-serving Admiralty secretary Sir Phillip Stephens, whose knowledge of navy business was unrivalled. Through him Banks easily obtained all that was required. In the years that followed the *Resolution* controversy Banks repaired his reputation within Admiralty circles, notably with each of Stephen's successors. A number of Admiralty officials and navy officers were elected fellows of the Royal Society during Banks's presidency, thereby strengthening ties between these organizations. Banks also acted to ensure that science was situated more centrally in naval concerns and, in particular, that British exploration voyages routinely included staff and equipment for the pursuit of science. His 1772 brush with the Admiralty probably served, then, to reinforce the earlier lessons of cultivating key officials in such bodies in order to promote his vision for science as an instrument of the state as well as a wider benefit to humanity (see Chambers, ed., *Indian and Pacific Correspondence*, vol. 1, letter 92 and the General Introduction). The present manuscript is a copy version in the hand of a clerk of which this is the first folio side.

139 'A GENERAL CHART Exhibiting the DISCOVERIES made by Capt$^{n.}$ *JAMES* COOK in this and his two preceeding VOYAGES; with the TRACKS of the SHIPS under his Command. By Lieut.$^{t.}$ Hen$^{y.}$ Roberts of HIS MAJESTY'S Royal Navy.'

By Henry Roberts. Engraved by William Palmer, published 1784.

The British Library, London. 973(2). 62.2 × 96 cm. Scale approx. 1:50,000,000.

This impressive map by Henry Roberts, with later annotation and colouring by an unknown hand, was published in 1784 to show the tracks of the lead vessels on all three of James Cook's Pacific voyages. These voyages revealed for the first time in reliable charts the geography of the Pacific more or less as it really is. The chart itself was included in the Atlas of the official account of the third Cook voyage. Cook's objectives during his third voyage were to survey the west side of North America, to establish how far it is from Asia and to investigate the practicability of a northern passage from the Pacific to Europe for reasons of communication and trade. Ice sheets confronted him at high northern latitudes and no inland route was found further south, but Cook did survey the Northwest Coast of America as instructed, and he was also the first European to visit and describe the islands of Hawai'i, where, following an enforced return to Kealakekua Bay for repairs, he met his death in February 1779. Cook's loss meant that the publication of an official voyage account would have to be overseen by others, thereby adding to the difficulties inherent in producing such large and complex works. Publication of each of Cook's previous voyage narratives had been dogged by problems due in part to the inexperience of the Admiralty in managing such ventures. Lord Sandwich only lighted on John Hawkesworth to edit the *Endeavour* voyage narrative through a recommendation from Charles Burney, and the choice proved an unhappy one. Then there was a clash over authorship of the second-voyage account between the Admiralty and the mission naturalists Johann Reinhold Forster and Johann Georg Adam Forster. Now Sandwich as first lord was engaged in arranging posthumous publication of yet another major account, an obligation to his fallen protégé that he evidently took seriously since he persisted in it even after losing office.

The arrangements show, however, that by this time there was in London rather more experience of such projects than there had been in the Hawkesworth days. For instance, the first and second volumes would be compiled from Cook's journals by Dr John Douglas, who had helped Cook to write the second-voyage account. James King, who commanded HMS *Discovery* on the way home, was in October 1780 put forward by Banks to the Admiralty as author of the third volume covering events after Cook's demise. Banks and the third-voyage artist John Webber would oversee the engravers employed to prepare an agreed selection of mission drawings for publication. In all 25 engravers were used, nearly half of whom had worked on material for previous official Cook-voyage accounts. Alexander Dalrymple, hydrographer to the East India Company, was to supervise the engraving of the charts and coastal views, with the exception of the general chart seen here. Completion of this chart and oversight of its engraving were given to Henry Roberts, who was also tasked with drawing fair copies of the voyage charts chosen for publication. A young Roberts had served as an able seaman on Cook's second circumnavigation, during which he copied the coastal views and charts of various officers as well as producing accomplished coastal views of his own. On the final Cook voyage Roberts served as master's mate, producing fewer coastal views than he had before along with some charts, but making fair copy charts from Cook's surveys and, importantly, constructing for Cook a general chart of the world with all his Pacific voyages on it. On returning home, Roberts updated this chart using the best authorities. Hence work on this remarkable visual statement of Cook's achievements as a navigator and mission commander lay with Roberts.

Notwithstanding the considerable range of expertise now available, preparation of this account gave rise to periodic disagreements between the various parties involved and, in the end, a delay before the work eventually emerged in 1784. In respect of the general chart the conflict primarily concerned who would complete this task, with Dalrymple wanting to make changes to the chart without, it seems, the sanction of Roberts. During the summer months of 1782 the disagreements boiled over, with Banks trying to assert control over the situation through the then first lord, Lord Keppel. This and the possibility that Dalrymple might assume full control of the general chart caused Roberts to threaten to dissociate himself from it, but happily applications to Lord Sandwich resolved the difficulty in favour of Roberts. Thus the general chart was published under his name with William Palmer as its engraver, and was included as plate 1 in the folio Atlas. The contretemps is detailed in contemporary correspondence, including that of James King, who sided with Roberts. Roberts and King were, however, less fortunate with the polar chart of Cook's North Pacific surveys, which was completed under Dalrymple with Banks's support and published in the official Atlas as plate 36. An independent version of this chart by Roberts and King appeared a month after the Atlas, and it included in print for the first time details of Samuel Hearne's exploration of the North American interior that Dalrymple had omitted (see David, *The Charts & Coastal Views of Captain Cook's Voyages*, vol. 3). Banks's library copy of the official three-volume account and its accompanying Atlas containing the larger illustrated scenes and views are at British Library, pressmarks 454.h.9–10 and 456.h.24.

The delay in publishing the account was due not only to various differences, but also to King being sent to the West Indies on convoy duty for a year late in 1781, after which he returned home in failing health. Mounting exasperation at the lack of progress led Banks to propose a minuted meeting at Sandwich's country home on 5 December 1783, where work on the charts (excepting the general chart, now under the direction of Roberts) was thoroughly reviewed by those involved and a clear plan for their completion hammered out. Banks also advised on the costs of engraving and printing the account, and the price at which it should be sold. In exchange for rights to publish a French translation of the account by Jean Nicholas Demeunier, he obtained from the Parisian writer and publisher Charles Joseph Panckoucke a supply of high-quality French paper on which to print the plates. He further arranged a German translation by the importunate younger Forster. Significantly, he intervened in the distribution of profits from the publication, ensuring that half went to James Cook's family, with the interest passing to Elizabeth Cook in her lifetime. A quarter went to James King as author of the final part of the account. An eighth went to the representatives of the late Charles Clerke, who commanded the mission after Cook, but died of consumption before its return. And one eighth was awarded to the *Resolution*'s master, William Bligh, less £100 deducted for the executors of Willam Anderson, the deceased surgeon whose journal was used for the natural history. Thus Banks safeguarded income for the immediate Cook family, and also remuneration for Bligh, who with good reason had complained that, although he had completed a large quantity of the survey work under Cook, virtually no recognition of the fact was made by Roberts in the published charts. Indeed, in the entire work one inset map only is attributed to Bligh. The first edition appeared on 4 June 1784 in 2,000 copies and sold out, it is said, in three days. Two subsequent editions were issued the next year. Shortly after the appearance of the first edition, the Royal Society Cook medal was issued as another token of the president's regard for James Cook's memory (see number 140 below).

The present map is interesting for showing in coloured blocks, added by an unknown hand following its publication, the different areas of Pacific and Eastern trade permitted to British chartered commercial companies in this period as well as the waters allowed to whalers operating in the southern oceans. Hand annotation at the foot of the chart provides a key to the coloured blocks and records parliamentary acts that defined what rights existed. The chart has been cropped as further written text is partially visible at its bottom edge. The surviving annotation reads as follows:

The red Stain denotes the Countries and Seas Comprehended in the East India Company's Charter –. The blue denotes the Countries and seas included in the South Sea Company's Charter the mixture of red and blue along the Northwest coast of America denotes the countries and Seas where the East India Company and the South Sea Company may both carry on a concurrent Trade

The yellow takes in the whole of the limits allowed by an Act of the 35th of George 3d Cap. 92 Sect. 19 for permitting Southern Whale Ships to pass to the Eastward of the Cape of Good Hope

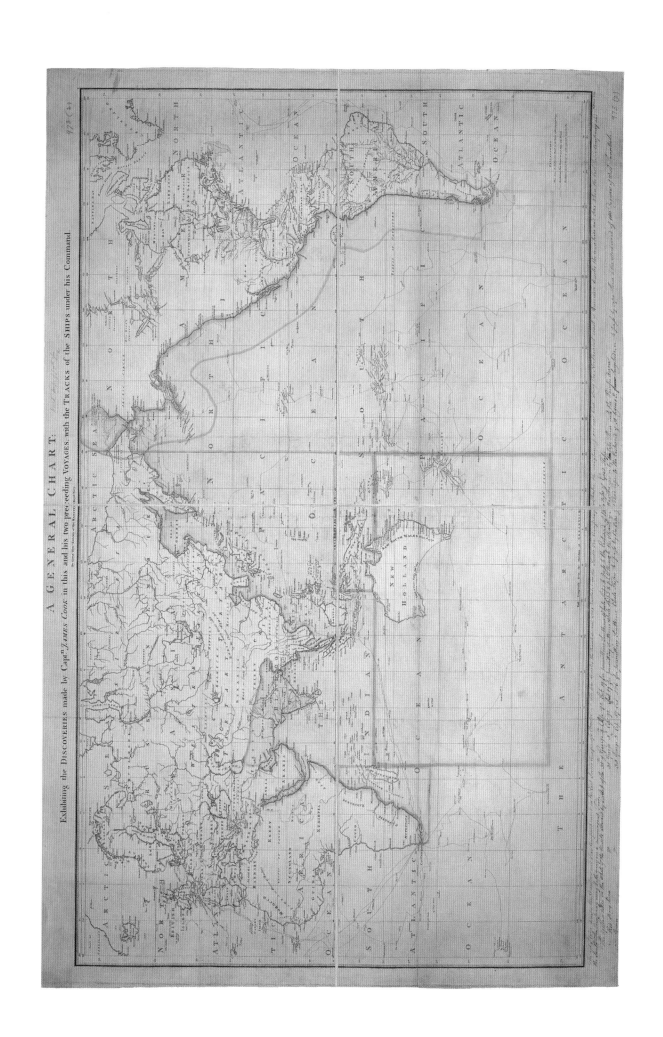

[The] Red Strong line [takes in the whole of the limits allowed by an Act of the] 35 George 3ᵈ Cap. 92 Sect. 19 for permitting Southern Whale Ships to pass through the Straits of Magelhanes or by /the/ Cape Horn into the Pacific Ocean

[The] Green [strong line takes in the whole of the limits allowed by an Act of the] 38 George 3ᵈ Cap. 57 Sect 5 & 6 for permitting Southern Whale Ships to pass by the Cape of Good Hope to the Eastward of 51 degrees E. /from/ Longitude, or to pass by Cape Horn to the Westward of 180 Degrees of West Longitude.

Major economic and strategic issues were raised by Cook's Pacific discoveries. Following his voyages, and particularly once settlements started to be established on the east coast of Australia from 1788, the vast monopolies of bodies like the East India Company were increasingly called into question. There was already pressure at home on the East India Company from a Pitt government keen to curb its powers. Pitt's government supported British whalers seeking to hunt in the southern oceans by offering them premiums, by pressing for relaxation of the terms under which company licences were granted to them to sail in restricted waters and by generally widening access to seas previously reserved to the trading companies. It also planned survey voyages to seek island bases at which whalers could refresh, and it held out to them opportunities to engage in fur trading on the Northwest Coast of America. These things ran counter to the monopolistic interests of both the South Sea Company and the East India Company, who naturally resisted such changes.

British activity of these kinds also had the effect of undermining Spanish power in the Pacific, notably in the case of British fur trading ventures on the Northwest Coast of America leading to the Nootka Crisis of 1790 in which war between Spain and Britain was narrowly avoided. Moreover, New South Wales offered strategic possibilities as a location from which to attack Spanish America, and it furnished a base where visiting whalers could refresh and refit (New Zealand and Hawai'i offered further such bases).

By the turn of the century the number and commercial strength of Britain's colonies in New South Wales had grown markedly. Larger ships could be built in them; these could be used to fish and trade further afield; some even started to arrive in London with seal oil, furs and other commodities, but were there impounded for having violated the East India Company's trading monopoly. As a result of these developments Banks intervened with authorities in the capital to try to shift the penal settlements on to a similar footing to that of other British colonies. He pressed, albeit with little success, for additional fishery and trading rights to be accorded them, bolstering his argument with the idea that this would serve to tie such settlements more firmly within Britain's global imperial network, and avoid a similar debacle to the one that overtook Britain's North American colonies (see Chambers, ed., *Indian and Pacific Correspondence*, esp. vols 7–8).

140 Royal Society medal of James Cook

By Lewis Pingo.
Private collection.
4.5 cm. Bronze.

Commissioned by the Royal Society under Joseph Banks as president, and funded by subscription, a commemorative medal of James Cook was completed in 1784 by the chief engraver at the Mint, Lewis Pingo. The medal was struck in gold, silver and bronze, and allotted to its subscribers. According to society records the number of medals in the first strike amounted to 13 gold, 289 silver and 500 bronze. After the gold medals had been distributed, a small number remained, and these were presented to British and Continental royalty. There was a surplus of proceeds from the issue of these medals and this allowed some more gold medals to be produced, one of which was given to Elizabeth Cook, widow of James.

The obverse of this medal shows a bust of Cook in profile wearing his captain's uniform. It commemorates him with the inscription, 'IAC. COOK OCEANI INVESTIGATOR ACERRIMVS' (the most acute explorer of the ocean). Beneath, in smaller lettering is, 'REG. SOC. LOND. SOCIO. SVO' (the Royal Society of London for its member). The reverse carries Britannia with an oar placed atop the globe. The inscription here reads, 'NIL INTENTATVM NOSTRI LIQUERE (our men left nothing unattempted) and AUSPICIIS GEORGII III (under the protection of George III).

Gold, silver and bronze examples of the Pingo medal survive in various public and private collections. Sarah Sophia Banks, sister of Sir Joseph Banks and a notable collector of ephemera, medals and coins, presented a gold version to the British Museum.

141   Bust, Sir Joseph Banks, *aet* 73

By Francis Leggatt Chantrey RA, commissioned by William Alexander in 1816. Plaster model exhibited at the Royal Academy, 1816 (953), and the finished marble exhibited there in 1818 (1105).

The Linnean Society, London. SC/1. Height 54 cm. White marble.

In 1816 Francis Chantrey, a rising star among British sculptors, started work on a bust of Sir Joseph Banks, probably the last likeness of the then aged naturalist to be produced during his lifetime. In a fine work Chantrey represented Banks in classical attire as a venerable patron of learning. William Alexander, first keeper of prints and drawings at the British Museum, commissioned the bust. Its plaster model was exhibited in 1816 at the Royal Academy, and a finished marble was shown at the academy in 1818. In the following year, Chantrey presented this bust to the Royal Society. Replicas were also produced. The present example was commissioned by the Linnean Society of London through a subscription of its members and it was delivered in 1822. Another replica is at Petworth House, West Sussex. In 1821 the British Museum and the Royal Society commissioned Chantrey to produce a full-length seated figure of Banks, who by then was dead. This now resides at the Natural History Museum, London.

Banks was an honorary fellow and founder member of the Linnean Society in 1788. He played a key role in the acquisition from Sweden of collections formerly belonging to Carl Linnaeus and his son, both by then deceased, that afterwards led to the establishment of the society. The sale letter for these collections was received at Spring Grove on 23 December 1783, and Banks passed it to his guest at the breakfast table, the young botanist James Edward Smith. At the time Banks was embroiled in acrimonious clashes at the Royal Society involving a number of rebellious fellows and his finances were in poor shape. His refusal to acquire Linnaeus's library and natural history collections, collections for which he had previously offered the same asking price of 1,000 guineas, is evidence of his difficulties. However, Banks's gesture enabled Smith to acquire them using money from his father, a palpable gain for British science and a sure loss for the Swedes, some of whom might well have tried to keep the collections in their country had Smith not been so prompt in shipping his prize. The collections arrived in London in 26 chests late in October 1784, comprising 19,000 sheets of preserved plants, 3,200 insects, 1,500 shells, 2,500 minerals, almost 3,000 books and about 3,000 letters and manuscripts. Smith, who went on to lead a distinguished career in botany, used them to establish the society of which he would be the first president.

142   Portrait, Sir Joseph Banks, *aet* 77

By Thomas Phillips RA. Commissioned by the Royal Horticultural Society, August 1820, completed 1821

The Royal Horticultural Society. Drawings Collection 62090-1001. Unframed 143 × 111.5 cm. Oil on canvas.

Thomas Phillips first painted Banks in 1808–9, when Banks was depicted in his presidential garb for the Spanish mathematician and astronomer José de Mendoza y Rios. In that portrait Banks is shown holding Humphry Davy's 1808 Bakerian lecture with an 1801 Mendoza paper on an improved reflecting circle on the table before him. Banks's white silk waistcoat is decorated with small flowers, and so the portrait suggests his botanical interests, but it also alludes to his regard for other sciences such as electrical chemistry (Davy) and astronomy (Mendoza).

The star and red sash of the Order of the Bath, an honour awarded to Banks in 1795, both appear. Phillips made three copies of the Mendoza portrait in about 1810. The copies are not all precisely the same in point of dress or detail, only one closely resembling the original (this was commissioned by the Duke of Somerset, and is now in the Caroline Simpson Collection, Sydney), but Banks is shown seated and facing the viewer in each. Five years later Phillips reproduced the 1809 portrait, this time for the Royal Society, with Banks again seen in his full presidential attire, and that version is still held by the society. There were a number of engravings after Phillips. The best of them, by Niccolò Schiavonetti in 1812, showed Banks as president and was dedicated to his great patron George III.

The present portrait is a final copy, commissioned in August 1820 by the Royal Horticultural Society following Banks's death. At Phillips's request the Royal Society supplied its painting for a model, and, as he had in some earlier versions, Phillips removed the presidential trappings. The resulting portrait shows Banks in his study, soberly dressed in grey. There is a pamphlet in his right hand and on the desk before him an illustration, here identified as based on a watercolour of the Spring Grove Peach by the botanical artist William Hooker, although the match is not an exact one (Phillips switched the peach cross section from one side of the sheet to the other, presumably to situate it closer to the viewer). The original Hooker watercolour of this peach was produced by order of the Royal Horticultural Society's council in the year of Banks's death, and it shows a fruit that was in 1816 named after Banks's Isleworth residence by the then president of the society Thomas Andrew Knight, Spring Grove being a location where Banks conducted various horticultural and other experiments. The illustration thus references the sitter's interests as well as his connection with the society that commissioned this portrait. Banks's cane and nearby glasses symbolize his age and infirmity, although his facial appearance is much the same as in earlier versions with no indication of his having grown older.

The commission from the Horticultural Society, as it was initially known, arose from the fact that Banks was a founder member of the society and had attended the first planning meeting for its establishment, held at James Hatchard's bookshop, Piccadilly, on 7 March 1804. In July he took up the post of vice-president. The idea for the society was that of John Wedgwood, son of the potter Josiah Wedgwood. Wedgwood wanted to encourage and diffuse practical improvements in the garden cultivation of ornamental and culinary plants and trees. In this sense the new society was not to be a scientific one like the Royal Society, whose position Banks jealously guarded. Instead it would emulate other existing societies that promoted practical improvements, such as the Bath and West of England Society, an agricultural society founded in 1777, and the Royal Society of Arts and Manufactures, which from 1754 sponsored artisan and technological innovations. At the initial meeting with Banks and Wedgwood were William Townsend Aiton, superintendent of Kew Gardens; James Dickson, the nurseryman; William Forsyth, superintendent of the royal gardens at St James's Palace and Kensington Palace; Charles Greville, the mineral collector and nephew of Sir William Hamilton and the botanist Richard Anthony Salisbury.

In the years that followed, Wedgwood's ideas were expanded upon. A membership by ballot and subscription was established; meetings were held, and an administration put in place; premiums and medals were issued for those horticultural improvements deemed worthy of them; a journal was published; a garden acquired; a library built up and research conducted. Banks served as a council member until he resigned in 1816. By that time a royal charter had been issued, in 1809, making the society the Royal Horticultural Society. The first transactions were published in 1807 (Banks contributed a number of papers to this journal) and the first medal was struck in 1808. The society still owns the present portrait of the most eminent of its founders. A print engraving by Samuel William Reynolds was twice published in 1822 and 1828 (see also the 1813 portrait of Banks that Phillips painted for the Corporation of Boston, in which Banks appears in his militia uniform holding a fen drainage plan, number 1).

143   Portrait, Joseph Banks, *aet* 29

By Benjamin West RA, 1771–2. Exhibited at the Royal Academy, 1773 (310).

The Collection: Art and Archaeology in Lincolnshire (Usher Gallery, Lincoln). LCNUG: 1989/9. Unframed 230 × 157.2 cm. Oil on canvas.

By Benjamin West probably commissioned by Robert Banks-Hodgkinson, Joseph Banks's uncle, this enormous portrait was apparently begun in December 1771, some five months after Banks's return from the *Endeavour* voyage. It shows Banks as a pioneering collector of Pacific natural history and ethnography (for which see the essay by N. Chambers in this volume, esp. page 218). During Banks's lifetime the portrait hung at Overton Hall, the Derbyshire residence that he inherited from Banks-Hodgkinson, and the only Banks property with a wall space large enough to accommodate such a canvas. Perhaps rather surprisingly for a work of this size, and one that was excellently engraved by John Raphael Smith in 1773 and so had been available in the print shops from Banks's youth, the portrait was thought to be lost from the mid-nineteenth century until it resurfaced at auction in the 1980s. It apparently remained at Overton after the hall was sold in 1829 to none other than the son of William Milnes, Banks's Overton steward of earlier years. Thereafter it stayed in Derbyshire among distant family relations of the Milnes and at other locations until at last it was put up for sale and eventually acquired by Lincolnshire County Council (see H.B. Carter in Gilbert, *The Banks/Cook Portfolio*, pp. 16–17).

NOTES

PAGES 21–6

Background to the Endeavour *Voyage*

1 John C. Beaglehole, ed., *The Journals of Captain James Cook on his Voyages of Discovery*, 3 vols in 4 parts (Cambridge: Hakluyt Society, 1955–69), vol. 1, p. 604.
2 *Ibid.*, vol. 1, p. cv.
3 *Ibid.*, vol. 1, pp. cclxxix–cclxxxiii.
4 Wayne Myers, 'Cook, the physician', in Antoinette and Robert Shalkop, eds, *Exploration in Alaska: Captain Cook commemorative lectures* (Anchorage: Cook Inlet Historical Society, 1980), pp. 59–88 on p. 63.
5 Randolph Cock, 'Precursors of Cook: the voyages of the *Dolphin* (1764–8) as precursors of Cook's voyages of exploration', *The Mariner's Mirror*, 85:1 (1999), pp. 30–52.
6 Ray Parkin, *H.M. Bark Endeavour: her place in Australian history* (Carlton, Victoria: Melbourne University Press, 1997), p. 71.
7 Derek Howse, 'Navigation and astronomy in the voyages', in Derek Howse, ed., *Background to Discovery: Pacific Exploration from Dampier to Cook* (Berkeley: University of California Press, 1990), pp. 160–83 on p. 165.
8 James Cook, 'An Observation of an Eclipse of the Sun at the Island of New-found-land. August 5, 1766', *Philosophical Transactions of the Royal Society*, LVII (1766), pp. 215–6.
9 William Windham to Lieutenant Burney, see Austin Dobson, ed., *Diary and Letters of Madame D'Arblay*, 6 vols (London, 1905), vol. 4, p. 378.
10 John C. Beaglehole, ed., *The Endeavour Journal of Joseph Banks 1768–1771*, 2 vols (Sydney: Trustees of the Public Library of New South Wales in association with Angus & Robertson, 1962), vol. 1, p. 22.
11 Harold B. Carter, *Sir Joseph Banks, 1743–1820* (London: British Museum (Natural History), 1988), p. 177.
12 Patrick O'Brian, *Joseph Banks: A Life* (London: Collins Harvill, 1987), p. 25.
13 John Ellis to Linnaeus, 19 August 1768, in J.E. Smith, ed., *A Selection from the Correspondence of Linnaeus and other Naturalists*, 2 vols (London, 1821), vol. 1, pp. 230–1.

PAGES 80–5

Dressing Up, Taking Over and Passing On

1 See Gilbert White, 'Correspondence, Sermon, Account-Book, Garden Kalendar, Animals and Plants, Geology, Roman-British Antiquities, &c.', in Thomas Bell, ed., *The Natural History and Antiquities of Selborne, in the County of Southampton*, 2 vols (London: John van Voorst, 1878), vol. 2, pp. 97–101. This is not the place to comment in detail on inaccuracies in Sheffield's account, but it should be noted that the coconut palm is not native to New Zealand, where a variety of different plant sources were used to make fibres, notably New Zealand flax (*Phormium tenax*).
2 Averil M. Lysaght, *Joseph Banks in Newfoundland and Labrador, 1766: His Diary, Manuscripts and Collections* (London: Faber and Faber, 1971), pp. 253–5.
3 Adrienne L. Kaeppler, *'Artificial Curiosities' being An exposition of native manufactures collected on the three Pacific voyages of Captain James Cook, R.N. at the Bernice Pauahi Bishop Museum January 18, 1978–August 31, 1978*. Bernice P. Bishop Museum Special Publication 65 (Honolulu: Bishop Museum Press, 1978), p. 128; Adrienne L. Kaeppler, 'Tracing the History of Hawaiian Cook-Voyage Artefacts in the Museum of Mankind', in T.C. Mitchell, ed., *Captain Cook and the South Pacific*, The British Museum Yearbook 3 (London: British Museum Publications for the Trustees, 1979), p. 171.
4 Jeremy Coote, *Curiosities from the Endeavour: A Forgotten Collection – Pacific Artefacts Given by Joseph Banks to Christ Church, Oxford after the First Voyage* (Whitby: Captain Cook Memorial Museum, 2004), p. 6.
5 Kaeppler, 'Artificial Curiosities', pp. 41, 171; Coote, *Curiosities from the Endeavour*, p. 6.
6 Karen Stevenson and Steven Hooper, 'Tahitian *Fau*: Unveiling an Enigma', in Judith Huntsman, ed., *Polynesian Art: Histories and Meanings in Cultural Contexts*, Special Issue, *The Journal of the Polynesian Society*, 116, no. 2 (June 2007), p. 201. The *fau* in the present volume was collected by Johann Reinhold Forster and his son Johann George on Cook's second voyage of 1772–1775; see Jeremy Coote, Peter Gathercole, and Nicolette Meister (with contributions by Tim Rogers and Frieda Midgley), '"Curiosities sent to Oxford": The Original Documentation of the Forster Collection at the Pitt Rivers Museum', *Journal of the History of Collections*, 7, no. 2 (2000), pp. 177–92.
7 Rüdiger Joppien and Bernard Smith, *The Art of Captain Cook's Voyages*, 4 vols (New Haven and London: Yale University Press, 1985), vol. 1: *The Voyage of the Endeavour, 1768–1771, with a Descriptive Catalogue of All the Known Original Drawings of Peoples, Places, Artefacts and Events and the Original Engravings Associated with them*, p. 75. See also Averil M. Lysaght, 'Banks's Artists and his *Endeavour* Collections', in T.C. Mitchell, ed., *Captain Cook and the South Pacific*, p. 25.
8 John Hawkesworth, ed., *An Account of the Voyages Undertaken by the Order of His Present Majesty for Making Discoveries in the Southern Hemisphere, and Successively Performed by Commodore Byron, Captain Wallis, Captain Carteret, and Captain Cook, in the Dolphin, the Swallow, and the Endeavour: drawn up from the journals which were kept by the several commanders, and from the papers of Joseph Banks, Esq*, 3 vols (London: W. Strahan and T. Cadell, 1773).
9 Volker Harms, 'Ein "Ancestor Panel," der Māoris von de ersten südsee-reise (1768–1771) James Cooks in der ethnographischen sammlung der Universität Tübingen entdeckt', *Baessler-Archiv*, 46, no. 2 (1998), pp. 429–41.
10 John C. Beaglehole, ed., *The Endeavour Journal of Joseph Banks 1768–1771*, 2 vols (Sydney: Trustees of the Public Library of New South Wales in association with Angus & Robertson, 1962), vol. 1, p. 421.
11 Amiria Salmond, 'Artefacts of Encounter: The Cook-Voyage Collections in Cambridge', in Jeremy Coote, ed., *Cook-Voyage Collections of Artificial Curiosities in Britain and Ireland, 1770–2014*, Museum Ethnographers Group Occasional Paper 5 (Oxford: Museum Ethnographers Group, [first issued 2011, forthcoming journal publication 2016]).
12 I draw on here previous discussions by

Adrienne L. Kaeppler; see, for example, 'Artificial Curiosities', pp. 39–42; 'From the South Seas to the World via London', in Jeremy Coote, ed., *Cook-Voyage Collections of Artificial Curiosities in Britain and Ireland, 1770–2014*, Museum Ethnographers Group Occasional Paper 5 (Oxford: Museum Ethnographers Group, [forthcoming, 2016]).

13 Jennifer Newell, 'Revisiting Cook at the British Museum', in Jeremy Coote, ed., *Cook-Voyage Collections of Artificial Curiosities in Britain and Ireland, 1770–2014*, Museum Ethnographers Group Occasional Paper 5 (Oxford: Museum Ethnographers Group, [forthcoming, 2016]).

14 Johann Christian Fabricius, *Briefe aus London vermischten Inhalts* (Dessau and Leipzig, 1784); cited Adrienne L. Kaeppler, 'Tracing the History of Hawaiian Cook-Voyage Artefacts in the Museum of Mankind', in T.C. Mitchell, ed., *Captain Cook and the South Pacific*, The British Museum Yearbook 3 (London: British Museum Publications for the Trustees, 1979), p. 174.

15 Amiria Salmond, 'Artefacts of Encounter'.

16 Stig Rydén, *The Banks Collection: An Episode in 18th-Century Anglo-Swedish Relations*, Statens Etnografiska Museum Monograph Series, no. 8 (Stockholm: Almqvist & Wiksell, 1965). The Statens Etnografiska Museum is the Ethnographical Museum of Sweden, in Stockholm.

17 Anon., 'A Collection of American Indian and Oceanic Art: The Property of The Trustees of The Warwick Castle Resettlement', in Sotheby & Co., *Catalogue of Primitive Art and Indian Sculpture: The Property of the The Trustees of The Warwick Castle Resettlement, The Property of R. St. Barbe Baker, Esq., The Property of Mrs. E. Meyerowitz, and Other Owners…/ Catalogue of African, Oceanic, Pre-Columbian and American Indian Art also Indian Sculpture…* (London: Sotheby and Co., 1969), pp. 66–77. [Catalogue of a sale held by Sotheby and Co., London, at the Royal Watercolour Society, London, on 8 December 1969.]

18 Anon., '136 A Tahitian Basalt Pounder…', in Christie's, *Art and Ethnography from Africa, North America and the Pacific Area* (London: Christie's, 1986), p. 63. [Catalogue of a sale held at Christie's, London, on 23 June 1986.]

19 Anon., 'Artificial Curiosities, from America, China, and the Newly-Discovered Islands in the South Seas', in Skinner and Co., *A Catalogue of the Portland Museum, Lately the Property of the Duchess Dowager of Portland, Deceased: Which Will Be Sold By Auction, By Mr. Skinner and Co. On Monday the 24th of April, 1786, and the Thirty-Seven Following Days, at Twelve O'Clock, Sundays, and the 5th of June, (the Day his Majesty's Birth-Day is Kept) Excepted; At her Late Dwelling-House, in Privy-Gardens, Whitehall; By Order of the Acting Executrix…* (London, Skinner and Co., 1786), pp. 58–9. [Catalogue of a sale held at Skinner and Co., London, from 24 April 1786.]

20 Anon., *A Descriptive Catalogue of the Rarities, in Mr. Greene's Museum at Lichfield* (Lichfield, 1773), pp. 15–16.

21 J. Broster, *Bibliotheca Llwydiana: A Catalogue of the Entire Library, (Near Ten Thousand Volumes) from the Presses of Caxton, Wynkyn de Worde, Pynson, &c. &c. &c. and Philosophical Apparatus, Late the Property of John Lloyd, Esq. L.L.D. Deceased, Fellow of the Royal, Antiquarian, and Linnæan Societies, and Late Member in Parliament for the County of Flint; Which Will Be Sold By Auction, Without Reserve, At His Seat at Wygfair, Near St. Asaph, Denbighshire, on Monday, January 15th, 1816, and Twelve Following Days, (Sundays, and Thanksgiving-Day on Thursday January the 18th, Excepted.) By Mr. Broster, of Chester* (Denbigh, 1816).

22 Jeremy Coote, 'An Interim Report on a Previously Unknown Collection from Cook's First Voyage: The Christ Church Collection at the Pitt Rivers Museum, University of Oxford', *Journal of Museum Ethnography*, no. 16 (2004), pp. 111–21; Coote, *Curiosities from the Endeavour*.

23 Harold B. Carter, *Sir Joseph Banks, 1743–1820* (London: British Museum (Natural History), 1988), p. 54.

24 The original letter is held in the Banks Correspondence in the Library of the Royal Botanic Gardens, Kew: B.C. I 32, and there is a copy in the Dawson Turner Collection at the Natural History Museum, London: DTC, I 49–50; Warren R. Dawson, *The Banks Letters: A Calendar of the Manuscript Correspondence of Sir Joseph Banks Preserved in the British Museum, the British Museum (Natural History) and other Collections in Great Britain* (London: British Museum (Natural History), 1958) pp. 318–19.

25 Māori scholar Ngahuia Te Awekotuku has argued that the word 'cleaver' would be more accurate than 'club' in giving a sense of the ways in which such hand-weapons were used to strike, jab, and slice in dance-like movements; see, for example, Ngahuia Te Awekotuku, 'Maori: People and Culture', in Dorota Czarkowska Starzecka, ed., *Maori Art and Culture* (London: British Museum Press, 1998), p. 38. The items exhibited at Lincoln were: from Tahiti a shark hook, a canoe bailer, a barkcloth beater, a chisel or gouge, a breast ornament or *taumi* and a noseflute; from New Zealand three belts, a cloak or *kaitaka*, a fish hook and five cleavers.

26 Jeremy Coote, 'Joseph Banks's Forty Brass Patus', *Journal of Museum Ethnography*, no. 20 (2008), pp. 49–68.

27 Beaglehole, ed., *The Endeavour Journal of Joseph Banks 1768–1771*, vol. 2, p. 15.

28 *Ibid*, vol. 1, p. 53.

29 *Ibid*, vol. 1, p. 437.

30 Lysaght points out that although the latter are 'invariably and unfortunately referred to as his servants' they were in fact 'four trained collectors'; see Lysaght, 'Banks's Artists and his *Endeavour* Collections', p. 11. Some Māori objects collected by Roberts have recently been identified in the collection of the Bowes Museum; Les Jessop, 'Cook-Voyage Collections in North-East England', in Coote, ed., *Cook-Voyage Collections of Artificial Curiosities in Britain and Ireland, 1770–2014*.

31 Harold B. Carter, 'Note on the Drawings by an Unknown Artist from the Voyage of HMS Endeavour', in Margarette Lincoln, ed., *Science and Exploration in the Pacific: European Voyages to the Southern Oceans in the 18th Century* (London: Boydell, 1988), pp. 133–4.

32 Anne Salmond, *The Trial of the Cannibal Dog: Captain Cook in the South Seas* (London: Allen Lane, 2003), p. 126.

33 Paul Tapsell, 'Footprints in the Sand: Banks's Māori Collection, Cook's First Voyage, 1768–71', in Michelle Hetherington and Howard Morphy, eds, *Discovering Cook's Collections* (Canberra: National Museum of Australia Press, 2009), pp. 92–111.

34 For a recent reassessment of the Banks–Parkinson dispute, see Noah Heringman's interesting discussion of early 'knowledge workers' in the overlapping fields of antiquarianism and natural history, in which Sydney Parkinson is presented as a collector and creator of knowledge in his own right; Noah Heringman, *Sciences of Antiquity: Romantic Antiquarianism, Natural History, and Knowledge Work* (Oxford: Oxford University Press, 2013).

PAGES 154–9

The Material History of the Endeavour

1. This is not the only stamp used by Banks to denote items from his library, but it is the one most common one.
2. Samuel Hearne, *A Journey from Prince of Wales's Fort in Hudson's Bay to the Northern Ocean, undertaken … for the discovery of Copper mines, a North West passage, &c. in the years 1769–1772* (Dublin, 1796). Banks's library copy, British Library pressmark 454.f.20.
3. Thomas Pennant, *Arctic Zoology*, 2 vols and supplement (London, 1784–7), pp. ix–x. Banks's library copy, British Library pressmarks G.2807 and G.2808.
4. Rüdiger Joppien and Neil Chambers, 'The Scholarly Library and Collections of Knowledge of Sir Joseph Banks', in Giles Mandelbrote and Barry Taylor, eds, *Libraries within the Library* (London: British Library, 2009), pp. 222–43.
5. Sarah Sophia Banks's collections at the British Library cover playbills, press cuttings, printed books on various subjects and a manuscript collection of heraldic displays. An important collection of her coins is held at the British Museum.
6. Philip Rowland Harris, *A History of the British Museum Library, 1753–1973* (London: British Library, 1998), p. 20.
7. The manuscript maps from Banks's collection held at the British Library are shared between the manuscript and maps departments.
8. These can now be found in the library's manuscript collections, BL Add. MS 4861, 4864, 2867, 4868, 4869, 4870 and 4884.
9. Philip Hatfield, *Lines in the Ice: exploring the roof of the world* (London: British Library, 2016).
10. Neil Chambers, *Joseph Banks and the British Museum: The World of Collecting, 1770–1830* (London: Pickering and Chatto, 2007). See also the essay by N. Chambers in this volume.
11. A number of manuscripts from the collections of James Cook were acquired by the British Museum at auction in 1868 and are also held in the manuscript collections. Another set of bound illustrations and charts at BL Add. MS 7085 are of uncertain provenance, but it is possible they were in the possession of John Hawkesworth during the publication of his edition of the *Endeavour* account, before being passed to the print seller Thomas Bulging and then sold to the British Museum in 1827. See Andrew David, *The Charts & Coastal Views of Captain Cook's Voyages*, 3 vols (London: the Hakluyt Society in association with the Australian Academy of the Humanities, 1988, 1992 and 1997), vol. 1, pp. lvi–lx.
12. BL Add. MS 15507, ff. 2–3.
13. BL Add. MS 7085 f. 3. There is also a single printed sheet of the map that was produced using Buchan's drawings among maps formerly belonging to Joseph Banks, British Library MAPS 181 m 1, vol. 3, pl. 69.
14. For example, see BL Add. MS 9345, ff. 46v–49, and BL Add. MS 9345, ff. 51v–52.
15. BL Add. MS 21593E.
16. BL MAPS 181 m 1.
17. BL MAPS 181 m 1, vol. 3, pl. 69.
18. David, *The Charts & Coastal Views of Captain Cook's Voyages*, vol. 1, p. xli.
19. BL Add. MS 15508, ff. 20–21.
20. John C. Beaglehole, ed., *The Endeavour Journal of Joseph Banks 1768–1771*, 2 vols (Sydney: Trustees of the Public Library of New South Wales in association with Angus & Robertson, 1962), vol. 1, pp. 281–3: '[T]heir chief amusement was carried on by the stem of an old canoe, with this before them they swam out as far as the outermost breach, then one or two would get onto it and opposing the blunt end to the breaking wave were hurried in with incredible swiftness.'
21. Banks is not unique in this area among the British Library's founding collectors. The manuscripts of Sir Hans Sloane detailing Caribbean flora and fauna are also a record of the ecological changes undergone on islands such as Jamaica.
22. BL Add. MS 21593C. For more on Tupaia's chart see, Harriet Parsons, 'British-Tahitian collaborative drawing strategies on Cook's *Endeavour* voyage', in Shino Konishi, Maria Nugent and Tiffany Shellam, eds, *Indigenous Intermediaries: new perspectives on exploration archives* (Canberra: Australia National University Press, 2015), pp. 147–67.
23. Philip Hatfield, *Canada in the Frame: Colonial Copyright and the Photographic Image* (London: McGill-Queens University Press, forthcoming 2017).
24. BL Add. MS 23920, f. 60 a–f.
25. BL Add. MS 23920, ff. 9, 10, 14 and 15.

PAGES 200–18

Exploring Collections from the Endeavour *Voyage*

1. Neil Chambers, 'Joseph Banks and Collections in the Age of Empire', in R.G.W. Anderson, M.L. Caygill, A.G. MacGregor and L. Syson, eds, *Enlightening the British: Knowledge, discovery and the museum in the eighteenth century* (London: The British Museum, 2003), pp. 99–113.
2. Roy A. Rauschenberg, ed., 'The Journals of Joseph Banks' Voyage up Great Britain's West Coast to Iceland the Orkney Isles July to October 1772', *Proceedings of the American Philosophical Society*, 117:3 (1973), pp. 186–226.
3. Stanfield Parkinson, ed., *A Journal of a Voyage to the South Seas, in his Majesty's Ship, the Endeavour. Faithfully transcribed from the Papers of the late Sydney Parkinson, draughtsman to Joseph Banks, Esq. on his late Expedition, with Dr. Solander, round the World* (London, 1773), reissued in 1784 with 'Explanatory Remarks on the Preface' by J. Fothergill, p. xi.
4. Guy L. Wilkins, 'A Catalogue and Historical Account of the Banks Shell Collection', *Bulletin of the British Museum (Natural History) Historical Series*, 1, no. 3 (1955), p. 74.
5. Rex Banks *et al.*, eds, *Sir Joseph Banks: A Global Perspective* (London, Royal Botanic Gardens, Kew, 1994), pp. 210–11. See also, British Museum (Natural History), *The History of the Collections Contained in the Natural History Departments of the British Museum*, 2 vols (London: British Museum, 1904 and 1906), vol. 2, p. 564.
6. Harold B. Carter, *Sir Joseph Banks, 1743–1820* (London: British Museum (Natural History), 1988), p. 95.
7. Thomas Bell, ed., *The Natural History and Antiquities of Selborne, in the County of Southampton*, 2 vols (London: John van Voorst, 1878), vol. 2, pp. 97–101.
8. Adm. 1/1609. For the Trinity College donation, see Trinity MS Add. a. 106, ff. 108r–9v and MS Add. a. 106, ff. 211r–212v.
9. BM CE 3/6 1632.
10. Neil Chambers, ed., *The Indian and Pacific Correspondence of Sir Joseph Banks 1768–1820*, 8 vols (London: Pickering and Chatto, 2008–13), vol. 1, letter 198, 18 August 1779.
11. BM CE 3/6 1743–4.
12. Jonathan King, *Artificial Curiosities from the Northwest Coast of America: native American artefacts in the British Museum collected on the third voyage of Captain James Cook and acquired through Sir

After the Endeavour: What Next for Joseph Banks?

Joseph Banks (London: British Museum Publications, 1981).
13 SL PN 1:18.
14 Chambers, ed., *The Indian and Pacific Correspondence of Sir Joseph Banks*, vol. 1, letter 239, 31 May 1782.
15 Stig Rydén, *The Banks Collection: An Episode in 18th-Century Anglo-Swedish Relations*, Statens Etnografiska Museum Monograph Series, no. 8 (Stockholm: Almqvist & Wiksell, 1965).
16 William J. Smith, 'A Museum for a Guinea', *Country Life*, 127, no. 3288 (10 March 1960); William J. Smith, 'Sir Ashton Lever of Alkrington, 1729–1788', *Lancashire and Cheshire Antiquarian Society Transactions*, 72 (1962); Adrienne L. Kaeppler, 'Tracing the History of Hawaiian Cook-Voyage Artefacts in the Museum of Mankind', in T.C. Mitchell, ed., *Captain Cook and the South Pacific*, The British Museum Yearbook 3 (London: British Museum Publications for the Trustees, 1979).
17 Peter J.P. Whitehead, 'Zoological Specimens from Captain Cook's Voyages', *Journal of the Society for the Bibliography of Natural History*, 5 (1969), no. 3, pp. 161–201, and 'A Guide to the Dispersal of Zoological Material from Captain Cook's Voyages', *Pacific Studies*, 2 (1978), no. 1, pp. 52–93; Adrienne L. Kaeppler, *Holophusicon. The Leverian Museum. An Eighteenth-Century English Institution of Science, Curiosity, and Art* (Altenstadt: ZKF Publishers, 2011).
18 Adrienne L. Kaeppler, 'To attempt some new discoveries in that vast unknown tract', in Michelle Hetherington and Howard Morphy, eds, *Discovering Cook's Collections* (Canberra: National Museum of Australia Press, 2009), pp. 63 and 77.
19 Edward J. Miller, *That Noble Cabinet: A history of the British Museum* (London: André Deutsch, 1973), pp. 226–7. William T. Stearn, *The Natural History Museum at South Kensington* (London: The Natural History Museum, 2008 reprint), pp. 283–4. See Professor Stearn's comments on the separation of Banks's herbarium and library in 1880, p. 284.
20 British Museum (Natural History), *The History of the Collections Contained in the Natural History Departments of the British Museum*, 2 vols (London: British Museum, 1904 and 1906), vol. 1, pp. 24–9.

1 John Gascoigne, 'Banks, Sir Joseph, baronet (1743–1820)', *Oxford Dictionary of National Biography* (Oxford: Oxford University Press, 2004), vol. 3, p. 692.
2 Vanessa Collingridge, *Captain Cook* (London: Ebury Press, 2002), p. 209; Richard Hough, *Captain James Cook* (London: Hodder & Stoughton, 1994), p. 177.
3 See James A. Home, ed., *The Letters and Journals of Lady Mary Coke*, 4 vols (Edinburgh: David Douglas, 1889–96), III, p. 435.
4 John C. Beaglehole, ed., *The Endeavour Journal of Joseph Banks 1768–1771*, 2 vols (Sydney: Trustees of the Public Library of New South Wales in association with Angus & Robertson, 1962), vol. 1, p. 51.
5 John C. Beaglehole, ed., *The Journals of Captain James Cook on his Voyages of Discovery*, 3 vols in 4 parts (Cambridge: Hakluyt Society, 1955–69), vol. 2, p. xxi.
6 Banks to the Comte de Lauraguais, 6 December 1771, published in Neil Chambers, ed., *The Indian and Pacific Correspondence of Sir Joseph Banks 1768–1820*, 8 vols (London: Pickering and Chatto, 2008–13), vol. 1, letter 32.
7 Quoted by Vanessa Collingridge, *Captain Cook*, p. 213.
8 Harold B. Carter, *Sir Joseph Banks, 1743–1820* (London: British Museum (Natural History), 1988), p. 99.
9 Beaglehole, ed., *The Journals of Captain James Cook on his Voyages of Discovery*, vol. 2, p. xxiv.
10 Ibid., p. xxvii.
11 Ibid., p. 3.
12 Ibid., p. 4.
13 Ibid., p. xxviii.

14 Ibid., p. 931.
15 Ibid., p. 6.
16 Ibid., p. 220.
17 Ibid., p. 7.
18 Ibid., p. 937.
19 Neil Chambers, ed., *The Indian and Pacific Correspondence of Sir Joseph Banks 1768–1820*, 8 vols (London: Pickering and Chatto, 2008–13), vol. 1, letter 97, 3 June 1772.
20 Beaglehole, ed., *The Journals of Captain James Cook on his Voyages of Discovery*, vol. 2, p. xxxi, note.
21 Ibid., pp. clxvii–clxx. These are Cook's instructions.
22 Beaglehole, ed., *The Journals of Captain James Cook on his Voyages of Discovery*, vol. 2, p. 643.
23 Uno von Troil, *Letters on Iceland* (Dublin, 1780), pp. 1–2.
24 Roy A. Rauschenberg, 'The Journals of Joseph Banks. Voyage up Great Britain's West Coast to Iceland and to the Orkney Isles July to October 1772', *Proceedings of the American Philosophical Society*, 117, no. 3 (1973), pp. 186–226.
25 The passport is to be found in the National Library of Australia, Canberra, Banks Papers, MS 9/3/118. It has been published in Rauschenberg, 'The Journals of Joseph Banks', pp. 217–18.
26 Claus Heide to Banks, 25 June 1772, BL Add. MS 8094, ff. 29–30.
27 Carter, *Sir Joseph Banks*, p. 123.
28 *The Scots Magazine*, Edinburgh, vol. 34 (November 1772), p. 637.
29 See Anna Agnarsdóttir, 'Sir Joseph Banks and the Exploration of Iceland', in Rex E.R. Banks, *et al.*, eds, *Sir Joseph Banks: A Global Perspective* (London: Royal Botanic Gardens, 1994), pp. 31–48.

BIBLIOGRAPHY AND SOURCES

Agnarsdóttir, A., 'Sir Joseph Banks and the Exploration of Iceland', in R.E.R. Banks *et al.*, eds, *Sir Joseph Banks: A Global Perspective* (London: Royal Botanic Gardens, 1994), pp. 31–48

Anderson, R.G.W., *Journal of the History of Collections*, 20, no. 1 (2008), pp. 151–2

Anon., *A Descriptive Catalogue of the Rarities, in Mr. Greene's Museum at Lichfield* (Lichfield, 1773)

Anon., *An Epistle from Mr. Banks, Voyager, Monster-hunter, and Amoroso, to Oberea, Queen of Otaheite. Transfused by A.B.C. Esq. Second Professor of the Otaheite, and of every other unknown Tongue. Enriched with the finest Passages of the Queen's Letter to Mr. Banks. Printed at Batavia, for Jacobus Opano*, 2nd edition (London: John Swan and Thomas Axtell, [1773]). Introduction dated 20 December 1773

Anon., 'Artificial Curiosities, from America, China, and the Newly-Discovered Islands in the South Seas', in *A Catalogue of the Portland Museum, Lately the Property of the Duchess Dowager of Portland, Deceased: Which Will Be Sold By Auction, By Mr. Skinner and Co. On Monday the 24th of April, 1786, and the Thirty-Seven Following Days, at Twelve O'Clock, Sundays, and the 5th of June, (the Day his Majesty's Birth-Day is Kept) Excepted; At her Late Dwelling-House, in Privy-Gardens, Whitehall; By Order of the Acting Executrix…* (London: Skinner and Co., 1786)

Anon., 'A Collection of American Indian and Oceanic Art: The Property of The Trustees of The Warwick Castle Resettlement', in *Catalogue of Primitive Art and Indian Sculpture: The Property of the The Trustees of The Warwick Castle Resettlement, The Property of R. St. Barbe Baker, Esq., The Property of Mrs. E. Meyerowitz, and Other Owners…/Catalogue of African, Oceanic, Pre-Columbian and American Indian Art also Indian Sculpture…* (London: Sotheby and Co., 1969)

Anon., '136 A Tahitian Basalt Pounder…', in *Art and Ethnography from Africa, North America and the Pacific Area* (London: Christie's, 1986)

Attenbrow, V., *Sydney's Aboriginal Past: Investigating the archaeological and historical records*, 2nd edition (Sydney: University of New South Wales, 2010)

Baird, I. and Ionescu, C., eds, *Eighteenth-Century Thing Theory in a Global Context: From Consumerism to Celebrity Culture* (Farnham, Surrey and Burlington, Vermont: Ashgate, 2013)

Banks, R.E.R. *et al.*, eds, *Sir Joseph Banks: A Global Perspective* (London: Royal Botanic Gardens, Kew, 1994)

Banks's Journal – Beaglehole, J.C., ed., *The Endeavour Journal of Joseph Banks 1768–1771*, 2 vols (Sydney: Trustees of the Public Library of New South Wales in association with Angus & Robertson, 1962)

Beaglehole, J.C., *The Life of Captain James Cook* (London: Hakluyt Society, 1974)

Bell, T., ed., *The Natural History and Antiquities of Selborne, in the County of Southampton*, 2 vols (London: John van Voorst, 1878)

Blake, R., *George Stubbs and the Wide Creation: Animals, people and places in the life of George Stubbs 1724–1806* (London: Chatto & Windus, 2005)

Bligh, W., *A Narrative of the Mutiny on board His Majesty's Ship Bounty; and the subsequent voyage of part of the crew, in the ship's boat, from Tofoa, one of the Friendly Islands, to Timor, a Dutch Settlement in the East Indies* (London: George Nicol, 1790)

Bligh, W., *A Voyage to the South Sea, undertaken by command of His Majesty, for the purpose of Conveying the Bread-Fruit Tree to the West Indies, in His Majesty's Ship the Bounty, commanded by Lieutenant William Bligh. Including an account of the Mutiny on Board the Said Ship, and the subsequent voyage of part of the crew, in the ship's boat, from Tofoa, one of the Friendly Islands, to Timor, a Dutch Settlement in the East Indies* (London: George Nicol, 1792)

British Library – Abbreviated, BL
    Additional Manuscripts, various volumes as cited in the main catalogue
    Printed library volumes, by author or editor, with pressmark for each one in the main catalogue
    Maps Department

British Museum – Abbreviated, BM
    Archives –
        Book of Presents
        Standing Committee Minutes
    Britain, Europe and Prehistory
    Prints and Drawings
    Department of Africa, Oceania and the Americas –
        Dalton registration slips
        Edge-Partington registration slips, NWC, TAH and NZ series
        Read registration slips

British Museum (Natural History), *The History of the Collections Contained in the Natural History Departments of the British Museum*, 2 vols (London: British Museum, 1904 and 1906)

Broster, J., *Bibliotheca Llwydiana: A Catalogue of the Entire Library, (Near Ten Thousand Volumes) from the Presses of Caxton, Wynkyn de Worde, Pynson, &c. &c. &c. and Philosophical Apparatus, Late the Property of John Lloyd, Esq. L.L.D. Deceased, Fellow of the Royal, Antiquarian, and Linnæan Societies, and Late Member in Parliament for the County of Flint; Which Will Be Sold By Auction, Without Reserve, At His Seat at Wygfair, Near St. Asaph, Denbighshire, on Monday, January 15th, 1816, and Twelve Following Days, (Sundays, and Thanksgiving-Day on Thursday January the 18th, Excepted.) By Mr. Broster, of Chester* (Denbigh, 1816)

Brown, P., *New illustrations of Zoology, containing fifty coloured plates of new, curious, and non-descript birds, with a few quadrupeds, reptiles and insects. Together with a short and scientific description of the same* (London: Printed for B. White, 1776)

Burney, F., *Diary and Letters of Madame D'Arblay*, 6 vols (London, 1905). Barrett, C., ed., preface and notes by Dobson, A.

Calaby, J.H., Mack, G. and Ride, W.D.L., 'The application of the generic name *Macropus* Shaw 1790 and of other names commonly referred to the grey kangaroo', *Memoirs of the Queensland Museum*, 14 (1962), pp. 25–31

Calaby, J.H., Mack, G. and Ride, W.D.L., 'The generic name *Macropus* Shaw, 1790 (Mammalia). Z.N.(S.) 1584', *Bulletin of Zoological Nomenclature*, 20 (1963), pp. 376–9

Captain Cook Memorial Museum, Whitby – Banks to Sandwich, [30 May 1772], WHICC 174.36.3

Carr, D.J., ed., *Sydney Parkinson: Artist of Cook's Endeavour Voyage* (London and Canberra: British Museum (Natural History) and Croom Helm Limited, 1983)

Carter, H.B., *Sir Joseph Banks (1743–1820): A guide to the biographical and bibliographical sources* (London: St Paul's Bibliographies and the British Museum (Natural History), 1987)

Carter, H.B., *Sir Joseph Banks, 1743–1820* (London: British Museum (Natural History), 1988)

Carter, H.B., 'Note on the Drawings by an Unknown Artist from the Voyage of HMS Endeavour', in Margarette Lincoln, ed., *Science and Exploration in the Pacific: European Voyages to the Southern Oceans in the 18th Century* (London: Boydell, 1988), pp. 133–4

Carter, H.B., Diment, J.A., Humphries, C.J. and Wheeler, A., 'The Banksian Natural History Collections of the *Endeavour* Voyage and their Relevance to Modern Taxonomy', in *History in the Service of Systematics: Papers from the Conference to Celebrate the Centenary of the British Museum (Natural History) 13–16 April, 1981*, Society for the Bibliography of Natural History, Special Publication 1 (London, 1981), pp. 62–8

Chambers, N., ed., *The Letters of Sir Joseph Banks: A Selection, 1768–1820* (London: Imperial College Press, 2000)

Chambers, N., 'Joseph Banks and Collections in the Age of Empire', in R.G.W. Anderson, M.L. Caygill, A.G. MacGregor and L. Syson, eds, *Enlightening the British: Knowledge, discovery and the museum in the eighteenth century* (London: The British Museum, 2003), pp. 99–113

Chambers, N., *Joseph Banks and the British Museum: The World of Collecting, 1770–1830* (London: Pickering and Chatto, 2007)

Chambers, N., ed., *The Scientific Correspondence of Sir Joseph Banks 1765–1820*, 6 vols (London: Pickering and Chatto, 2007)

Chambers, N., ed., *The Indian and Pacific Correspondence of Sir Joseph Banks 1768–1820*, 8 vols (London: Pickering and Chatto, 2008–13)

Cilento, R., 'Sir Joseph Banks, F.R.S., and the naming of the Kangaroo', *Notes and Records of the Royal Society*, 26, no. 2 (1971), pp. 157–61

Cock, R., 'Precursors of Cook: the voyages of the *Dolphin* (1764–8) as precursors of Cook's voyages of exploration', *The Mariner's Mirror*, 85:1 (1999), pp. 30–52

Collingridge, V., *Captain Cook* (London: Ebury Press, 2002)

Cook, J., 'An Observation of an Eclipse of the Sun at the Island of New-found-land. August 5, 1766', *Philosophical Transactions of the Royal Society*, LVII (1766), pp. 215–6

Cook, J., *A Voyage to the Pacific Ocean. Undertaken by the command of His Majesty, for making Discoveries in the Northern Hemisphere. To determine the Position and Extent of the West Side of North America; its Distance from Asia; and the Practicability of a Northern Passage to Europe...*, 3 vols (London: Printed by W. and A. Strahan for G. Nicol, and T. Cadell, 1784)

Cook's Journals – Beaglehole, J.C., ed., *The Journals of Captain James Cook on his Voyages of Discovery*, 3 vols in 4 parts (Cambridge: Cambridge University Press for the Hakluyt Society, 1955–69)

Coote, J., 'An Interim Report on a Previously Unknown Collection from Cook's First Voyage: The Christ Church Collection at the Pitt Rivers Museum, University of Oxford', *Journal of Museum Ethnography*, no. 16 (2004), pp. 111–21

Coote, J., *Curiosities from the Endeavour: A forgotten collection, Pacific Artefacts Given by Joseph Banks to Christ Church, Oxford after the First Voyage* (Whitby: Captain Cook Memorial Museum, 2004)

Coote, J., 'Joseph Banks's Forty Brass Patus', *Journal of Museum Ethnography*, no. 20 (2008), pp. 49–68

Coote, J., ed., *Cook-Voyage Collections of Artificial Curiosities in Britain and Ireland, 1770–2014*, Museum Ethnographers Group Occasional Paper 5 (Oxford: Museum Ethnographers Group, [forthcoming unseen, 2016])

Coote, J. *et al.*, '"Curiosities sent to Oxford": The Original Documentation of the Forster Collection at the Pitt Rivers Museum', *Journal of the History of Collections*, 7, no. 2 (2000), pp. 177–92

Culture and Media, Department of, Case 13 2012/13: 'Two paintings by George Stubbs, *The Kongouro from New Holland* (The Kangaroo) and *Portrait of a Large Dog* (The Dingo)'

Curtis, W., *Flora Londinensis*, 2 vols (London, [1775] 1777–1798). Including 432 hand-coloured plates issued in six fascicles

Dalrymple, A., *An Historical Collection of the Several Voyages and Discoveries in the South Pacific Ocean* (London, 1770 and 1771)

Dampier, W., *New Voyage Round the World* (London: James Knapton, 1697)

Darwin, C., *Narrative of the Surveying Voyages of His Majesty's Ships Adventure and Beagle, between the years 1826 and 1836, describing their examination of the southern shores of South America, and the Beagle's Circumnavigation of the Globe*, vol. 3 (London: Henry Colburn, 1839)

David, A., *The Charts & Coastal Views of Captain Cook's Voyages*, 3 vols (London: the Hakluyt Society in association with the Australian Academy of the Humanities, 1988, 1992 and 1997)

Dawson, W.R., *The Banks Letters: A Calendar of the Manuscript Correspondence of Sir Joseph Banks Preserved in the British Museum, the British Museum (Natural History) and other Collections in Great Britain* (London: British Museum (Natural History), 1958)

Derbyshire Record Office – Banks to Perrin, 1 December 1768, DRO D239 M/F 15882

Diment, J.A. *et al.*, *Catalogue of the Natural History Drawings commissioned by Joseph Banks on the Endeavour Voyage 1768 –1771...*, Bulletin of the British Museum (Natural History) Historical Series, vol. 11 (1984), Part 1: Botany – Australia

Diment, J.A. *et al.*, *Catalogue of the Natural History Drawings Commissioned by Joseph Banks on the Endeavour Voyage 1768 –1771...*, Bulletin of the British Museum (Natural History) Historical Series, vol. 12 (1987), Part 2: Botany – Brazil, Java, Madeira, New Zealand, Society Islands and Tierra del Fuego

Di Piazza, A., and Pearthree, E., 'A New Reading of Tupaia's Chart', *Journal of the Polynesian Society*, 116, no. 3 (2007), pp. 321–40

Dixson Library, Sydney

Dolan, B., *Josiah Wedgwood: Entrepreneur to the Enlightenment* (London: Harper Perennial, 2005)

Donovan, E., *An Epitome of the Natural History of the Insects of New Holland, New Zealand, New Guinea, Otaheite, and other islands in the Indian, Southern, and Pacific Oceans...* (London, 1805)

Driessen, H.A.H., 'Outriggerless Canoes and Glorious Beings', *The Journal of Pacific History*, 17 (1982), pt 1, pp. 3–28

Driessen, H.A.H., 'Outriggerless Canoes and Glorious Beings Revisited', *The Journal of Pacific History*, 19 (1984), pt 4, pp. 248–57

Druett, J., *Tupaia: Captain Cook's Polynesian Navigator* (Santa Barbara, California: Praeger, 2011)

Duff, R., ed., *No Sort of Iron: Culture of Cook's Polynesians* (Christchurch: Art Galleries and Museums' Association of New Zealand, 1969)

Duyker, E., *Nature's Argonaut: Daniel Solander 1733–1782, Naturalist and Voyager with Cook and Banks* (Melbourne: Miegunyah/Melbourne University Press, 1998)

Duyker, E. and Tingbrand, P., eds, *Daniel Solander: Collected Correspondence 1753–1782* (Melbourne: Miegunyah/Melbourne University Press, 1995)

Edge-Partington, J., *An Album of the Weapons, Tools, Ornaments, Articles of Dress etc. of the Natives of the Pacific Islands. Drawn and Described from examples in public & private collections in England by James Edge-Partington*, First Series, 2 vols (London: Issued for private circulation by James Edge-Partington and Charles Heape, 1890)

Edge-Partington, J., *An Album of the Weapons, Tools, Ornaments, Articles of Dress etc. of the Natives of the Pacific Islands. Drawn and Described from examples in public & private collections in England by James Edge-Partington*, Second Series (London: Issued for private circulation by J. Edge-Partington and Charles Heape, 1895)

Edge-Partington, J., *An Album of the Weapons, Tools, Ornaments, Articles of Dress etc. of the Natives of the Pacific Islands. Drawn and Described from examples in public & private collections in Australasia by James Edge-Partington*, Third Series (London: Issued for private circulation by J. Edge-Partington and Charles Heape, 1898)

Edwards, D., *A Natural History of Uncommon Birds* (London, 1743–51)

Etnografiska Museet, Stockholm.
A collection of artefacts that Banks presented to the Swedish friend of Daniel Solander, Johan Alströmer, during his visit to London from 1777–8 that included items from the first and second Cook voyages. In 1848 the collection was donated by the Alströmer family to the Royal Swedish Academy of Sciences, from whence it later passed to the Etnografiska Museet, Stockholm

Farington, J., *The Farington Diary*, Grieg, J., ed., 3 vols, 2nd edition (London: Hutchinson & Co., 1922, 1923 and 1924)

Ferdon, E.N, *Early Tahiti As The Explorers Saw It, 1767–1797* (Tuscon: The University of Arizona Press, 1981)

Fletcher, H.R., *The Story of the Royal Horticultural Society, 1804–1968* (Oxford: Oxford University Press, 1969)

Flower, W.H., *Catalogue of specimens…in the Museum of the Royal College of Surgeons of England*, vol. 2 (London: Royal College of Surgeons, 1884)

Forster, J.G.A., *A voyage round the world, in His Britannic Majesty's sloop Resolution*, 2 vols (London: B. White; J. Robson; P. Elmsly; G. Robinson, 1777)

Forster, J.G.A., *A Letter to the Right Honourable the Earl of Sandwich, First Lord Commissioner of the Board of Admiralty, &c, from George Forster, F.R.S.* (London: G. Robinson, 1778)

Forster, J.R., and Forster, J.G.A., *Observations made during a voyage round the world, on physical geography, natural history, and ethic philosophy* (London: printed for G. Robinson, 1778)

Fox, G.T., ed., *Synopsis of the Newcastle Museum, late the Allan, formerly the Tunstall, or Wycliffe Museum…* (Newcastle: T. and J. Hodgson, 1827)

Gascoigne, J., *Joseph Banks and the English Enlightenment: Useful Knowledge and Polite Culture* (Cambridge: Cambridge University Press, 1994)

Gascoigne, J., 'Banks, Sir Joseph, baronet (1743–1820)', *Oxford Dictionary of National Biography* (Oxford: Oxford University Press, 2004), vol. 3, p. 692

Gascoigne, J., *Captain Cook, Voyager Between Worlds* (London: Continuum Books, 2007)

Gathercole, P., 'Lord Sandwich's Collection of Polynesian Artefacts', in Lincoln, M., ed., *Science and Exploration in the Pacific: European voyages to the southern oceans in the eighteenth century* (Woodbridge, Suffolk: Boydell Press in association with the National Maritime Museum, 1998), pp. 103–115

Gerard, J., *The Herball or Generall Historie of Plantes* (London: John Norton, 1597)

Gilbert, P., ed., Carter, H.B., David, A.C.E., *The Banks/Cook Portfolio* (London: Hill House Publishers, 1990)

Goldsmith, O., *An History of the Earth, and Animated Nature* (London: printed for J. Nourse, 1774), 8 vols

Gray, J.E., *List of the Specimens of Mammalia in the Collection of the British Museum* (London, 1843)

Green, C., and Cook, J., 'Observations made, by appointment of the Royal Society, at King George's Island in the South Sea; by Mr. Charles Green, formerly Assistant at the Royal Observatory at Greenwich, and Lieut. James Cook, of his Majesty's Ship the *Endeavour*', *Philosophical Transactions of the Royal Society*, vol. 61 (1771), pp. 397–421

Groves, E.W., 'Notes on the Botanical Specimens Collected by Banks and Solander on Cook's First Voyage, together with an Itinerary of Landing Localities', *Journal of the Society for the Bibliography of Natural History*, 4:1 (1962), pp. 57–62

Harms, V., 'Ein "Ancestor Panel," der Māoris von de ersten südsee-reise (1768–1771) James Cooks in der ethnografischen sammlung der Universität Tübingen entdeckt', *Baessler-Archiv*, 46, no. 2 (1998), pp. 429–41

Harris, P.R., *A History of the British Museum Library, 1753–1973* (London: British Library, 1998)

Hatfield, P., *Lines in the Ice: Exploring the Roof of the World* (London: British Library, 2016)

Hatfield, P., *Canada in the Frame: Colonial Copyright and the Photographic Image* (London: McGill-Queens University Press, forthcoming 2017)

Hawkesworth, J., ed., *An account of the voyages undertaken by the order of His Present Majesty for making discoveries in the southern hemisphere, and successively performed by Commodore Byron, Captain Wallis, Captain Carteret, and Captain Cook, in the Dolphin, the Swallow and the Endeavour: drawn up from the journals which were kept by the several commanders, and from the papers of Joseph Banks, Esq*, 3 vols (London: W. Strahan and T. Cadell, 1773)

Hearne, S., *A Journey from Prince of Wales's Fort in Hudson's Bay to the Northern Ocean, undertaken … for the discovery of Copper mines, a North West passage, &c. in the years 1769–1772* (Dublin, 1796)

Heringman, N., '"Peter Pindar," Joseph Banks, and the Case Against Natural History', *Wordsworth Circle*, 35, no. 1 (2004), pp. 21–9

Heringman, N., *Sciences of Antiquity: Romantic Antiquarianism, Natural History, and Knowledge Work* (Oxford: Oxford University Press, 2013)

Hetherington, M., and Morphy, H., eds, *Discovering Cook's Collections* (Canberra: National Museum of Australia Press, 2009)

Hoare, M.E., *The Tactless Philosopher: Johann Reinhold Forster (1729–98)* (Melbourne: The Hawthorn Press, 1975)

Home, J.A., ed., *The Letters and Journals of Lady Mary Coke*, 4 vols (Edinburgh: David Douglas, 1889–96)

Hooker, W. et al., *Drawings of Fruit*, vols 1–10 (1815–1830), Royal Horticultural Society, Lindley Library, London

Hooper, S., *Pacific Encounters: Art & Divinity in Polynesia 1760–1860* (London: The British Museum Press, 2006)

Hough, R., *Captain James Cook* (London: Hodder & Stoughton, 1994)

Howse, D., 'The Principal Scientific Instruments Taken on Captain Cook's Voyages of Exploration 1768–80', *The Mariner's Mirror*, 65:2 (1979), pp. 119–35

Howse, D., ed., *Background to Discovery: Pacific Exploration from Dampier to Cook* (Berkeley: University of California Press, 1990)

Howse, D., and Hutchinson, B., 'The Saga of the Shelton Clocks', *Antiquarian Horology* 6:5 (1969), pp. 281–98

Hunterian Museum and Art Gallery, University of Glasgow

International Commission on Zoological Nomenclature (ICZN), 'Opinion 760. *Macropus* Shaw, 1790 (Mammalia): Addition to the official list together with the validation under the plenary powers of *Macropus giganteus* Shaw, 1790', *Bulletin of Zoological Nomenclature*, 22 (1966), pp. 292–95

Iredale, T., and Troughton, E. Le G., 'Captain Cook's Kangaroo', *Australian Zoologist*, 3, no. 8 (1925), pp. 311–16

Iredale, T., and Troughton, E. Le G., 'The identity of Cook's Kangaroo', *Records of the Australian Museum*, 20, no. 1 (1937), pp. 67–71

Iredale, T., and Troughton, E. Le G., 'The actual identity of Captain Cook's kangaroo', *Proceedings of the Linnaean Society of New South Wales*, 87 (1962), pp. 177–84

Jackson, S.M. and Groves, C.P., *Taxonomy of Australian Mammals* (Collingwood, Victoria: CSIRO Publishing, 2015)

Jessop, L., 'The Exotic Artefacts from George Allan's Museum, and other 18th Century Ethnographic Collections surviving in The Hancock Museum, Newcastle upon Tyne', *Transactions of the Natural History Society of Northumbria*, 63, part 3 (June 2003)

Jessop, L., 'Cook-Voyage Collections in North-East England', in Jeremy Coote, ed., *Cook-Voyage Collections of Artificial Curiosities in Britain and Ireland, 1770–2014*, Museum Ethnographers Group Occasional Paper 5 (Oxford: Museum Ethnographers Group, [forthcoming, 2016])

Joppien, R., 'Sir Joseph Banks and the World of Art in Great Britain', in R.E.R. Banks *et al.*, eds, *Sir Joseph Banks: A Global Perspective* (London: Royal Botanic Gardens, 1994), pp. 87–103

Joppien, R. and Chambers, N., 'The Scholarly Library and Collections of Knowledge of Sir Joseph Banks', in Mandelbrote, G. and Taylor, B., eds, *Libraries Within the Library* (London: British Library, 2009), pp. 222–43

Joppien, R. and Smith, B., *The Art of Captain Cook's Voyages*, 4 vols (New Haven: Published for the Paul Mellon Centre for Studies in British Art by Yale University Press, 1985–7)

Kaeppler, A.L., 'Cook Voyage Provenance of the "Artificial Curiosities" of Bullock's Museum', *Man: The Journal of the Royal Anthropological Institute*, New Series, 9, no. 1 (March 1974), pp. 68–92

Kaeppler, A.L., *'Artificial Curiosities' being An exposition of native manufactures collected on the three Pacific voyages of Captain James Cook, R.N. at the Bernice Pauahi Bishop Museum January 18, 1978–August 31, 1978*, Bernice P. Museum Special Publication 65 (Honolulu: Bishop Museum Press, 1978)

Kaeppler, A.L., 'Tracing the History of Hawaiian Cook-Voyage Artefacts in the Museum of Mankind', in T.C. Mitchell, ed., *Captain Cook and the South Pacific*, The British Museum Yearbook 3 (London: British Museum Publications for the Trustees, 1979), pp. 167–97

Kaeppler, A.L., 'To attempt some new discoveries in that vast unknown tract', in M. Hetherington and H. Morphy, eds, *Discovering Cook's Collections* (Canberra: National Museum of Australia Press, 2009), pp. 58–77

Kaeppler, A.L., *Holophusicon. The Leverian Museum. An Eighteenth-Century English Institution of Science, Curiosity, and Art* (Altenstadt: ZKF Publishers, 2011)

Kaeppler, A.L. *et al.*, *James Cook and the Exploration of the Pacific* (London: Thames & Hudson, 2009)

Kaye, I., 'Captain James Cook and the Royal Society', *Notes and Records of the Royal Society*, 24, no. 1 (1969), pp. 7–18

Kerr, D., *Census of Alexander Shaw's Tapa Cloth Book, 1787* (Dunedin, New Zealand: Donald Kerr, 2013)

King, D.S., *Missionaries and Idols in Polynesia* (San Francisco and London: Beak Press in association with Paul Holberton Publishing, 2015)

King, J.C.H., *Artificial Curiosities from the Northwest Coast of America: native American artefacts in the British Museum collected on the third voyage of Captain James Cook and acquired through Sir Joseph Banks* (London: British Museum Publications, 1981)

Langdon, R., 'The European Ships of Tupaia's Chart', *The Journal of Pacific History*, 15, pt 4 (1980), pp. 225–32

Langdon, R., 'Of Time, Prophecy, and the European Ships of Tupaia's Chart', *The Journal of Pacific History*, 19, pt 4 (1984), pp. 239–47

Lanyon-Orgill, P.A., *Captain Cook's South Sea Island Vocabularies* (London, 1979)

Lincoln, M., ed., *Science and Exploration in the Pacific: European voyages to the southern oceans in the eighteenth century* (Woodbridge, Suffolk: Boydell Press in association with the National Maritime Museum, 1998)

Linnaeus, C., *Systema Naturæ per Regna Tria Naturæ, secundum Classes, Ordines, Genera, Species, cum Characteribus, Differentiis, Synonymis, Locis*, 2 vols, 10th edn (Stockholm: Laurentius Salvius, 1758, 1759)

Linnaeus, C., *Species Plantarum, Exhibentes Plantas Rite Cognitas, ad Genera Relatas, cum Differentiis Specificis, Nominibus Trivialibus, Synonymis Selectis, Locis Natalibus, secundum Systema Sexuale Digestas*, 2 vols, 2nd edn (Stockholm: Laurentius Salvius, 1762, 1763)

Lysaght, A.M., 'Captain Cook's kangaroo', *The New Scientist*, 14 March 1957, pp. 17–19

Lysaght, A.M., 'Some Eighteenth Century Bird Paintings in the Library of Sir Joseph Banks (1743–1820)', *Bulletin of the British Museum (Natural History) Historical Series*, 1 (1959), no. 6

Lysaght, A.M., *Joseph Banks in Newfoundland and Labrador, 1766: His Diary, Manuscripts and Collections* (London: Faber and Faber, 1971)

Lysaght, A.M., 'Banks's Artists and his *Endeavour* Collections', in T.C. Mitchell, ed., *Captain Cook and the South Pacific*, The British Museum Yearbook 3 (London: British Museum Publications for the Trustees, 1979), pp. 9–80

MacGregor, N., *A History of the World in 100 Objects* (London: The British Museum, Allen Lane imprint of Penguin Books, 2010)

Mackay, D., *In the Wake of Cook: Exploration, Science & Empire, 1780–1801* (London: Croon Helm Ltd, 1985)

Mandelbrote, G., and Taylor, B., eds, *Libraries Within the Library* (London: British Library, 2009)

Marquardt, K.H., 'Do We Really Know the *Endeavour*?', *The Great Circle*, 8, no. 1 (April 1986), pp. 27–32

Marsden, E., ed., *A Brief Memoir of the life and writings of the late W. Marsden...* (London: J.L. Cox, 1838)

Marsden, W., 'Remarks on the Sumatran Languages, by Mr. Marsden. In a Letter to Sir Joseph Banks, Bart. President of the Royal Society', *Archaeologia*, 6 (January 1782), pp. 154–8

Marsden, W., *The History of Sumatra...*, 2nd edition (London, 1784)

Marsden, W., *A Catalogue of Dictionaries, Vocabularies, Grammars and Alphabets* (London, 1796)

Marsden, W., *A Dictionary of the Malayan Language* (London, 1812)

Marsden, W. *Miscellaneous Works of William Marsden* (London: J.L. Cox and Son, 1834)

Marshall, J.B., 'The handwriting of Joseph Banks, his scientific staff and amanuenses', *Bulletin of the British Museum (Natural History) Botany Series*, 6 (1986), no. 1

Maskelyne, N., *The Nautical Almanac and Astronomical Ephemeris, for the Year 1768. Published by Order of the Commissioners of Longitude* (London: W. Richardson and S. Clark, 1767)

Maskelyne, N., *The Nautical Almanac and Astronomical Ephemeris, for the Year 1769. Published by Order of the Commissioners of Longitude* (London: W. Richardson and S. Clark, 1768)

McNeely, I.F., 'Wilhelm von Humboldt and the World of Languages', *Ritsumeikan Studies in Language and Culture*, 23, no. 2 (October 2011), pp. 129–47

Megaw, J.V.S., 'Captain Cook and bone barbs at Botany Bay', *Antiquity*, 43 (1969), pp. 213–16

Megaw, J.V.S., 'More eighteenth-century trophies from Botany Bay', *Mankind*, 8 (3) (June 1972), pp. 225–6

Megaw, J.V.S., 'Something old, something new: further notes on the Aborigines of the Sydney district as represented by their surviving artefacts, and as depicted in some early European representations', in J. Specht, ed., *F.D. MaCarthy, Commemorative papers (Archaeology, Anthropology, Rock Art). Records of the Australian Museum*, Supplement 17 (1993), pp. 25–44

Megaw, J.V.S., 'There's a hole in my shield...: A textual foot-note', *Australian Archaeology*, 38 (1994), pp. 35–7

Miller, D.P., and Reill, P.H. eds, *Visions of empire: voyages, botany, and representations of nature* (Cambridge: Cambridge University Press, 1996)

Miller, E., *That Noble Cabinet: A history of the British Museum* (London: André Deutsch, 1973)

Mitchell Library, Sydney – Abbreviated, ML

Mitchell, T.C., ed., *Captain Cook and the South Pacific*, British Museum Yearbook 3 (London: British Museum Publications, 1979)

Morrison-Scott, T.C.S. and Sawyer, F.C., 'The Identity of Captain Cook's Kangaroo', *Bulletin of the British Museum (Natural History) Zoology Series*, 1, no. 3 (1950), pp. 45–50

Müller, J.S., *Illustratio Systematis Sexualis Linnæi*, 3 vols (London, [1770–]1777). Including 108 hand-coloured plates and a duplicate uncoloured set

Museum of Archaeology and Anthropology, University of Cambridge –
- Sandwich Collection – a gift of Pacific objects from James Cook to his patron, Lord Sandwich. Passed by Sandwich in October 1771 to Trinity College, Cambridge, and currently deposited at the museum
- Pennant Collection – including a mix of artefacts from Cook's and perhaps other voyages, some of which may have been gifts from Banks. Passed to the museum by the earl and countess of Denbigh in 1912 and 1913
- Widdicombe Collection – artefacts from the Pacific voyages of Cook, in particular the third voyage, many of which were originally acquired by the Holdsworth family of Devon at the 1806 sale of the Leverian museum. Objects from the *Endeavour* voyage are unlikely. Obtained by the museum, 1921–2

Myers, W., 'Cook, the physician', in Antoinette and Robert Shalkop, eds, *Exploration in Alaska: Captain Cook commemorative lectures* (Anchorage: Cook Inlet Historical Society, 1980), pp. 59–88

National Archives, Kew
   Admiralty Papers

National Maritime Museum, Greenwich, London – Abbreviated, NMM
   Admiralty Collection, Ship Plans
      Caird Library
      Charts and Maps
      Oil Paintings
      Weapons, Ordnance

National Museum of Scotland, Edinburgh – Department of Science and Technology

Natural History Museum, London – Abbreviated, NHM
   Botany Library, *Endeavour Florilegium* and drawings –
      Dawson Turner Collection (abbreviated DTC), Botany Library
      General Herbarium
      General Library, Rare Books
      Department of Zoology, birds and shells
      Zoology Library, *Endeavour* drawings and Solander Slips

O'Brian, P., *Joseph Banks: A Life* (London: Collins Harvill, 1987)

Oliver, D.L., 'Food Plants. Breadfruit', *Ancient Tahitian Society*, 3 vols (Honolulu: University Press of Hawaii, 1974), vol. 1, *Ethnography*, pp. 234–44

Owen, R., *Descriptive Catalogue of the Osteological Series contained in the Museum of the Royal College of Surgeons of England*, vol. 1 Pisces, Reptilia, Aves, Marsupialia (London: Taylor and Francis, 1853)

Parkin, R., *H.M. Bark Endeavour: her place in Australian history* (Carlton, Victoria: Melbourne University Press, 1997)

Parkinson, Stanfield, ed., *A Journal of a Voyage to the South Seas, in his Majesty's Ship, the Endeavour. Faithfully transcribed from the Papers of the late Sydney Parkinson, draughtsman to Joseph Banks, Esq. on his late Expedition, with Dr. Solander, round the World* (London, 1773). Reissued in 1784 with 'Explanatory Remarks on the Preface' by J. Fothergill

Parsons, H., 'British-Tahitian collaborative drawing strategies on Cook's *Endeavour* voyage', in Konishi, S., Nugent, M. and Shellam, T., eds, *Indigenous Intermediaries: new perspectives on exploration archives* (Canberra: Australia National University Press, 2015), pp. 147–67

Pennant, T., *Tour in Scotland and Voyage to the Hebrides 1772*, 2 vols (Chester: John Monk, 1774 and London: Benjamin White, 1776)

Pennant, T., *History of Quadrupeds*, 2 vols (London: B. White, 1781)

Pennant, T., *Arctic Zoology*, 2 vols and supplement (London, 1784–7)

Phelps, S., *Art and Artefacts of the Pacific, Africa and the Americas: The James Hooper Collection* (London: Hutchinson & Co., 1976),

Pitt Rivers Museum, University of Oxford
   Banks Collection – donated by Joseph Banks probably directly to Christ Church, Oxford, at an unknown date prior to mid-January 1773 and now deposited at the Pitt Rivers Museum.
   Forster Collection – donated by J.R. Forster to the University of Oxford in 1776 and now held by the Pitt Rivers Museum

Ponzi, F., *18th-century Iceland: A Pictorial Record from the Banks and Stanley Expeditions* (Iceland: Almenna bókafélagið, 1987)

Poole, W.E., '*Macropus giganteus*', *Mammalian Species*, no. 187 (1982), pp. 1–8

Ramsbottom, J., 'Note: Banks's and Solander's Duplicates', *Journal of the Society for the Bibliography of Natural History*, 4:3 (1963), p. 197

Rauschenberg, R.A., 'The Journals of Joseph Banks. Voyage up Great Britain's West Coast to Iceland and to the Orkney Isles July to October 1772', *Proceedings of the American Philosophical Society*, 117, no. 3 (1973), pp. 186–226

Raven, H.C., 'The Identity of Captain Cook's Kangaroo', *Journal of Mammalogy*, 20, no. 1 (1939), pp. 50–7

Reilly, R. and Savage, G., *Wedgwood: the Portrait Medallions* (London: Barrie & Jenkins Ltd, 1973)

Ride, W.D.L. and Calaby, J.H., in 'Comments on the Proposed Stablilization of the *Macropus* Shaw 1790. Z.N.(S) 1584', *Bulletin of Zoological Nomenclature*, 21 (1964), pp. 250–55

Ritvo, H., *The Platypus and the Mermaid and Other Figments of the Classifying Imagination* (Cambridge, Massachusetts: Harvard University Press, 1997)

Rolfe, W.D.I., 'A Stubbs drawing recognized', *The Burlington Magazine*, December 1983, pp. 738–41

Rose, R.G., *A Material Culture of Ancient Tahiti*, 9 vols. Unpublished PhD thesis. Harvard University, Cambridge, Massachusetts (1971)

Rott, H.W., ed., *George Stubbs 1724–1806: Science into Art* (Munich, 2012)

Royal Society, London – Abbreviated, RS
   Letters and Papers
      Miscellaneous Manuscripts
      Philosophical Transactions, Executive Secretary Sequence

Rydén, S., *The Banks Collection: An Episode in 18th-Century Anglo-Swedish Relations*, Statens Etnografiska Museum Monograph Series, no. 8 (Stockholm: Almqvist & Wiksell, 1965)

Salmond, A., *Two Worlds: First meetings between Maori and Europeans, 1642–1772* (Auckland: Viking, 1991)

Salmond, A., *Between Worlds: Early exchanges between Maori and Europeans, 1773–1815* (Auckland: Viking, 1997)

Salmond, A., *The Trial of the Cannibal Dog: Captain Cook in the South Seas* (London: Allen Lane, 2003)

Salmond, A., 'Artefacts of Encounter: The Cook-Voyage Collections in Cambridge', in Jeremy Coote, ed., *Cook-Voyage Collections of Artificial Curiosities in Britain and Ireland, 1770–2014*, Museum Ethnographers Group Occasional Paper 5 (Oxford: Museum Ethnographers Group, [first issued 2011, forthcoming journal publication 2016])

School of Oriental and African Studies, London – Marsden Collection, MS 12153

Science Museum, London – Astronomy Collection

The Scots Magazine, Edinburgh, 34 (November 1772)

Scott-Waring, J., *An Epistle from Oberea, Queen of Otaheite, to Joseph Banks, Esq. Translated by T.Q.Z. Esq. Professor of the Otaheite Language in Dublin, and of all the Languages of the undiscovered Islands in the South Sea; And enriched with Historical and Explanatory Notes* (London: J. Almon, 1774 [1773]). Introduction dated 20 September 1773

Scott-Waring, J., *A Second Letter from Oberea, Queen of Otaheite, to Joseph Banks, Esq; Translated from the Original, Brought over by his Excellency Otaipairoo, Envoy Extraordinary and Plenipotentiary from the Queen of Otaheite, to the Court of Great Britain, Lately arrived in his Majesty's Ship the Adventure, Capt. Furneaux. With some curious and entertaining Anecdotes of this celebrated Foreigner before and since his Arrival in England; Together with explanatory Notes from the Queen's former Letter, and from Dr. Hawkesworth's Voyages* (London: Printed by T.J. Carnegy for E. Johnson, [1774])

Shaw, A., *A catalogue of the different specimens of cloth collected in the three voyages of Captain Cook, to the Southern Hemisphere; with a particular account of the manner of the manufacturing the same in the various islands of the South Seas; partly extracted from Mr Anderson and Reinhold Forster's observations, and the verbal account of some of the most knowing of the navigators: with some anecdotes that happened to them among the natives* (London, 1787)

Skelton, R.A., ed., *The Journals of Captain James Cook on his Voyages of Discovery; Charts & Views drawn by Cook and his Officers and reproduced from the original manuscripts* (Cambridge: Cambridge University Press for the Hakluyt Society, 1969)

Smith, A., 'Archaeology, local history and community in French Polynesia', *World Archaeology*, 42, no. 3 (2010), pp. 367–80

Smith, B., *European Vision and the South Pacific, 1768–1850: A study in the History of Art and Ideas* (Oxford: Clarendon Press, 1960)

Smith, J.E., Sir, ed., *A Selection from the Correspondence of Linnæus and other Naturalists, from the original manuscripts*, 2 vols (London, 1821)

Smith, V., *Intimate Strangers: Friendship, Exchange and Pacific Encounters* (Cambridge: Cambridge University Press, 2010)

Smith, W.J., 'A Museum for a Guinea', *Country Life*, 127, no. 3288 (10 March 1960)

Smith, W.J., 'Sir Ashton Lever of Alkrington, 1729–1788', *Lancashire and Cheshire Antiquarian Society Transactions*, 72 (1962)

Sorrell, M., 'A zebra, a tigress and a cheetah: New light on George Stubbs's exotic animal subjects', *The Britisih Art Journal*, 15, no. 1 (2014), pp. 99–109

Starzecka, D., ed., *Maori Art and Culture* (London: Published for the Trustees of the British Museum by British Museum Press, 1998)

Starzecka, D., Neich, R., Pendergrast, M., *The Māori Collections of the British Museum* (London: The British Museum, 2010)

Stearn, W.T., *The Natural History Museum at South Kensington* (London: The Natural History Museum, 2008 reprint)

Stearn, W.T., Wilson, S. and Buchanan, H., *Great Flower Books, 1700–1900* (London: H.F. & G. Witherby Ltd, 1990)

Steinheimer, F.D., 'A hummingbird nest from James Cook's *Endeavour* voyage, 1768–1771', *Archives of Natural History*, 30, no. 1 (2003), pp. 163–5

Stevenson, K., and Hooper, S., 'Tahitian Fau – Unveiling an Enigma', in Judith Huntsman, ed., *Polynesian Art: Histories and Meanings in Cultural Contexts*, Special Issue, *The Journal of the Polynesian Society*, 116, no. 2 (June 2007), pp. 181–211

Stubbs, G., *The Anatomy of the Horse* (London: J.Purser for the author, 1766)

Suárez, T., *Early Mapping of the Pacific* (Periplus Editions, 2004)

Sutro Library, California – Abbreviated, SL

Tanner, J., *From Pacific Shores: Eighteenth-century Ethnographic Collections at Cambridge* (Cambridge, 1999)

Tapsell, P., 'Footprints in the Sand: Banks's Māori Collection, Cook's First Voyage, 1768–71', in M. Hetherington and H. Morphy, eds, *Discovering Cook's Collections* (Canberra: National Museum of Australia Press, 2009), pp. 92–111

Taylor, B., *Stubbs* (London: Phaidon Press Ltd, 1975)

Trinity College, Cambridge – Trinity MS Add. a. 106

Troil, U., von, *Letters on Iceland* (Dublin, 1780)

Troughton, E. Le G. and McMichael, D.F., in 'Comments on the Proposed Stablilization of the *Macropus* Shaw 1790. Z.N.(S.) 1584', *Bulletin of Zoological Nomenclature*, 21 (1964a), pp. 255–8

Troughton, E. Le G. and McMichael, D.F., in 'Comments on the Proposed Stablilization of the *Macropus* Shaw 1790. Z.N.(S.) 1584', *Bulletin of Zoological Nomenclature*, 21 (1964b), pp. 329–31

Tullberg, T.F.H., *Linnéportrait, vid Uppsala universitets minnefest pa tvahundraarsdagen af Carl von Linnés födelse* (Stockholm: Aktiebolaget Ljus, 1907)

Turnbull, D., 'Cook and Tupaia, a Tale of Cartographic *Méconnaissance*?', in Lincoln, M., ed., *Science and Exploration in the Pacific: European voyages to the southern oceans in the eighteenth century* (Woodbridge, Suffolk: Boydell Press in association with the National Maritime Museum, 1998), pp. 117–131

Webb-Johnson, A., 'The George Adlington Syme Oration: Surgery in England in the Making', *The Australian and New Zealand Journal of Surgery*, 9 (1939), pp. 10–30

Weld, C.W., *A History of the Royal Society, with Memoirs of the Presidents*, 2 vols (London, 1848)

Wheeler, A., 'Catalogue of the Natural History Drawings commissioned by Joseph Banks on the Endeavour Voyage 1768 –1771…', *Bulletin of the British Museum (Natural History) Historical Series*, 13 (1986), Part 3: Zoology

White, J., *Journal of a voyage to New South Wales* (London: J. Debrett, 1790)

Whitehead, P.J.P., 'Zoological Specimens from Captain Cook's Voyages', *Journal of the Society for the Bibliography of Natural History*, 5, no. 3 (1969), pp. 161–201

Whitehead, P.J.P., 'A Guide to the Dispersal of Zoological Material from Captain Cook's Voyages', *Pacific Studies*, 2, no. 1 (1978), pp. 52–93

Wilkins, G.L., 'A Catalogue and Historical Account of the Banks Shell Collection', *Bulletin of the British Museum (Natural History) Historical Series*, 1 (1955), no. 3

Willis, R.J., 'The earliest known Australian bird painting: a Rainbow Lorikeet, *Trichoglossus haematodus moluccanus* (Gmelin) by Moses Griffith, painted in 1772', *Archives of Natural History*, 15, no. 3 (1988), pp. 323–9

Younger, R.N., *Kangaroo, Images Through the Ages* (Hawthorn, Victoria: Century Hutchinson Australia Pty Ltd, 1988)

PHOTOGRAPHIC CREDITS

We gratefully acknowledge the copyright for the images obtained from the following

Boston Guildhall, Boston Borough Council

The British Library, London

© The Trustees of the British Museum

© Captain Cook Memorial Museum, Whitby

© The Collection: Art and Archaeology in Lincolnshire
(Usher Gallery, Lincoln)

Derbyshire Record Office

By permission of the trustees of the Goodwood Collection

Museum of Archaeology and Anthropology, University of Cambridge

Museum of the Royal College of Surgeons (Hunterian Museum)

The National Archives

Courtesy of the Trustees of the Natural History Museum, London.

Courtesy of the Trustees of the Natural History Museum, Tring

© National Maritime Museum, Greenwich, London

© National Museums of Scotland, Edinburgh

© Pitt Rivers Museum, University of Oxford

Private collection

Royal Collection Trust/© Her Majesty Queen Elizabeth II 2014

RHS, Lindley Library

The Royal Society, London

Science Museum/Science & Society Picture Library